Advances in
HYDROSCIENCE

VOLUME 5

Contributors to Volume 5

HERMAN BOUWER
CHESTER C. KISIEL
J. R. PHILIP
YIN-CHAO YEN

Advances in
HYDROSCIENCE

Edited by
VEN TE CHOW
UNIVERSITY OF ILLINOIS
URBANA, ILLINOIS

Volume 5 - 1969

ACADEMIC PRESS   New York and London

COPYRIGHT © 1969, BY ACADEMIC PRESS, INC.
ALL RIGHTS RESERVED
NO PART OF THIS BOOK MAY BE REPRODUCED IN ANY FORM,
BY PHOTOSTAT, MICROFILM, OR ANY OTHER MEANS, WITHOUT
WRITTEN PERMISSION FROM THE PUBLISHERS.

ACADEMIC PRESS, INC.
111 Fifth Avenue, New York, New York 10003

United Kingdom Edition published by
ACADEMIC PRESS, INC. (LONDON) LTD.
Berkeley Square House, London W.1

LIBRARY OF CONGRESS CATALOG CARD NUMBER 64-17306

PRINTED IN THE UNITED STATES OF AMERICA

List of Contributors

HERMAN BOUWER, U.S. Water Conservation Laboratory, U.S. Department of Agriculture, Phoenix, Arizona

CHESTER C. KISIEL, Hydrology and Water Resources Office, University of Arizona, Tucson, Arizona

J. R. PHILIP, Division of Plant Industry, Commonwealth Scientific and Industrial Research Organization, Canberra, Australia

YIN-CHAO YEN, U.S. Army Terrestrial Sciences Center, Hanover, New Hampshire

Foreword

This volume of *Advances in Hydroscience* contains articles in three major areas of hydroscience: stochastic hydrology, subsurface flow, and solid-state hydrology.

In the area of stochastic hydrology, C. C. Kisiel provides a comprehensive review of the time series of hydrologic data. Stochastic hydrology is rapidly gaining recognition as an important and basic subject in water resources development because it helps to achieve efficient and economic designs of water resources systems through the use of analytical or simulation models. In this article, hydrologic data are treated as statistical data which can be arranged in a sequence in accordance with the time of occurrence. This is logical because the hydrologic data reflect the histories of the hydrologic phenomena. The sequence so treated is mathematically known as a *time sequence* or *time series*. In statistical mathematics, recent literature on time series analysis is abundant. By treating the hydrologic data as time series, a host of new mathematical tools is therefore available for the stochastic analysis of the data, thus creating a new era in the field of scientific hydrology.

Two articles on subsurface flow are included in this volume. H. Bouwer describes the advances in various scientific disciplines and their application of the analysis and prediction of seepage losses from open channels. This information is invaluable for the design of channels used in irrigation, drainage, and other branches of hydraulic engineering where open channels are employed as a means of conveyance for water.

J. R. Philip presents an authoritative review of the mathematical-physical approach to the study of water movement in unsaturated material which has been developed in the last fifteen years. This theoretical development has been achieved largely by soil physists who are concerned with agronomic or ecological aspects of hydrology, but it will be highly appreciated by engineers and applied hydrologists who may find it useful in planning and designing water projects dealing with subsurface water.

The term *solid-state hydrology* is coined by the editor in order to draw a parallel with the well-accepted term solid-state physics. Solid-state hydrology deals with the science of water in its solid state, such as ice, snow, and frost. Y.-C. Yen's article is a lucid summary of recent scientific development on some properties of snow concerning its process of metamorphism and sintering and its thermal and radiation properties. This summary should be welcomed by workers in the field of snow and ice research.

VEN TE CHOW

Urbana, Illinois
December, 1968

Contents

List of Contributors v

Foreword vii

Contents of Previous Volumes xi

TIME SERIES ANALYSIS OF HYDROLOGIC DATA
CHESTER C. KISIEL

    I. Introduction 1
   II. Conceptual Classification of Stochastic Processes 26
  III. Generalized Harmonic Analysis 30
  IV. Generating Processes 50
   V. Linear Systems 57
  VI. Estimation of Correlation and Spectral Density Functions 68
 VII. Examples of Correlograms and Power Spectra in Hydrology 86
VIII. Importance of Time Dependence in Hydrologic Analysis and Synthesis 99
  IX. Summary 109
      Symbols 110
      References 112

THEORY OF SEEPAGE FROM OPEN CHANNELS
HERMAN BOUWER

   I. Introduction 121
  II. Hydrodynamics of Seepage 122
 III. Measurement of Soil Hydraulic Conductivity 163
  IV. Summary and Conclusions 169
     Symbols 169
     References 170

RECENT STUDIES ON SNOW PROPERTIES
YIN-CHAO YEN

   I. Introduction 173
  II. Metamorphism and Intergranular Bonds 174
 III. Thermal Properties of Snow 183
  IV. Radiation Properties of Snow 201
     Symbols 212
     References 213

THEORY OF INFILTRATION
J. R. PHILIP

- I. Introduction 216
- II. Water Transfer in Unsaturated Soils 217
- III. Mathematical Preliminaries 229
- IV. Exact Solutions of Absorption Equations 235
- V. Exact Solutions of Infiltration Equations 248
- VI. Linearized Solutions 258
- VII. Delta-Function Solutions 266
- VIII. Comparison Technique for Estimating Integral Properties of Solutions 270
- IX. Exact Solutions of More Complicated Problems 275
- X. Physics of One-Dimensional Infiltration 279
- XI. Effects of Geometry and Gravity on Infiltration 284
- Symbols 289
- References 291

Author Index 297

Subject Index 302

Contents of Previous Volumes

VOLUME 1

Sonar
BRADFORD A. BECKEN

Hydroelasticity
S. R. HELLER, JR.

Statistical Hydrodynamics in Porous Media
ADRIAN E. SCHEIDEGGER

New Contributions to Hydroballistics
F. S. BURT

Hydraulics of Wells
MAHDI S. HANTUSH

AUTHOR INDEX—SUBJECT INDEX

VOLUME 2

Tsunamis
W. G. VAN DORN

Chemical Geohydrology
WILLIAM BACK AND BRUCE B. HANSHAW

Hydrodynamics of the Dolphin
MAX O. KRAMER

Hydromechanics of Inland Navigation
SHU-T'IEN LI

Technical Development in Ground Water Recharge
PAUL BAUMANN

AUTHOR INDEX—SUBJECT INDEX

VOLUME 3

Viscous Resistance of Ships
F. H. TODD

Magnetohydrodynamics of Channel Flow
SHIH-I PAI

Hydrodynamics of Blood Flow
E. O. ATTINGER

Biological Treatment of Waste Water
W. WESLEY ECKENFELDER, JR.

Sea Water Conversion
EVERETT D. HOWE

Linearized Steady Theory of Fully Wetted Hydrofoils
TETSUO NISHIYAMA

Evaporation Retardation by Monolayers
WYNDHAM J. ROBERTS

Dynamic Programming in Water Resources Development
NATHAN BURAS

AUTHOR INDEX—SUBJECT INDEX

CONTENTS OF PREVIOUS VOLUMES

VOLUME 4

Hovering Craft over Water
N. HOGBEN

Soil Moisture Theory
E. C. CHILDS

The Theory of Wind-Generated Waves
O. M. PHILLIPS

Flow in the Zone of Aeration
R. W. STALLMAN

Hydraulic Jumps
N. RAJARATNAM

Stochastic Reservoir Theory
E. H. LLOYD

Storm Surges
CHARLES L. BRETSCHNEIDER

AUTHOR INDEX—SUBJECT INDEX

TIME SERIES ANALYSIS OF HYDROLOGIC DATA

CHESTER C. KISIEL

Hydrology and Water Resources Office
University of Arizona, Tucson, Arizona

I.	Introduction	1
	A. General	1
	B. Hydrologic Models in General	2
	C. Nature of Hydrologic Time Series	12
II.	Conceptual Classification of Stochastic Processes	26
III.	Generalized Harmonic Analysis	30
	A. Periodic Time Series	33
	B. Transient Time Series	39
	C. Random Time Series	40
IV.	Generating Processes	50
V.	Linear Systems	57
VI.	Estimation of Correlation and Spectral Density Functions	68
	A. Autocorrelation and Variance Spectral Functions	68
	B. Cross-Correlation and Covariance Spectra Calculations	77
	C. Tests of Significance	79
	D. Limitations of Correlation and Power Spectrum Methods in Hydrology	84
VII.	Examples of Correlograms and Power Spectra in Hydrology	86
VIII.	Importance of Time Dependence in Hydrologic Analysis and Synthesis	99
IX.	Summary	109
	Symbols	110
	References	112

I. Introduction

A. General

Current water resources development, domestic and international, requires the best engineering procedures for the evaluation, prediction, and control of water resources. Lack of adequate data, in part, hampers fulfillment of these objectives. On the other hand, dissatisfaction with traditional methods of hydrologic analysis has led research investigators to draw upon developments of the past 30 to 40 years in mathematical statistics, probability theory, and communications theory. The result has been the evolution of a field of specialization called *stochastic hydrology*. According to the Committee on Surface-Water Hydrology [1] of the American Society of Civil Engineers:

Stochastic hydrology is defined as the manipulation of statistical characteristics of hydrologic variables to solve hydrologic problems, on the basis of the stochastic properties of the variables. A stochastic variable is defined as a *chance* variable or one whose value is determined by a probability function.

Stochastic hydrology is operational in the sense that its ultimate objective is to predict more efficient and economic designs of water resources systems through analytical or simulation models. In this article, stochastic hydrology is taken to encompass the analysis of hydrologic time series and the construction of predictive stochastic models of the rainfall-runoff process. It is the objective of this article to review basic concepts, recent developments, and their potential application in stochastic hydrology, with emphasis on time series analysis and synthesis (model building).

B. Hydrologic Models in General

The hydrologic cycle represents the entire process of circulation and redistribution of water by the atmosphere and the earth. The process constantly moves toward a balance between water on the earth and atmospheric moisture. On land, the balance is represented by the following simple statement of the law of mass conservation:

$$P = R + E + \Delta S \tag{1}$$

in which P is the precipitation, R is runoff, E is evaporation, and ΔS is the change in storage. For each basin, Eq. (1) represents a mass balance between inflow, outflow, and a change in storage within the basin. The entire cycle is quite complex in the sense that cause-effect is not unidirectional in all instances; *feedback* (two-way causality) is an inherent property of the hydrologic cycle by virtue of evaporation.

The cycle in a basin system is schematically illustrated in the flow chart of Fig. 1. The basin *system* is a black box that transforms the rainfall signal into a smoothed streamflow. In the words of the communications engineer, the basin *filters* out the high-frequency components of the input in that it *smooths* out the amplitude of the rainfall. Even in the absence of human activity, *the transformation system* of a *virgin* watershed is quite complex. Human activity on a watershed could either simplify or complicate the transforming system. As larger surface areas become impervious and suburban areas develop, one facet of the hydrologic cycle, infiltration, is virtually eliminated as a modifying factor. However, increased runoff rates and erosional problems may make the system even more complex. Our problem in hydrology is to *identify* the nature of the input, state, and nature of the basin system and nature of the output under all conditions.

Time Series Analysis of Hydrologic Data

The study of the hydrologic cycle has intensified in recent years because of the accelerating interaction of man and his environment and because of increased realization that more efficient and *prevention-oriented* use of a relatively *fixed* water resource is essential to our civilization. The approach to hydrologic study is varied and reflects differing philosophies: analytical or empirical, classical deterministic or stochastic, macroscopic or microscopic, field experimentation or laboratory experimentation, brute manpower or computer simulation. In truth, the best approach to a better understanding of the hydrologic process is a synthesis of all of these alternatives under a realization that each has its limitations. Actually, the process of model building is a cycle that represents a continuous interplay between analysis, synthesis, and data collection (see Fig. 2). Furthermore, it is multidirectional, as shown in Table I.

The major divisions of hydrologic study are given by Amorocho and Hart [4] as *physical science research* (*physical hydrology*) and *system investigations* (*parametric* and *stochastic hydrologies*). They further subdivide the topics and methods of hydrologic study, as shown in Table II.

Mathematical models in hydrology (as in all science and engineering) are suitable abstractions of complex physical phenomena. The model should be both sufficiently complete in its description as to produce useful results and sufficiently limited in its complexity as to be manageable. Einstein[1] has made the following comments apropos to model building:

> Our experience hitherto justifies us in believing that Nature is the realization of the simplest conceivable mathematical ideas. I am convinced that we can discover, by means of purely mathematical constructions, the concepts and laws connecting them with each other, which furnish the key of the understanding of natural phenomena. Experience may *suggest* the appropriate mathematical concepts, but they most certainly *cannot* be deduced from it. *Experience remains, of course, the sole criterion of the physical utility of a mathematical construction.* But the creative principle resides in mathematics. In a certain sense, therefore, I hold it true that pure thought can grasp reality, as the ancients dreamed.
>
> It seems that the human mind has first to construct forms independently before we can find them in Nature. Kepler's marvelous achievement is a particularly fine example of the truth that knowledge cannot spring from experience alone but only from the comparison of the inventions of the intellect with observed facts.

To some, the preceding quotation may seem superfluous, but long-time familiarity with deterministic "laws of nature" may cause one to forget that

---

[1] The quotations are given by von Schelling [4a].

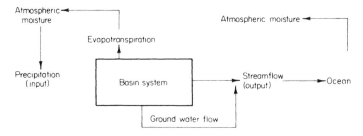

FIG. 1. Hydrologic cycle.

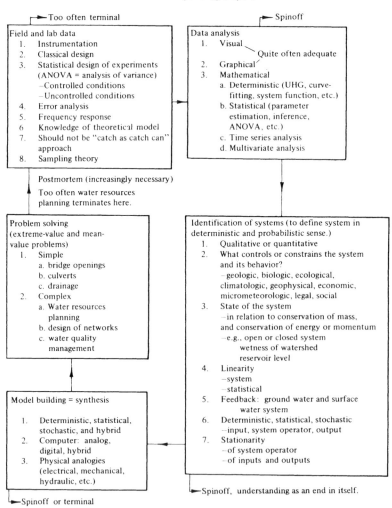

FIG. 2. Cycle of model building.

Time Series Analysis of Hydrologic Data

TABLE I

MODEL APPROACH TO THE STUDY OF THE HYDROLOGIC SYSTEM

Mathematical model	Physical model	Computer model
Deterministic	Electrical	Electronic analog: active,
Probabilistic	Mechanical	passive
Stochastic (dynamic)	Hydraulic:	Digital:
Nonstochastic (static)	flumes,	numerical (noncreative use)
Hybrid	fluid amplifiers,	nonnumerical (creative use):
Deterministic-	laboratory catchments	inferential analysis,
probabilistic	(see Chow, and Amorocho and Hart [3])	mathematical logic
		Hybrid (analog-digital)

TABLE II

TOPICS AND METHODS OF HYDROLOGIC STUDY[a]

Physical hydrology	Systems investigations
Meteorology and climatology	Parametric methods
Energy conversion and transfer	Correlation analysis
Biosystems	Partial system synthesis
Soil physics	with linear analysis
Watershed hydraulics and hydromechanics	General system synthesis
Geomorphology	General nonlinear analysis
Geology	Stochastic methods
Hydrochemistry and geochemistry	Markov chains
	Time series analysis and synthesis
	Monte Carlo methods

[a] Adapted from Amorocho and Hart [4].

these models are a fabrication of the human mind and based on axioms, in some instances, which can be shown to be neither true nor false. The intellectual processes involved in the development of predictive models, be they deterministic or probabilistic, are well illustrated in Fig. 3, first proposed in the context of electrical engineering.

Table III gives possible models of the rainfall, runoff, and basin *transformation system*. Lee [10], Marshall [13], and Davenport and Root [14], among many others, treat the transformation of periodic, aperiodic (transient), and random signals by a linear, time-invariant system. Chow [9]

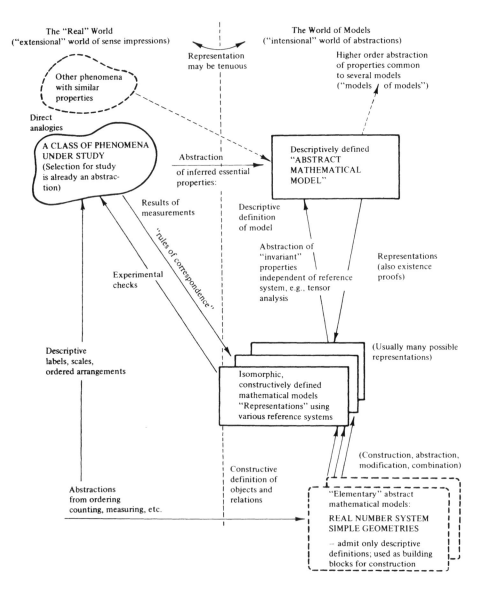

FIG. 3. Methodology of applied mathematics. From lecture notes by Korn [4b].

Time Series Analysis of Hydrologic Data

TABLE III

CONCEPTUAL CLASSIFICATION OF THE HYDROLOGIC SYSTEM

I. Input (rainfall)
   A. According to causality and chance
      1. Deterministic
      2. Quasi-deterministic
      3. Probabilistic
         a. Stochastic
            (1) Pure-random sequence
               (a) Stationary process
               (b) Nonstationary process
            (2) Nonpure-random sequence
               (a) Stationary process
               (b) Nonstationary process
         b. Nonstochastic
            (1) Frequency analysis
            (2) Order statistics
            (3) Probability of discrete events
      4. Hybrid: combination of deterministic and probabilistic, for example, the Stanford Watershed Model and its modification [5, 6]
   B. According to temporal and spatial distribution
      1. Uniform
      2. Nonuniform

II. Transformation system
   A. According to causality and chance: as in the first column (see Ishihara and Takasao [7])
   B. According to linearity and time-invariance of basin response (after Ishihara and Takasao [7])
      1. Linear time-invariant (see Cheng [8], Chow [9], Lee [10])
      2. Linear time-variant (see Stubberud [11], Chow [9])
      3. Nonlinear time invariant (see Amorocho and Orlob [12], Marshall [13])
      4. Nonlinear time variant
   C. According to parametric form (see Amorocho and Hart [4])
      1. Lumped-parametric form (see Cheng [8])
         (ordinary linear differential equations)
      2. Distributed-parameter
         (partial differential equations)
   D. According to openness of system (*all* the outflow emerges at lower end of catchment)
      1. Open: determined largely by subsurface geology of the catchment
      2. Closed: see Amorocho and Hart [3]
   E. According to geomorphic properties
   F. According to storage of energy
      1. Reactive (energy stored as in a reservoir)
      2. Nonreactive (dissipation of energy as by friction forces)

III. Output (runoff or evaporation)
   A. According to causality and chance as for input

Input → Watershed transformation system → Output

presents a review of such systems, e.g., the unit hydrograph and the instantaneous unit hydrograph, as well as possible conceptual models of linear, time-variable, and nonlinear systems in hydrology. Stubberud [11] discusses linear time-variant systems. Dawdy and Matalas [15] review some of the stochastic models in hydrology. Amorocho and Orlob [12], Minshall [16], Ishihara and Takasao [7], and Chow [9] report on the nonlinearity of most watersheds. Both Ishihara and Takasao [7] and Chow [17] assert that the hydrologic process in nature is nonstationary-stochastic and produces a non-pure-random sequence in most instances.

A further breakdown of potential hydrologic models is achieved by considering *nonlinear systems*. Nonlinear systems do not possess *one* or more of the properties of a linear system, namely, superposability and proportionality. Such systems cannot be accurately described by linear algebraic or linear differential equations. The signal processes (in terms of runoff *within* the basin) that occur between input and output are *nonlinear processes*. The *integrated basin response* at the gaging station is the end effect of many individual signal processes distributed within the basin. According to Marshall [13], in most

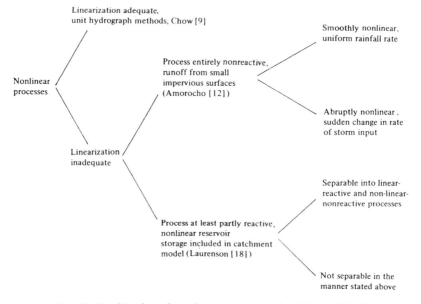

FIG. 4. Classification of nonlinear processes (after Marshall [13]).

cases nonlinear systems will not satisfy the superposition principle, will not exhibit proportionality between the amplitudes of input and output, and will generate, in the output, not only frequencies contained in the input but new frequencies related to these. An exception to the foregoing is the fact that some nonlinear systems are approximately in accordance with the proportionality principle. As seen in Fig. 4, a linear model may be an adequate approximation for some nonlinear processes. Otherwise, a process is classified according to the capability of the basin system to store energy as in natural or artificial reservoirs. In a *nonreactive process*, no appreciable energy can be stored in the system; all energy delivered to the system is dissipated by friction forces except that in the streamflow. The *reactive system* has at least one component capable of storing energy. From Fig. 4, we note that not all components of the basin transformation system need be nonlinear. Indeed, this notion is implicit in the runoff routing models proposed by Laurenson [18] and Rockwood [19].

We may identify three approaches to the construction of models of hydrologic phenomena. In the *first*, one may mold one's professional knowledge and conceptual interpretation of the phenomenon into a suitable framework (deterministic or probabilistic) by purely rational considerations involving the interplay of inductive and deductive reasoning. The resulting model, however, may be too complex and thus mathematically intractable. As a result, the desired model is forced into tractable forms of plausible *deterministic*[2] *models* (e.g., linear differential equations) or of *stochastic models* (e.g., the Poisson or the Markov process). From actual data of the physical process, one determines the required parameters of the model. Since the model is based to some degree on reality, it is generally the most accurate, the most adaptable to *comparable* situations, the most reliable when extrapolated beyond the observed ranges of parameters, and the most likely to give insights into the structure of the physical process. The latter advantage may result in an improved model. However, for probabilistic models, the procedure of model building and judging is not quite the same. According to Cramer [21]:

> In all cases when an element of *randomness* appears in the phenomena, the mathematical model will include some *probabilistic* concepts. Certain quantities will be regarded as random variables, the probability distributions of which must be more or less specified in the course of building the model. The process of verifying the accuracy of the model, and of using it for analysis, prediction, and decision-making, will then assume *statistical* character.

The second and more pragmatic approach to model building is to adopt a convenient mathematical model with little or no attempt to justify the choice on the basis of knowledge of the physical process. In its crudest form, the approach involves the fitting of empirical curves to observed data either by eye or by statistical methods to obtain the curve of best fit to the postulated

---

[2] The meaning here is in the narrow sense of the word, that is, determinism is synonymous with causation. However, Bunge [20] points out that causation is only one category of determinism.

deterministic model. When the goal is an empirical description of the phenomenon, the approach may prove quite satisfactory and valuable for decision-making. In addition to traditional methods of curve-fitting [22], multivariate statistical methods are useful if the underlying assumptions are not violated and if the predictor equations are not extended beyond the limits of the data. These include the following methods of analysis as reviewed by Wallis [23]: *regression, principal component, varimax, oblimax, key cluster,* and *object.* Their predictive ability in hydrology improves as more is known about the hydrologic system. Recently, Benson [24] has reviewed the dangers of *spurious correlation* in hydrology and hydraulics. Although fruitful inductions may be obtained from functional plotting and statistical methods of curve fitting, the subsequent models must be used with caution, since several mathematical models with entirely different implications may provide apparently equally good fits to observed data.

The third and final approach (so-called *second-order approach*) is of major value in the construction of mathematical models of *hydrologic time series.* Use is made of the concepts of generalized *harmonic analysis* [25–33]. Observed data are treated directly to produce efficient estimates of spectral representations (*variance spectra*) or of moment functions (*correlograms*) of the stochastic processes. When one lacks sufficient knowledge, either of the physical phenomena generating the observations or of the theory of stochastic processes necessary to model these phenomena, spectral analysis of the second moment functions provides clues as to the major sources of variation in the phenomenon. It enables us to detect trends, periodicities, and persistence linkages within a single time series or between multistation time series. Tukey [34] indicates that "bi-spectral analysis will enable us to learn much about *nonlinearities*, even when data are obtained at places and times rather remote from those where the nonlinear interaction took place." Frenkiel and Schwartzchild [35] were able to show from a large number of spectral analyses that there were two different sources for the fluctuations in atmospheric turbulence: a low-frequency component due to solar radiation, and a high-frequency component due to thermal instability of the atmosphere. This finding has been the basis for a more detailed physical theory. In oceanography, the spectral approach has been used to study the relationship between mean sea levels and atmospheric variables [36] and to detect low-frequency ocean waves from disturbances (storms) 10,000 miles away [37]. With respect to meteorology and climatology, Mitchell has reviewed the fruits of the second-order approach in seeking out periodicities and persistence in meteorological time series [38]. Yevjevich [39, 40], Matalas [41, 42], and Julian [43] have used the technique in hydrology. In the area of water pollution control, Wastler [44], the Public Health Service [45], and Gunnerson [46] report on the use of individual and cross-power spectral techniques to ascertain the influence of streamflow, tidal fluctuations, and solar radiation on the levels of biochemical oxygen demand (BOD), dissolved oxygen, and algae in an estuary. Tukey [34], in reviewing the history of the second-order approach, details the possibilities of higher-order moments (e.g., *cross-spectra, cross-bispectra, cesptrum, trispectrum*) in future data analysis. Tukey states: "We must realize that wise choice of what is to be analyzed is a natural part of trying *to let the data speak for themselves*—something which each of us owes to his data as part of careful and painstaking analysis." To re-emphasize, the end objective of these empirical spectral analyses is to improve on existing models and to suggest new models for more efficient prediction in hydrology.

More information on model building is given by Amorocho and Hart [4], Bunge [20], Cramer [21], Wallis [23], Granger [28], Whittle [47], Cornell [48], Snyder [49], Doob [50], and Kinsman [51].

Notwithstanding the divergent paths taken to formulate mathematical models, caution is necessary in their *ultimate* interpretation. In judging the model, the user should not take seriously its more delicate results as direct reflections of reality. For example, Doob [50] cites the original stochastic model of Brownian motion, the Wiener process, which implied that the paths of the individual particles are continuous, but their first and second time derivatives (velocities and accelerations) do not exist. A later refinement, the Ornstein-Uhlenbeck process, smoothed out the velocities but led to no particle accelerations. However, once these unnatural properties of these models were overlooked, their predictions proved valuable. Analogous considerations apply to hydrologic models.

Models are subject to two types of errors. In the first instance, the model is not sufficiently complex, or an insufficient number of variables have been incorporated. Some of the neglected variables may be contributing a large percentage of the systematic variation in the dependent variable. In the second instance, the model's parameters vary randomly or systematically. If random, their analysis should be according to statistical sampling theory. If systematic, better experimental design or more careful testing may result in more accurate forecasts. Furthermore, in view of the number of parameters which are incorporated into component hydrologic models, such as the Stanford watershed model [5], sensitivity of the solution or final design to the range of values of a parameter which may be reasonably encountered in nature becomes an important consideration. The same solution may be obtained for a variety of combinations of the parameters. As a consequence, *nonuniqueness* is an important property of conceptual models in hydrology; *sensitivity analysis* is an important activity in model analysis [52, 53].

Evaluation of models may be quantitative (objective), qualitative (subjective), or statistical. In the first instance, a simple comparison of prediction and reality suffices. Shapiro and Ward [54] indicate that many investigations of atmospheric models, both theoretical and experimental, where only a few degrees of freedom are allowed, show pronounced tendencies for periodic fluctuations when, in fact, these do not exist in the real atmosphere. When data are inadequate, evaluation of models is difficult and subjective, as for example with the nonlinear regression model for peak flow as a function of physiographic characteristics of the basin [55]. Statistical tests of significance are the basis of inferring the most appropriate models from nonsequential data. However, in general, the theory of statistical inference is not well developed as yet for testing spectra or correlograms against proposed models of the stochastic process [28, 33]. But, unlike the scientist who seeks to describe natural phenomena, the engineer has a more pragmatic test for a model inasmuch as decisions on the utility of a model are a consequence of its application. If the model results in near-optimum decisions (as judged by experience), it may be adequate for engineering purposes.

C. NATURE OF HYDROLOGIC TIME SERIES

1. Hydrologic Processes

A *process* is any phenomenon that shows a continuous change in time, in an area, in a volume, or along a line. Inasmuch as change in both space and time is an inherent characteristic of most hydrologic phenomena, we may speak of *hydrologic processes*. It is well known that rainfall and runoff vary temporally and spatially over a drainage basin, that infiltration rates and capacities vary similarly, that the erosional process is not the same throughout the basin, that the distribution of conservative and nonconservative diffusants (tracers or pollutants) in a fluid volume (atmosphere, stream, or earth) is not only nonuniform but also time-dependent, and that the width and cross-sectional area of a stream vary in time and with distance. In short, the hydrologic process occurs in a dynamic system and should be so treated when constructing mathematical models. The time-dependent hydrologic process is of most concern to hydrologists.

The hydrologic process $\{X(t): t \in T\}$ unfolds in time with a definite sequence of occurrence of events as, for example, with flood hydrographs and rainfall hyetographs. The resulting *time series* may be *discrete* or *continuous*. The continuous record of streamflows, over a period of years, is a continuous time series in a very real sense; whereas the continuous record artificially transformed into a sequence of mean daily flows, mean monthly flows, or mean annual flows is a discrete time series. The mean values of flow are assumed to occur instantaneously at discrete time. Events in nature which occur approximately at discrete time are low flows in a stream, instantaneous maximum streamflows (floods), periods of no flow into or out of a reservoir, and periods of no rain. Definition of the time scale is important when constructing mathematical models of the hydrologic process.

2. Rank and Sequence of Hydrologic Events

The events that constitute a hydrologic time series may be analyzed according to *rank* of the magnitude of the event and according to *sequence* of occurrence of the event. Ranking of hydrologic events ignores their order of occurrence. It forms the basis of the traditional methods of frequency analysis which have been adequately described by Chow [17]. These methods assume the pertinence of classic probability theory of independent events to hydrologic phenomena. The resulting flow-duration curves, empirical frequency graphs, and functions have been used to predict both the magnitude of streamflow or rainfall and the *number* of times, on the average, that a given flow or rainfall will be equalled or exceeded in the future [56,–58]. In contrast to ranking, the study of the sequence of occurrence presumes that past events in the time series may influence the magnitudes of present and future events in the same series, or in a second time series. Thus, possible *linkage*, or *dependence*, between successive (concurrent) values of the same time series (concurrent time series) is not ignored. For example, yesterday's streamflow is strongly related to today's streamflow; a storm pattern that *persists* for several days in a given geographic locale obviously results in rainfall-runoff patterns that are highly conditioned by past events; the sun *causes* the diurnal variation of air temperature, and of water temperature and concentration of dissolved oxygen in a river; the oceanic tide is responsible for the semidiurnal fluctuations of gage heights in a tidal estuary and, in part, for the fluctuating of ground water levels along the sea coast. In general, the linkage that might exist between successive values of a hydrologic time series has been ignored when predicting the magnitudes and the frequency of occurrence of given events by traditional methods.

3. Homogeneity in Time and Space

The hydrologic time series, at a point in space, represents a *sample* from a population of all possible values of the phenomenon. Each event in the series must have the same description. The series is *homogeneous in time* if each event has the same chance to occur at all times. However, this is not the case with a dependent time series, as illustrated in the previous paragraph. Nonhomogeneity in time, at a single point, may result from natural, quasi-natural, or human activity. Types of natural changes include trend, periodicity, persistence, and catastrophic events (earthquakes, landslides) as shown in Fig. 5. Quasi-natural changes pertain to shifting rating curves at a gaging station (due to erosion of the channel), to the effect of trees, buildings, and urban areas, on the rainfall patterns at a nearby rain gage, or to the effect of changing shade conditions on temperature measurements at a point. Man-made changes include the effect of dams on evaporation losses, streamflow,

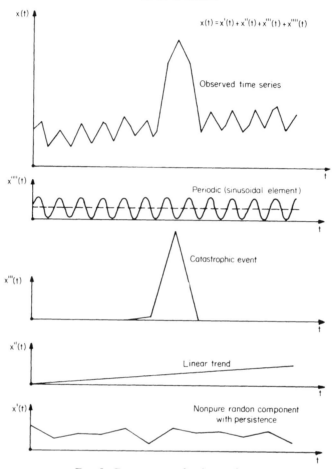

FIG. 5. Components of a time series.

and ground water levels, the effect of dessication of swamps and bogs on the local hydrology, the effect of watershed management and agricultural practices on evapo-transpiration, erosion, and runoff patterns, or the effect of air pollutants and concrete surfaces on the meteorology of a region. Lack of homogeneity indicates that samples of $\{X(t): t \in T\}$ are not chosen from the same population. Increasingly, human activity compromises the homogeneity of hydrologic (or climatological) data over large geographic areas such that these alterations are scarcely distinguishable from natural climatic changes. For a more detailed discussion of time homogeneity, see the literature [17, 59, 60].

Spatial homogeneity of hydrologic time series exists in the statistical sense when two or more time series possess statistical parameters that are not significantly different from each other, i.e., each time series has been drawn from the same population. Evidently, two rain gages located at two different elevations on the slope of a mountain are drawn from two different meteorological populations. On the other hand, two rain gages located only a few feet apart will record similar time series that are *space dependent*; both gages should be considered as one gage when calculating the mean rainfall over a basin. Therefore, in evaluating *statistical hydrologic homogeneity* (runoff and rainfall), spatially independent stations should be used.

The determination of *probable maximum precipitation* (PMP) by means of storm transposition and maximization techniques assumes meteorological homogeneity over a large geographic area with similar topographic features [61, 62]. In the regional analysis of the frequency of flood flows, as practiced by the U.S. Geological Survey [63, 64], statistical tests of significance are employed to establish hydrologically homogeneous areas. Other discussions of the spatial variation of hydrologic (meteorologic) variables are given by Panofsky and Brier [29], Holloway [30], Hely [65], Caffey [66], and the International Association of Scientific Hydrology [67].

4. Deterministic and Stochastic Processes

The variation within hydrologic processes may be *deterministic* or *stochastic*. In a *deterministic process*, a definite relation exists between the hydrologic variable and time, as in the equations for flood routing and the instantaneous unit hydrograph. The functional equation defines the process for the entire time of its existence; each successive observation of the phenomenon in time does *not* represent new information on the process. This is in contrast to a *stochastic process*, which evolves, entirely or in part, according to a random mechanism. The process may be *pure random* or *nonpure* random; the probability law from which the variate $x(t_i)$ is sampled at time t_i may be time-independent (the probability law does not change in time) or time-dependent. Figs. 6–10 give examples of these processes. If the stochastic process[3]

---

[3] This is the formal representation of a stochastic process in which the set T is the region of time over which the process is defined and the time parameter t *belongs* to T, the index set of all possible values. It is most important not to confuse this symbol with the random vector $X(t)$, which is a simple random variable with a distribution function, $F(x) = P[X(t) \leq N]$, as shaded in Figs. 6–8. In actual practice, a sample function (or some time history record of finite length, such as a 3-year sequence of rainfall or runoff at a single point) may be thought of as the observed result of a single experiment (3-year sample of rainfall data). The possible number of such experiments represents a sample space. The value x_i of the random variable $X(t_i)$ is the sampled value from the available population at time t_i. Figure 9 illustrates how the sampled value at time t_i differs as additional sample functions are taken. Figures 6–8 illustrate how the stochastic process $\{X(t): t \in T\}$ is a family or collection of random variables. The time parameter t may have a finite and discrete range: $t = 0, 2, \ldots, n$ an infinite and discrete range: $t = 0, 2, \ldots, \infty$; an infinite and discrete range: $t = 0, 1, 2, \ldots, \infty$ or an infinite and continuous range: $0 \leq t \leq \infty$.

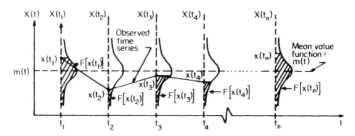

Fig. 6. Pure random process (stationary with time-independent probability law).

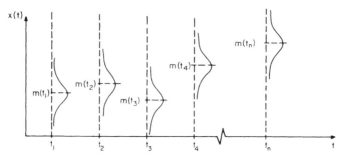

Fig. 7. Nonstationary stochastic process with variable mean value function (time-independent probability law).

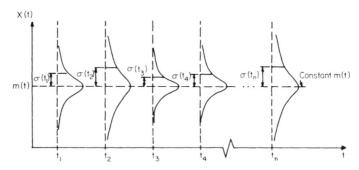

Fig. 8. Nonstationary stochastic process with time-varying variance (time-dependent probability law).

$\{X(t): t \in T\}$ is entirely random, the values $x(t_1)$, $x(t_2)$, ..., $x(t_n)$ from the populations $X(t_1)$, $X(t_2)$, ..., $X(t_n)$ in the time series are *internally independent* among themselves and, thus, constitute a *random sequence*. The displacement in time of a single particle undergoing Brownian motion in a fluid field is the classic example of a random time series. If the process is nonpure random, the values $x_1, x_2, \ldots, x_n$ are dependent and coupled among themselves. The resulting nonrandom time sequence does not fluctuate as rapidly

Time Series Analysis of Hydrologic Data

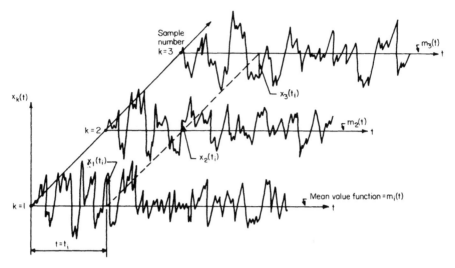

FIG. 9. Typical sample functions of a stationary stochastic process.

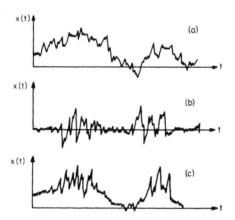

FIG. 10. Examples of nonstationary time traces: (a) time-varying mean value; (b) time-varying variance; (c) time-varying mean and variance.

as a random time sequence because of the regularity introduced by stochastic dependence (in a sense, deterministic behavior). From the preceding discussion, we note that a spectrum of different processes is possible in nature, ranging from a pure deterministic process to a pure random process.

Thus, we see that a stochastic process is more complex than its deterministic analog, the function, and its one-dimensional analog, the random variable. As previously defined, a stochastic process develops in time under the control of probability laws. The result is a time series called a *sample function* (see Fig. 9). In contrast to a deterministic function, the value x, at any time of a stochastic process, is not predictable with certainty. As an example, the partial sum

$$S_i = 1 + 2 + 3 + \cdots + t \tag{2}$$

is known for any time t, whereas the partial sum

$$S_n = x(t_1) + x(t_2) + \cdots + x(t_n) \tag{3}$$

can be described only in probabilistic terms at any time t_n, since the $X(t_i)$'s are independent, identically distributed random variables. The value of S_n has a stochastic dependence on the value of $S_1, S_2, \ldots,$ and S_{n-1}. The simple stochastic process of Eq. (3) is of interest in the determination of the reservoir storage-yield relation from an historical record [68–75] and in the distribution of precipitation or runoff *amounts* for periods of increasing length [76, 77].

In engineering hydrology, we are interested in the *dynamic response*, or streamflow $Y(t)$, of a basin to a *stochastic forcing function* (rainfall input) at several times, $X(t_1), X(t_2), \ldots, X(t_n)$. The probabilities of large peaks of input reinforcing or cancelling each other's effects on the basin response depends on the degree of *correlation* (*joint dependence*) between the input random variables $X(t_1), X(t_2), \ldots, X(t_n)$ which have a multidimensional *joint distribution function* $F[x(t_1), \ldots, x(t_n)]$. The basin response as well may be defined in terms of its own joint distribution function $F[y(t_1), \ldots, y(t_n)]$. A general description of the stochastic process for either the input or the output may be made by stating (for the input, e.g.) for all integers, n, the multidimensional *joint distribution function*, $F[x(t_1), \ldots, x(t_n)]$, of the n random variables $X(t_1), X(t_2), \ldots, X(t_n)$ for all times $t_1, t_2, \ldots, t_n$ in T. The joint distribution function for a stationary pure-random process where $x(t_1), x(t_2), \ldots, x(t_n)$ are independent of each other is given by

$$F[x(t_1), x(t_2), \ldots, x(t_n)] = F[x(t_1)]F[x(t_2)] \cdots F[x(t_n)]$$
$$F[y(t_1), y(t_2), \ldots, y(t_n)] = F[y(t_1)]F[y(t_2)] \cdots F[y(t_n)] \tag{4}$$

In practice, we may state the general model of the n-dimensional distribution, as, for example, the multivariate normal distribution, and specify the parameters necessary to define the distribution as functions of time. For a multivariate normal distribution, these parameters are the *mean value function* $m(t)$, the *variance* $\sigma^2(t)$, and the *covariance kernel*[4] $K(s, t)$ and are given, respectively, by

---

[4] In the context of the *same* time series, the covariance kernel is also called the *autocovariance function*. The autocovariance function is identical to the autocorrelation function $E[X(s)X(t)]$ when the expected values $E[X(s)]$ and $E[X(t)]$ are zero.

$$m(t) \equiv E[X(t)] \equiv \int_{-\infty}^{\infty} x(t)\, dF[X(t)] \tag{5}$$

$$\sigma^2(t) = \text{var}[X(t)] \tag{6}$$

$$K(s, t) \equiv \text{cov}[X(s), X(t)] \equiv E[X(s)X(t)] - E[X(s)]E[X(t)] \tag{7}$$

where $E[X(t)]$ is the *expected* value of $X(t)$ or the *average* over the *ensemble* of all possible sample functions (as shown in Fig. 9), cov[ ] signifies covariance, and $X(s)$ is the value of the process at an earlier time s. Alternate and more complete descriptions of a stochastic process are given by Parzen [78] and Prabhu [79]. More details on the multivariate normal distribution may be found in Johnson and Leone [80] or Wilks [81].

Frequently, the only information available on a stochastic process is its mean value function, variance, and covariance kernel. Although these completely specify the normal or "Gaussian" stochastic process, higher-order moments may be required for other processes. Nonetheless, the mean value function and the covariance kernel transmit much information on many processes. When the sample functions possess a "common" shape, the mean value function determines that shape. In this special case, *the mean value function may be equivalent to the deterministic model of the same phenomenon.* Such correspondence between the deterministic and stochastic models is a potential guide in the construction of stochastic models of similar physical phenomena.

The covariance kernel, as the variance of a random variable, is a second moment that depicts the "degree" or "strength" of correlation (or dependence) between the values of the process at any two times, s and t. A special case of Eq. (7) results when $s = t$, since the kernel reduces to the variance function of the process:

$$K(t, t) = E[X^2(t)] - E^2[X(t)] = \text{var}[X(t)] \tag{8}$$

Directly related to the product-moment correlation coefficient (of traditional statistics) is the *correlation kernel*, varying between plus one and minus one. Its definition, in terms of the covariance kernel, is

$$\rho(s, t) = \frac{\text{cov}[X(s), X(t)]}{\{\text{var}[X(s)]\,\text{var}[X(t)]\}^{1/2}} = \frac{K(s, t)}{[K(s, s)K(t, t)]^{1/2}} \tag{9}$$

The dimensionless correlation kernel is a measure of the linear predictability (stochastic dependence) of one value x_t of the process at time t given another value x_s of the process at time s. If $\rho(s, t)$ is zero, then the stochastic process $\{X(t): t \geq 0\}$ is composed of random variables $X(t_1), X(t_2), \ldots, X(t_n)$ that are *internally independent* in time, and the sequence is said to be *decoupled*. Recall that such a process has been called a *pure random sequence*. Internal independence is emphasized, since the same time series could still be *externally dependent* (*cross-correlated* with) on another time series—for example, runoff amounts or rates in relation to rainfall amounts (or rates), low flows in relation to ground water levels, runoff rates on main stream in relation to runoff rates on a nearby tributary. The notions of internal and external dependence are relevant to later discussion of the notions of autocorrelation and cross-correlation as used in generalized harmonic analysis of periodic, aperiodic (transient), and random functions [10].

In a *stationary* time series, the probability law is *invariant* to shifts of the time axis. The mean, variance, and the one-dimensional dustribution function $F(x)$ of the process are functionally independent of time, as shown in Fig. 6. An infinite number of *sample functions* may be obtained from the process, as shown in Fig. 9. If it were possible to sample the flows at a stream gaging station N times over the same 50-year period, N different time series (called *sample records for finite time series*) would be available for analysis. Each such sample record has an equally likely chance of occurring if the process is stationary. Of course, repetitious sampling of flows is not possible, but, if the flows in that stream remain time-homogeneous and stationary, the next 50-year record would constitute a sample record whose mean and variance are just as likely to occur as those in the previous 50-year record. It is this inherent statistical variability of the mean and variance which has been neglected in the traditional method of Rippl for estimating the reservoir storage-yield relation [70, 71, 73]. This same concept of statistical variability forms the basis of regional flood-frequency analyses as employed by the U.S. Geological Survey [63]. Stationarity is a key assumption in the empirical analysis and in the formulation of models of hydrologic time series.

A *strictly* stationary process exists formally if, for all integers n, for all times $t_1, t_2, \ldots, t_n$ and all time shifts of magnitude h, the random vectors $\{x(t_1), x(t_2), \ldots, x(t_n)\}$ and $\{x(t_1 + h), x(t_2 + h), \ldots, x(t_n + h)\}$ are distributed identically. This signifies $\{X(t, \zeta): t \in T, \zeta \in Z\}$ and $\{X(t + h, \zeta): (t + h) \in T, \zeta \in Z\}$ have the same distribution functions, so that

$$F_\zeta[x(t_1), \ldots, x(t_n)] = F_\zeta[x(t_1 + h), \ldots, x(t_n + h)] \qquad (10)$$

It is very difficult to verify stationarity in the strict sense. However, in practice, *covariance stationarity* or *stationarity in the wide sense* provides the desired simplicity of analysis. Under this simplification, the covariance kernel, $K(s, t)$, becomes independent of absolute time and a function of the absolute difference, $|t - s|$, only. Conceptually, the absolute difference, $|t - s|$, equals τ, a *lead* or lag time. The time series is shifted upon itself by an amount τ to ascertain the degree of memory of the past. Figure 11 illustrates the concept. By taking products of concurrent values of the shifted and unshifted time series, e.g., $x(t_i)x(s_i)$, one obtains the desired autocorrelation function.

Time Series Analysis of Hydrologic Data

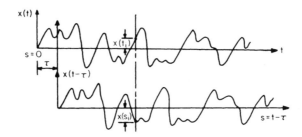

FIG. 11. Time shifting and autocorrelation.

It is generally agreed that most hydrologic time series are nonstationary, nonpure random sequences. Thus, the mean and variance tend to vary with time, the distribution function is time-dependent, and the values x_i of the time series tend to be coupled. Factors contributing to this complexity are human activity over the basin, catastrophic events, meteorological oscillations, and erosional processes. Kazmann [82] writes of *hydrologic-nonequilibrium* as the constant condition of a river basin. Hurst [83], Alexander [84], Leopold [85], Yevjevich [40], and Julian [43] present data on the nonpure random nature of streamflow sequences. The strength of this internal correlation depends on the *time interval* upon which the time series is built. Mean daily flows tend to have a high degree of internal dependence, mean monthly flows less so, and mean annual flows even less. As one smooths the higher frequency fluctuations by arithmetic averaging over longer periods of time, the *smoothed time series* exhibits less and less variability (variance) and covariability (covariance). Wet years tend to *cluster* with wet years, and dry years cluster with dry years, more so than in a sequence of random numbers. From a practical standpoint, greater reservoir capacity than predicted by the Rippl method is required because of this *cluster* effect. The foregoing considerations are quite important in building stochastic models in hydrology for use in water resources planning.

Hydrologic phenomena vary not only in time but also in space. If Z represents the set of points in space where rainfall or runoff may be observed and ζ_0 denotes a specific element of Z, then the stochastic process at ζ_0 is formally expressed as $\{X(t, \zeta_0): t \in T, \zeta_0 \in Z\}$. The objective of the time series analysis at the point ζ_0 is to infer the probability law of $\{X(t, \zeta): t \in T, \zeta \in Z\}$ for every point in the hydrologic or meteorologic regime Z. Matalas [86] cites as an example the occurrence of summer convective storms in the semi-arid regions of the Southwestern United States. These local storms have a high intensity and short duration, and occur more or less randomly over a wide area. The sequence of storm events in time at a point ζ_0 in this area represents a sample record from the stochastic process $\{X(t, \zeta_0): t \in T, \zeta_0 \in Z\}$. Each sample time series obtained over this region is just as likely as the next time series if environmental conditions do not change with time and space. Similarly, the historical record of streamflows at a gaging station is only one sample from the stochastic process defined previously. If it were possible to sample the same stream under identical environmental conditions for the same period in history, another sample time series is just as likely to have occurred. This time series would have statistical properties similar to the original if stationarity prevailed.

All possible sample functions of $\{X(t, \zeta_0): t \in T, \zeta_0 \in Z\}$ represent the ensemble of the process. However, in most hydrologic studies, only one member of an ensemble is available. In Fig. 9, it is observed that two or more types of averages may be obtained if all members (sample functions) of the ensemble were available. Each sample function has a mean value, the *time average*; in addition, at a specific time t_i the average of the $x_k(t_i)$ values for the kth sample over the entire ensemble is an *ensemble* average. The process is said to be *ergodic* if the time average for any member equals the corresponding ensemble average at any point in time. Thus, the ensemble average may be inferred from the time average if ergodicity were established. If the ensemble members correspond to points in space, then the space averages equal time averages. In hydrology this is really impossible, as the ensemble of sample functions is not available. Nonetheless, the notion of ergodicity is implicit in the construction of isohyetal or isopluvial maps wherein space averages are used in place of time averages. In the case of the summer convective storms in semi-arid regions, the probability of a storm at any point in the region is likely to be small. Hydrologists are quite familiar with rain gages that have never experienced the full impact of a storm, only its peripheral discharges. Lengthy records are required if meaningful time averages are to be realized. Pooling of time series at various gaging sites in the region would give an estimate of the space average if the assumption of ergodicity were valid. The concept is, in a sense, implicit in the *station-year method*. Furthermore, its applicability to semi-arid lands is credible since, according to Sellers [87], the region is climatologically homogeneous, at least to the extent that monthly precipitation departures from the long-period normals for both winter and summer are usually of the same sign throughout. On the other hand, the validity of the ergodic hypothesis in the study of the spatial variation of annual runoff and peak floods would be open to question in view of the many processes that transform rainfall to runoff. The concept is implicit, at least over limited geographic areas, in the statistical definition of regions of hydrologic homogeneity in order to predict flood frequencies for ungaged watersheds [63, 64], in the construction of areal maps of mean annual runoff [65], and in the extension of streamflow records by correlation methods [59, 88, 89]. In reality, hydrologic time series are neither stationary nor ergodic. Stationarity is a realistic assumption if the net effect of natural and human causes of environmental changes is small; ergodicity is not so physically realizable. Both concepts, however, are useful in hydrologic analysis and synthesis.

5. Causality and Chance (Determinism and Probability)

In general, solutions to problems of engineering hydrology, such as determination of peak flow, total hydrograph of runoff, or reservoir storage, have been based on deterministic models. The use of probability concepts in hydrology for model building is relatively recent. Many deterministic models have withstood the test of time, e.g., overland flow models, flood routing methods, and other equations of open channel flow; others have fallen by the wayside, e.g., empirical formulas for waterway area; still others, although unsatisfactory in many instances, continue to be used, e.g., the rational formula, storm maximization models, and the unit hydrograph methods. Any model, be it empirical or rational in origin, is merely an approximation of nature. Nature cannot be ordered into a theoretical straightjacket. Nonetheless, man continues to search for more perfect representations of the natural condition so as to minimize experimentation and to conserve resources. It is increasingly felt that stochastic models give a *realistic* conception of the rainfall-runoff process [90]. The following discussion is intended as an aid toward a better understanding of the stochastic approach to model building.

Papoulis [91], in distinguishing between determinism and probability, gives the following example:

Example 1. A projectile, thrown from the origin 0, has an initial velocity v_0 that forms an angle θ with the x axis. Its trajectory crosses the x axis at point B as in Fig. 12. We need the distance $0B$.

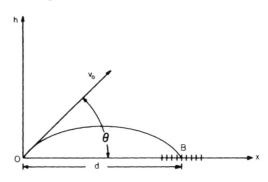

Fig. 12. Trajectory of projectiles.

From Newton's second law of motion, we can easily show that

$$d = (v_0^2/g) \sin 2\theta$$

At first glance, we presume that this model is a causal relation with no room for a probabilistic interpretation. However, upon recalling the assumptions underlying the model, namely, no air friction and no variation of g, the gravitational acceleration, and upon recognizing the *inexactness* in the measurement of the initial angle and velocity of the projectile, we see that the exact location of B is not determined by the equation. If we ignore all unknown parameters except the uncertainty in the angle θ, then the error in θ results in an error in the distance d, an example of error propagation [92]. Accordingly, if many projectiles with the "same" initial velocity (with certain errors) are ejected from the origin, the location of B varies about a mean value of d with respect to the origin. If the angle θ is a random variable, then the measured value of d fluctuates randomly about a mean value. In effect, the question being answered is not "What is the exact point of impact?," but "What is the probability that the object will impact at any given location on the horizontal axis?" Thus, the true value of the distance d is elusive, whereas the mean value d, based on *measurable* quantities, is consistent with the real world. Thus, the same phenomenon can be subjected to either a deterministic or probabilistic analysis.

Some may object to the foregoing interpretation. Papoulis continues: "The reader might argue that angle θ has *obviously a definite value*, even though we do not know it. If we knew θ, we should be in a position to predict *exactly* the impact point. The phenomenon is thus inherently deterministic, and probabilistic considerations are necessary only because of our ignorance. Our answer is that the physicist is not concerned with what *is* true, but only with what he can measure. Such explanations are therefore outside the sphere of his scientific interests."

In a similar vein, Kinsman [51] emphasizes that it is the role of the geophysicist (hydrologist) to tell a most *likely* story (not the true story) based on the recorded sense *appearances* of the phenomenon. In practical terms, the decision as to whether or not physical data are deterministic or random is usually based upon the ability to reproduce the data by controlled experiments. Laboratory experiments on various components of the hydrologic cycle facilitate the decision *only* with respect to the component model (saturated flow in the ground, infiltration, evaporation, open channel hydraulics). The decision becomes more difficult as the component models are synthesized in the form of a complete representation of the hydrologic process. Aside from errors of measurement (sense impressions) in the field or the laboratory, the predictions may vary as a result of major or minor discrepancies between the model and the prototype. The deterministic approach postulates that the major sources of variation are known *a priori*; whereas the stochastic approach accepts the inherent variability of the process about a mean value function, i.e., the deterministic component. Therefore, the deterministic process is a *degenerate* case of a stochastic process in which the error of forecasting is zero; hence in such a process a knowledge of the past is sufficient to predict the future.

Pertinent to an understanding of the interrelationship between causality, chance, and natural laws is the following excerpt from a philosophical discussion by Bunge [20]:

> Causation and chance are often regarded as polar concepts, and not without some reason as long as it is not forgotten that bulky masses of chance events may average out to nearly "causal lines" and that, *vice versa*, the interplay of numerous independent and individually determinate entities always results in chaotic situations.
>
> For example, the motion of a single particle in a smoke stream is like the almost random walk of a drunkard, yet the stream as a whole has a well-defined pattern, even when it degenerates into small vortex motions .... Random behavior at the microscopic or individual level may result in ordered behavior at a macroscopic or collective level, and orderly individual behavior may fit a random collective pattern. Moreover, chance has its peculiar laws. Thus, smoke particles will move away from a given point to an average distance proportional to the square root of the time elapsed....
>
> Perfect randomness is probably as much of an abstraction as perfect causation is, and yet neither is in our minds *alone*....
>
> *From complete randomness to strict causation there is a continuum of degrees of correlation.*[5]

In the deterministic approach, given certain key variables in the phenomenon and given certain values of these variables, we shall predict, and hopefully observe, a particular *state* or value of the process. However, in the stochastic approach, we neglect innumerable factors in the natural process and predict the likelihood with which each of the possible states will be observed. In determining the stochastic distribution of the possible *states* (effects), we consider *indirectly* and *collectively*, as a chance phenomenon, those *sources of variation* (causes) which are not known explicitly, which are not estimable, which are simply not known, and which are *indeed random*.

In either the deterministic or stochastic approach, the ultimate utility is in its ability to produce acceptable engineering predictions and decisions.

---

[5] Correlation in the context of this quote is that given by the dimensionless correlation kernel, Eq. (9).

Neither approach is capable of predicting the future stream of events with perfect certainty. Their pathways for achieving their goals are not identical. It is increasingly felt that the stochastic approach answers questions not heretofore answerable by traditional methods. According to Fiering [90], to write stochastic models off as poor substitutes for knowledge is to completely misunderstand the nature of the environment. Consider, for instance, the dynamic response of a basin system to a sequence of rainfall inputs. Deterministic methods are, in general, restricted to the calculation of the response history (streamflow hydrograph at a gaging station) to idealized storm inputs (uniform rainfall). The conceptual models of the instantaneous unit hydrograph assume an instantaneous input of rainfall over the entire basin (see Nash [93]). In contrast, the stochastic model of the input relation, given certain states of basin wetness, gives to the engineer the promise of making completely general statements concerning the probability of occurrence of *any* magnitude of maximum response of the catchment to *all* rainfall inputs. According to Cornell [48], a stochastic model, in general, is concerned not only with the *central tendency values* predicted by deterministic models, but also with the inherent and *unexplained variation* observed in physical phenomena. It evaluates this variation and thus gives a measure of the sensitivity of the idealized phenomenon to uncertainties. The stochastic approach through spectral analysis of variance may even indicate the primary sources of this variation, leading to their investigation and possibly to their better description and prediction either in the deterministic or stochastic frame of reference.

II. Conceptual Classification of Stochastic Processes

Essentially, three classes of stochastic processes encompass the specific processes of current interest in hydrology: counting process, transition type of process, and processes characterized by first and second moment functions (e.g., time series models).

The counting process is one that counts the occurrence of simple events of a specified type. Examples include the Poisson process (Parzen [78]), renewal processes (Cox [94]), and queueing models (Cox and Smith [95]). In formulating a stochastic analogue of the Stanford watershed model, Bagley [6] used a modified Poisson probability distribution of daily rainfall amounts to simulate input to a basin whose soil moisture reservoir (queue) was modeled according to queueing theory and Markov chains. Queueing theory was first applied to reservoirs by Moran [96] and later extended by Langbein [97], Fiering [98], and Lloyd [99]. Chow [9] has reviewed their findings.

Time Series Analysis of Hydrologic Data

The second class includes processes that develop in time as a series of transitions of a system from state to state. The process is specified by the probabilities of transition from one state to another and by the degree of dependence upon its past history. If, for a discrete time parameter process, the future of the process is independent of the past and present, then classical probability theory applies. If the future depends only on the present, then a first-order Markov process exists. If the future depends on some fixed, finite number of past states (values), then a higher-order Markov process exists. Applications to hydrology include synthesis of rainfall amounts by Pattison [100], prediction of the probability of sequences of wet and dry days by Weiss [101], probabilistic specification of different states of soil moisture content by Bagley [6], extension of rainfall records by Chow and Ramaseshan [102], and the augmentation of streamflow records by Julian [43], Brittan [103], Thomas and Fiering [104], Yagil [105], Beard [106], and Fiering [107, 108].

The third class of processes is based on empirical investigation of the first and second moment functions of actual time series. Time series analysis involves the estimation and reconstruction of the properties of the underlying generating process from the sample. The approach is similar to drawing a random sample from a population (e.g., an urn filled with numbered balls) and then estimating the properties of the population from the sample. The longer the historical time series, the better the estimates of its parameters, assuming stationarity. Let us assume that an observed time series

$$\{x(t); t = 1, 2, \ldots, N\}$$

is a single sample from a generating process $\{X(t); t > 0\}$. The latter simply indicates the manner in which the time series is formed for every moment of time; however, because of its stochastic nature, the process *cannot* predict the *actual* value of the series at any moment. Examples of simple generating processes for univariate stationary[6] time series are

$$X(t) = A \cos \omega_1 t + \xi(t) \qquad (11)$$

$$X(t) = \xi(t) + \alpha\xi(t-1) \qquad (12)$$

$$X(t) + \beta X(t-1) = \xi(t) \qquad (13)$$

$$X(t) = p(t) + q(t)\xi(t) \qquad (14)$$

where $\xi(t)$ is a random (noise), independent (uncorrelated) series, and $p(t)$ and $q(t)$ are polynomials in t. Equation (12) represents a one-step moving average process in which α is a weighting coefficient on past values. Equation

---

[6] Stationarity of the time series indicates that the generating mechanism (including chance elements) is of the same nature at each point in time.

(11) defines a linear cyclic process in which A is the amplitude and ω_1 is the fundamental angular frequency (radians per time unit). It could represent the seasonal variation of rainfall or runoff during a year, or the seasonal variation of temperature as a basis of describing the evaporative process in time. Equation (13) represents a *simple autoregressive process* (also called a *first-order Markov process* or *simple persistence*), wherein the present value $X(t)$ is determined by the previous value $X(t-1)$. More complex forms of Eqs. (11)–(13), respectively, are

$$X(t) = \sum_{i=1}^{m} A_i \sin(\omega_i t + \theta_i) + \xi(t) \tag{15}$$

$$X(t) = \sum_{i=0}^{m} \alpha_i \xi(t-i) \tag{16}$$

$$X(t) = \sum_{i=1}^{m} \beta_i X(t-i) + \xi(t) \tag{17}$$

$$X(t) + \sum_{i=1}^{m} \beta_i X(t-i) = \sum_{i=0}^{m} \alpha_i \xi(t-i) \tag{18}$$

where ω_i is the angular frequency of the cycle, θ_i the phase angle of the ith cycle, and α_i is a weighting factor in the *moving average process* defined by Eq. (16). Matalas [42] has applied model [Eq. (16)] to relate effective annual precipitation to annual runoff. The present value $X(t)$ of the autoregressive process [Eq. (17)] is formed from a linear sum of past values of the series and an independent noise term $\xi(t)$ that is unconnected with the past. The coefficients β_i of this process are a measure of the strength of association of the past value with the present value. These β_i's, related to the *serial correlation coefficients*, progressively get smaller as the time interval between past and present values increases. A time series generated by Eq. (17) will show oscillations or fluctuations of varying length; the average length depends on the β's. Equation (18), called a *linear regressive model*, is a mixture of terms from Eqs. (16) and (17). The spectral approach to be introduced in a later section allows us to discriminate between different generating mechanisms that combine to produce a univariate stationary time series. Properties of the models [Eqs. (11)–(18)] will be presented in a later section.

In the foregoing models of univariate time series, no trend components are included. *Trend* represents an increasing or decreasing tendency of a time series. Flows on the Colorado River at Lees Ferry exhibit a decreasing trend. Trend in a time series may show up as a changing mean or variance, as with the intensity of turbulent fluctuations at a point in the atmosphere over a diurnal cycle. A more subtle type of trend is a change in the importance of one component, such as a decrease in seasonal variation, or a change in the amount that the present value is correlated to the previous value of the series.

Time Series Analysis of Hydrologic Data

The following is a simple example of a nonstationary time series:

$$Y(t) = m(t) + X(t) \tag{19}$$

This model generates a time series only with trend in the mean $m(t)$. $X(t)$ is a stationary series, and $m(t)$ may be a linear relation, a polynomial, or a sequence of widely separated discontinuities. Granger [28] and Bendat and Piersol [109] present spectral methods for the study of nonstationarity in time series.

When two or more time series are related, then *multivariate* time series models are of interest. Hamon and Hannan [36], Hannan [27], Jones [110], and Fiering [108] have presented techniques for obtaining optimum linear predictor equations from sample data.

Hamon and Hannan employ *cross-spectral methods* (to be outlined in a later section), not only to use optimally the data for regression analysis but also to allow a penetrating examination of the data to check against departures from hypothesis in the linear regression models. Briefly, the method gives greatest weight to the angular frequencies where signal level $X(t)$ is highest relative to the noise level $\xi(t)$. These models are, respectively, *simple regression*, *multiple regression*, and *lagged regression*, which are given below:

$$Y(t) = \alpha + \beta X(t) + \xi(t) \tag{20}$$

$$Y(t) = \alpha + \sum_{i=1}^{m} \beta_i X(i, t) + \xi(t) \tag{21}$$

$$Y(t) = \sum_{k_1}^{k_2} \beta_i X(i, t) + \xi(t) \tag{22}$$

If the simple bivariate regression model [20] is valid, the population regression β should be the same in all frequency bands of the frequency-domain representation of the two time series. Furthermore, since no lagged values are employed in Eq. (20), $Y(t)$ and $X(t)$ should be either in phase (β positive) or in antiphase (β negative). The random component $\xi(t)$ should be stationary and independent of $X(t)$. Equation (21) is a general model for more than two time series. For example, the precipitation at two stations in a basin may be related in this manner to the runoff. Hamon and Hannan illustrate the method by relating mean sea level, atmospheric pressure, and two components of wind stress. Equation (22) introduces the effects of the past history of the X variates on $Y(t)$. However, the choice of the number of positive (k_2) and negative (k_1) lags is uncertain and arbitrary without *a priori* knowledge of the pertinent interrelationships. More involved schemes have been proposed by Jones [110]. Theoretical developments of the past few years clearly indicate the interrelationship of multivariate analysis to spectral analysis.

III. Generalized Harmonic Analysis

In this section, the theory of generalized harmonic analysis is summarized in increasing order of complexity for periodic, transient, and random functions. The theory is developed for linear, time-invariant systems. More details may be found elsewhere (Lee [10], Bendat and Piersol [109]), the latter being the most comprehensive presentation of the subject to date which is suitable for "beginners."

The theory is pertinent to later discussions of lag-time domain and frequency domain analysis of hydrologic time series, formulation of time series models and development of their theoretical correlation and spectral functions, frequency response characteristics of hydrologic measuring systems, properties of linear systems for single and multiple inputs, and optimum linear predictors (optimum realizable unit hydrographs). It illustrates the interrelationship between parametric and stochastic hydrology. The theory is relevant to the analysis and synthesis of short-term time series on the order of hours or days and to the study of long-term time series on the order of months and years.

It is presented in the context of continuous time series. The subsequent computational equations must consider the fact that hydrologic time series may be either discrete or continuous. Streamflow may be discrete or continuous, depending on whether the stream is perennial or intermittent. Precipitation is essentially discrete in its natural occurrence, depending on time scale. For either kind of phenomenon, the manner in which the data are measured and subsequently published determines the discreteness of the sequence. This aspect of hydrologic data analysis is important inasmuch as the original variance of the "true" continuous sequence is steadily reduced through smoothing by natural attenuation, instrument response characteristics, arithmetic averaging into average rates and volumes, and truncation effects of a finite time series.

The concepts are, likewise, pertinent to spatial distributions as in the study of stream meanders [111, 112], of ground roughness [113], and of the amplitudes of sand waves [114].

As emphasized previously, an important use of generalized harmonic analysis is as a data analysis tool. The spectral analysis of a data record is analogous to the decomposition of natural light into a rainbow (a spectrum) which results when *white* light is passed through a prism (see Fig. 13a). If the source of light is an artificial incandescent element, the spectrum consists of a finite number of bright lines of unequal intensity (energy), as shown by Fig. 13b. The extension of the foregoing concept to transient and random functions leads to the notion of a continuous spectrum of frequencies, rather than a discrete line spectrum. However, the property of the empirical or

Time Series Analysis of Hydrologic Data

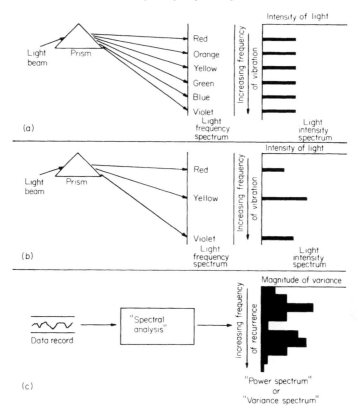

FIG. 13. Physical analogy to spectral analysis (from Wastler [44]): (a) prismatic resolution of a beam of sunlight; (b) prismatic resolution of a beam of light; (c) resolution of a data record by spectral analysis.

theoretical signal being analyzed is not its amplitude, but its variance [see Eq. (8)]. The variance is a measure of the dispersion of observations about a mean value; in a sense, it gives the average "intensity" of fluctuation of the phenomenon about a statistically steady mean value. In a single trace, the total variance is decomposed to find the frequency bands (periods) in which the variance contribution is statistically significant (see Fig. 13c). The resulting *variance*[7] (or *power*) *spectrum* is a sample spectrum that can be interpreted to better understand the physics of the generating process. Figure 14 shows typical spectra for a horizontal straight line, a trend line, and sine

---

[7] The term "power" is commonly seen in the literature, since the practical development of generalized harmonic analysis was initially undertaken by the communications engineer. The terms *variance* or *covariance* spectra are more descriptive. The units of the variance will depend on the units of the quantity measured.

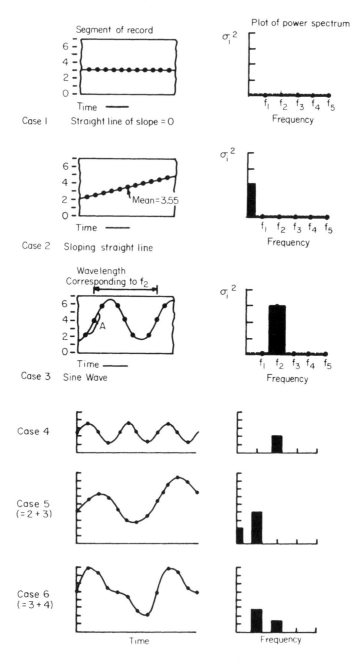

FIG. 14. Typical spectra obtained from several types of time series (adapted from Wastler [44], Gunnerson [46]).

waves. Note the concentration of variance in the zero frequency band for the nonstationary trend line and the concentration of variance about the discrete frequency of the sine wave. If the data record were not long enough, spectral analysis would not reveal whether the line A was part of a true trend line or part of a sine wave. Actually, the concept of a variance spectrum is designed to evaluate the nature of the nonrandomness in a data record in a manner that presupposes minimal *a priori* knowledge, if any, is available of the nonrandomness (stochastic linear dependence), trends, or even oscillations that might be present. Subsequently, generating processes are postulated on the basis of this circumstantial evidence, coupled with a reasonable understanding of the physics of the hydrologic system. These models are used to simulate the hydrologic process so as to facilitate prediction of mean-value and extreme-value conditions in hydrology [43, 71, 73, 100, 102, 104–108]. Hence, the analysis leads both to a better understanding of causality, indeterminism, and randomness in the system and to operational predictions for design of systems and sampling of the space-time continuum [46].

A. Periodic Time Series

Classical harmonic analysis involves the decomposition of oscillatory, periodic, almost periodic, and transient phenomena into an infinite number of sinusoidals with varying frequencies or periods. Its objective has been to seek out deterministic regularities and, in turn, to resynthesize the original phenomenon.

A summary of periodicities and oscillations of interest in hydrology is given in Table IV. The empirical or theoretical time series $x(t)$ may be expanded into a Fourier series according to the following formula:

$$x(t) = (a_{x0}/2) + \sum_{n=1}^{\infty} (a_{xn} \cos 2\pi n f_1 t + b_{xn} \sin 2\pi n f_1 t) \qquad (23)$$

in which $f_1 = T_1^{-1}$ is the fundamental *circular* frequency, inversely related to the fundamental period T_1 and to the fundamental *angular* frequency ω_1 by $f_1 = \omega_1/2\pi$, n is an integer representing the order of the harmonic, a_{xn} and b_{xn} are the harmonic coefficients, and $a_{x0}/2$ is the mean value of $x(t)$. The *harmonic coefficients* are defined, respectively, by

$$a_{xn} = (2/T_1) \int_0^{T_1} x(t) \cos 2\pi n f_1 t\, dt, \quad n = 0, 1, 2, \ldots \qquad (24)$$

$$b_{xn} = (2/T_1) \int_0^{T_1} x(t) \sin 2\pi n f_1(t)\, dt, \quad n = 1, 2, 3, \ldots \qquad (25)$$

Alternately, Eq. (23) may be expressed as

$$x(t) = (a_{x0}/2) + \sum_{n=1}^{\infty} c_{xn} \cos(2\pi n f_1 t - \theta_{xn}) \qquad (26)$$

TABLE IV

PERIODICITIES AND OSCILLATIONS OF INTEREST IN HYDROLOGY

1. Semidiurnal
   a. Tidal fluctuations along coastal beaches and in estuaries [44]
   b. Atmospheric tides in relation to precipitation variations [115]
2. Diurnal
   a. Solar radiation; air temperature
   b. Water temperature in rivers
   c. Concentration of dissolved oxygen in rivers and estuaries
   d. Evapotranspiration
   e. Water levels in wells (capillary pressure)
   f. Photosynthetic activity of algae in rivers, lakes, reservoirs, and estuaries
3. Semimonthly
   a. Semilunar variation of precipitation totals [116]
4. Annual
   a. Solar radiation; air temperature
   b. Precipitation totals
   c. Streamflow volumes
   d. Evapotranspiration
5. Biennial
   a. Mean zonal wind speed and temperature in the stratosphere with period ranging between 22 and 30 months [117]
6. Others
   a. Free oscillations of lakes [118]

where $c_{xn} = (a_{xn}^2 + b_{xn}^2)^{1/2}$, and θ_{xn} = phase angle = $\tan^{-1}(-b_{xn}/a_{xn})$. In words, Eq. (26) states that complex periodic data consist of a stationary mean value component, $a_{x0}/2$, and an infinite number of sinusoidal components (harmonics) that have amplitudes c_{xn} and phases θ_{xn} (see Fig. 15).

Pertinent to later generalizations is the following *time domain* representation in terms of complex exponentials as given by Lee [10]:

$$x(t) = \sum_{n=-\infty}^{\infty} F_x(n) \exp(jn\omega_1 t) \qquad (27)$$

where $j = \sqrt{-1}$, $F_x(n) = \frac{1}{2}(a_{xn} - jb_{xn})$ is the complex amplitude spectrum of the periodic function $x(t)$, and also the Fourier transform of $x(t)$:

$$F_x(n) = T_1^{-1} \int_{-T_1/2}^{T_1/2} x_1(t) \exp(-jn\omega_1 t) \, dt \qquad (28)$$

It is the *frequency domain* representation of $x(t)$ and results in a discrete line spectrum as in Fig. 15. Graphically, $F_x(n)$ may be shown in terms of a_n, b_n, and θ_n as in Fig. 16.

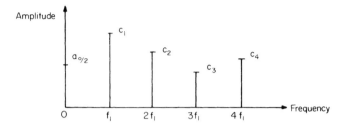

FIG. 15. Amplitude spectrum for complex periodic data.

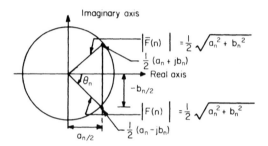

FIG. 16. Polar representation of complex amplitude spectrum $F(n)$.

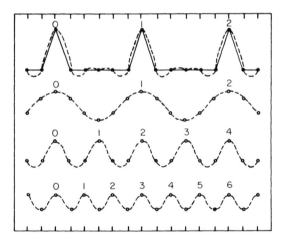

FIG. 17. A spike-shaped periodic function (solid line) with period equal to six data sampling intervals and its resolution into three harmonic components. Approximation to sum of harmonics is shown as a dashed curve superimposed on the function (adapted from Mitchell [38]).

The conjugate of $F_x(n)$ is obtained simply by exchanging $\exp(-jn\omega_1 t)$ with $\exp(jn\omega_1 t)$ in Eq. (28):

$$\bar{F}_x(n) = T_1^{-1} \int_{-T_1/2}^{T_1/2} x(t) \exp(jn\omega_1 t)\, dt \tag{29}$$

If the shape of the time series is that of a true periodicity (sine wave), then the entire fluctuation of the series would be contained in one harmonic (i.e., one sine and one cosine term whose period corresponds to the original function), and only one line at f_1 would be shown in Fig. 15. On the other hand, when the shape of the function is not sinusoidal, then at least two and at most $N/2$ harmonics are required to approximate the function (where N is the total number of data intervals in the period T_1). Figure 17 illustrates resynthesis of a spike-shaped periodic function with basic period of six data intervals by the sum of its three harmonic components. Note the closeness of agreement, particularly at the data points.

In connection with real hydrologic time series, the notions of frequency and period are an abstraction, and the representation of the series by a sum of trigonometric terms is a mathematical artifact. Consider the spike-shaped periodic function in Fig. 17 as it raises a fundamental question about the interpretation of the results of harmonic analysis when applied to real periodic functions in nature. Do the second and third harmonics in the illustration have any *physical* meaning? They certainly do not! Nonetheless, when properly used, the mathematical abstraction in terms of a sine wave language can be a powerful tool in data analysis even when the functions are not basically periodic, but transient, nonpure random, and random.

Consider two periodic time series, $x(t)$ and $y(t)$, each with identical fundamental period T_1. Displace $y(t)$ by a lag time $\tau = k\,\Delta t$, that is, by k discrete time intervals. The cross-correlation function $R_{xy}(\tau)$ is given by

$$R_{xy}(\tau) = T_1^{-1} \int_0^{T_1} x(t) y(t+\tau)\, dt = \sum_{n=-\infty}^{\infty} [\bar{F}_x(n) F_y(n)] \exp(jn 2\pi f_1 \tau) \tag{30}$$

in which $\bar{F}_x(n)$ is the conjugate of $F_x(n)$, and $F_y(n)$ is the complex amplitude spectrum of $y(t)$. The Fourier transform of $R_{xy}(\tau)$ is the cross-power spectrum function $S_{xy}(f)$:

$$S_{xy}(n) = \bar{F}_x(n) F_y(n) = T_1^{-1} \int_0^{T_1} R_{xy}(\tau) \exp(-jn 2\pi f_1 \tau)\, d\tau \tag{31}$$

The cross-correlation function in terms of the original harmonic coefficients in the Fourier series synthesis of $x(t)$ and $y(t)$ is

$$R_{xy}(\tau) = \frac{a_{x0} a_{y0}}{4} + \tfrac{1}{2} \sum_{n=1}^{\infty} c_{xn} c_{yn} \cos(n 2\pi f_1 \tau + \theta_{yn} - \theta_{xn}) \tag{32}$$

in which

$$c_{xn} = (a_{xn}^2 + b_{xn}^2)^{1/2}, \quad c_{yn} = (a_{yn}^2 + b_{yn}^2)^{1/2}$$

Similarly, the cross-spectral function is

$$S_{xy}(n) = \tfrac{1}{2} \sum_{n=1}^{\infty} c_{xn} c_{yn} [\cos(\theta_{yn} - \theta_{xn}) + j \sin(\theta_{yn} - \theta_{xn})] \quad (33)$$

If $x(t)$ is displaced in time rather than $y(t)$, then Eqs. (32) and (33) hold if subscripts x and y are interchanged. The relations between $R_{xy}(\tau)$ and $R_{yx}(\tau)$ and between $S_{xy}(n)$ and $S_{yx}(n)$ are

$$R_{xy}(\tau) = R_{yx}(-\tau) \quad (34)$$

$$S_{xy}(n) = \bar{S}_{yx}(n) \quad (35)$$

Consider the special case when $x(t) = y(t)$. The autocorrelation function, $R_{xx}(\tau)$, and the power spectrum function, $S_{xx}(n)$, are defined:

$$R_{xx}(\tau) = T_1^{-1} \int_0^{T_1} x(t)x(t+\tau)\,dt = \sum_{n=-\infty}^{\infty} [\bar{F}_x(n)F_x(n)] \exp(jn2\pi f_1 \tau) \quad (36)$$

$$S_{xx}(n) = \bar{F}_x(n)F_x(n) = T_1^{-1} \int_0^{T_1} R_{xx}(\tau) \exp(-jn2\pi f_1 \tau)\,d\tau \quad (37)$$

In terms of the harmonic coefficients, these functions are

$$R_{xx}(\tau) = (a_{x0}^2/4) + \tfrac{1}{2} \sum_{n=1}^{\infty} c_{xn}^2 \cos n 2\pi f_1 \tau \quad (38)$$

$$S_{xx}(n) = c_{xn}^2/2 \quad (39)$$

We note from Eq. (36) that the autocorrelation function is even

$$R_{xx}(\tau) = R_{xx}(-\tau) \quad (40)$$

and is a maximum at $\tau = 0$ (see Fig. 18); but the cross-correlation function $R_{xy}(\tau)$ is, in general, neither odd nor even

$$R_{xy}(\tau) = R_{yx}(-\tau) \quad (41)$$

and is not necessarily a maximum at $\tau = 0$. If $\tau = 0$, then $R_{xx}(0)$ is the mean square value of $x(t)$, i.e., $E\{[x(t)]^2\}$. Equation (7) may be written as

$$K_{xx}(\tau) = R_{xx}(\tau) - m_x^2 \quad (42)$$

In a similar way for two periodic functions, we define a cross-covariance kernel as

$$K_{xy}(\tau) = R_{xy}(\tau) - m_x m_y \quad (43)$$

If the values of $x(t)$ and $y(t)$ are standardized by subtracting m_x and m_y, respectively, from each value of $x(t)$ and $y(t)$, then $K_{\dot{x}\dot{x}}(\tau) = R_{\dot{x}\dot{x}}(\tau)$, and $K_{\dot{x}\dot{y}}(\tau) = R_{\dot{x}\dot{y}}(\tau)$ for the transformed series $\dot{x}(t)$ and $\dot{y}(t)$. When $\tau = 0$, we see that the covariance kernel $K_{\dot{x}\dot{x}}(0) = \text{var}[\dot{x}(t)]$. The variance of the transformed series is $\sum_{n=1}^{\infty} c_{\dot{x}n}^2/2$. From $S_{\dot{x}\dot{x}}(n) = c_{\dot{x}n}^2/2$, we see that the power spectrum

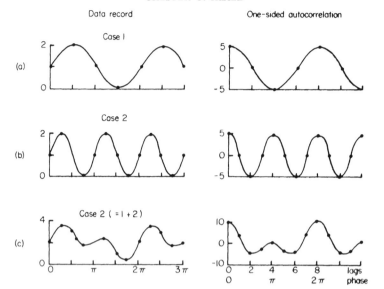

FIG. 18. Autocorrelation functions of simple sine waves (from Gunnerson [46]).

function gives the portion of the variance explained by the nth harmonic (see Fig. 14, case 6).

In Eq. (9), we defined a *correlation kernel* $\rho(\tau)$ for $\{X(t): t \in T\}$ as

$$\rho_{xx}(\tau) = \frac{K_{xx}(\tau)}{[K_{xx}(0)K_{xx}(0)]^{1/2}} = \frac{K_{xx}(\tau)}{\text{var}[x(t)]} \quad (44)$$

It is also called the *normalized cross-covariance function*. For standardized values,

$$\rho_{\dot{x}\dot{x}}(\tau) = \frac{R_{\dot{x}\dot{x}}(\tau)}{\text{var}[\dot{x}(t)]} \quad (45)$$

and is termed the autocorrelation coefficient. Note that $\rho_{xx}(0) = 1$. For two periodic functions, we define a *cross-correlation coefficient* $\rho_{xy}(\tau)$ as

$$\rho_{xy}(\tau) = \frac{K_{xy}(\tau)}{[K_{xx}(0)K_{yy}(0)]^{1/2}} = \frac{K_{xy}(\tau)}{\{\text{var}[x(t)]\,\text{var}[y(t)]\}^{1/2}} \quad (46)$$

which for standardized observations becomes

$$\rho_{\dot{x}\dot{y}}(\tau) = \frac{R_{\dot{x}\dot{y}}(\tau)}{[R_{\dot{x}\dot{x}}(0)R_{\dot{y}\dot{y}}(0)]^{1/2}} \quad (47)$$

that is, $m_{\dot{x}}$ and $m_{\dot{y}}$ are zero for the standardized sequences $\dot{x}(t)$ and $\dot{y}(t)$.

From Eq. (38), we see that the phase information is not inherent in the autocorrelation function. However, for two time series of a periodic nature, the respective phase information is preserved as seen in Eq. (32). In addition,

Time Series Analysis of Hydrologic Data

all periodic time series with the same amplitude and period, but different phase angles, have the same autocorrelation function that oscillates symmetrically and with period T_1 as a cosine function (see Fig. 18a). If, however, at least one sine wave of unequal period is combined with the original wave, the autocorrelation function of Fig. 18a has an oscillatory movement rather than the periodicity of Figs. 18a and 18b. It should be recognized that the autocorrelation function would have the same form if the original data, transformed data, or normalized data are employed. In the first instance, the amplitude of the graph is displaced upward by m_x^2 [see Eq. (42)]. $R_{xx}(\tau)$ for the original and transformed data are naturally dimensional, whereas $\rho_{xx}(\tau)$ fluctuates between plus and minus one and is dimensionless. A negative autocorrelation for a periodic function indicates that the displaced periodic function is out of phase with the original function.

B. Transient Time Series

The Fourier series representation of a periodic function over a sequence of discrete frequencies (harmonics) may be generalized to the continuous frequency domain by allowing T to approach infinity:

$$\lim_{\substack{T_1 \to \infty \\ nf_1 \to f}} nf_1 = n/T_1 = df \tag{48}$$

The *transient function*, therefore, is a periodic function with infinite periods. As a result, the transform pair relations Eqs. (27) and (28) become

$$x(t) = \int_{-\infty}^{\infty} F_x(f) \, e^{j2\pi ft} \, df \tag{49}$$

$$F_x(f) = \int_{-\infty}^{\infty} x(t) \, e^{-j2\pi ft} \, dt \tag{50}$$

$F(f)$ is a continuous function of the circular frequency and is, in general, complex. For Eqs. (49) and (50) to be valid, $\int_{-\infty}^{\infty} |x(t)| \, dt$ and $\int_{-\infty}^{\infty} |y(t)| \, dt$ must be finite.

For two transient time series (for example, flood waves), the cross-correlation function and cross-energy density spectrum are given as

$$R_{xy}(\tau) = \int_{-\infty}^{\infty} x(t) y(t + \tau) \, dt = \int_{-\infty}^{\infty} S_{xy}(f) \, e^{j2\pi f\tau} \, df \tag{51}$$

$$S_{xy}(f) = \int_{-\infty}^{\infty} R_{xy}(\tau) \, e^{-j2\pi f\tau} \, d\tau \tag{52}$$

If $x(t) = y(t)$, then Eqs. (51) and (52) define, respectively, the autocorrelation function and the energy density spectrum for a single time series. When $\tau = 0$, Eq. (51) for the single time series defines the total energy for the series.

Note that to take mean values of the integral in Eq. (51) would negate its existence, since the averaging interval would be infinite time. Except for the fact that both the autocorrelation and cross-correlation functions are transient, their symmetry properties are the same as those for periodic functions, that is, $R_{xx}(\tau) = R_{xx}(-\tau)$ and $R_{xy}(\tau) = R_{yx}(-\tau)$.

Short-term transients of interest to hydrologists include flood pulses, storms, and ground water response to surface inputs. Long-term transients are referred to as trends. However, these are difficult to detect, except on a circumstantial basis through spectral analysis, to any degree of certainty, as an apparent trend may be part of a low-frequency oscillation or a chance effect. Matalas [86] points out that tree ring growth in semi-arid regions, expressed as widths of annual rings, tends to vary about a monotonically decreasing nonlinear trend line, an inherent botanical property of growth. Variations about this line may be the result of variations in precipitation and temperature. Analysis of the standardized variations about the trend line *may* contain information of value to long-term water resources planning in marginal lands [119].

C. RANDOM TIME SERIES

As illustrated in Fig. 9, each realization of the random process $\{X(t): t \in T\}$ possesses an identical autocorrelation function over the ensemble:

$$R_x(\tau) = \lim_{T \to \infty} T^{-1} \int_0^T x(t)x(t+\tau)\,dt = \int_{-\infty}^{\infty} S_x(f)\,e^{j2\pi f \tau}\,df \qquad (53)$$

Its Fourier transform $S_x(f)$ is the two-sided power spectral density function:

$$S_x(f) = \int_{-\infty}^{\infty} R_x(\tau)\,e^{-j2\pi f \tau}\,d\tau \qquad (54)$$

As $R_x(\tau) = R_x(-\tau)$, Eq. (53) defines a real-valued even function, and Eqs. (53) and (54) may be written as a cosine transform pair:

$$R_x(\tau) = \int_{-\infty}^{\infty} S_x(f) \cos 2\pi f \tau\,df \qquad (55)$$

$$S_x(f) = \int_{-\infty}^{\infty} R_x(\tau) \cos 2\pi f \tau\,d\tau \qquad (56)$$

When $\tau = 0$, $R_x(0)$ is the mean square value; when the sequence is standardized, the new series has zero mean and $R_x(0) = \text{var}[\dot{x}(t)]$, which is the area under the spectral density function over its entire frequency range $(-\infty, \infty)$ as shown in Fig. 19. $S_x(f)$ has units of variance per unit circular frequency, whereas $S_x(f)\,df$ is the infinitesimal variance in the frequency band df. The

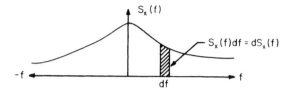

FIG. 19. Nature of spectral density function.

normalized spectral density function for a random function is analogous to the continuous probability density function, just as the power spectrum function for a periodic function is analogous to the *discrete* probability mass function. Note that the integral of the normalized spectral density function over its entire range must be one, as is the case with a probability distribution function.

The existence of negative frequencies is a mathematical artifact and not a physical reality. As a result, it is common to write a one-sided spectral density function $G_x(f) = 2S_x(f)$, where f varies only over $(0, \infty)$, and is otherwise zero and which is defined as

$$G_x(f) = \lim_{\Delta f \to 0} \lim_{T \to \infty} [(\Delta f)T]^{-1} \int_0^T x^2(t, f, \Delta f)\, dt \qquad (57)$$

In Eq. (57), an estimate of $G_x(f)$ depends on the frequency bandwidth Δf and sampling interval T. Some authors [109] express $G_x(f)$ as

$$G_x(f) = 2 \int_{-\infty}^{\infty} R_x(\tau) e^{-j2\pi f \tau}\, d\tau = 4 \int_0^{\infty} R_x(\tau) \cos 2\pi f \tau\, d\tau \qquad (58)$$

which differs by a factor of 2 with the computational algorithms put forth by others [25, 28].

In fact, it is possible for the "beginner" to become confused with notation and scale factors. Errors by factors of 2, π, $\pi^{1/2}$, and 4π are possible. Table V

TABLE V

COMMONLY USED TRANSFORM PAIRS[a]

$f_1(t) = \int_{-\infty}^{\infty} F_1(f) e^{j\omega t}\, df$	$F_1(f) = \int_{-\infty}^{\infty} f_1(t) e^{-j\omega t}\, dt$
$f_2(t) = \frac{1}{2} \int_{-\infty}^{\infty} F_2(f) e^{j\omega t}\, df$	$F_2(f) = 2 \int_{-\infty}^{\infty} f_2(t) e^{-j\omega t}\, dt$
$f_3(t) = \int_{-\infty}^{\infty} F_3(\omega) e^{j\omega t}\, d\omega$	$F_3(\omega) = (2\pi)^{-1} \int_{-\infty}^{\infty} f_3(t) e^{-j\omega t}\, dt$
$f_4(t) = (2\pi)^{-1/2} \int_{-\infty}^{\infty} F_4(\omega) e^{j\omega t}\, d\omega$	$F_4(\omega) = (2\pi)^{-1/2} \int_{-\infty}^{\infty} f_4(t) e^{-j\omega t}\, dt$
$f_5(t) = (2\pi)^{-1} \int_{-\infty}^{\infty} F_5(\omega) e^{j\omega t}\, d\omega$	$F_5(\omega) = \int_{-\infty}^{\infty} f_5(t) e^{-j\omega t}\, dt$
$f_6(t) = \frac{1}{2} \int_{-\infty}^{\infty} F_6(\omega) e^{j\omega t}\, d\omega$	$F_6(\omega) = \pi^{-1} \int_{-\infty}^{\infty} f_6(t) e^{-j\omega t}\, dt$

[a] Adapted from Streets [120].

as given by Streets [120] presents six different forms that are used for transform pairs. These are used randomly in three different situations: (a) system weighting functions, transfer functions; (b) aperiodic signals, energy spectral densities; (c) correlation functions, power spectral densities. In addition, four different forms are used for spectral densities (see Korn and Korn [121], Eqs. (10), (17), (18)). These are related by

$$\Gamma(f) = 2\gamma(f) = 2\pi G(\omega) = 4\pi g(\omega) \tag{59}$$

as seen in Table VI, which presents the notation used in 16 reference sources on the subject. Factors of 2 and 2π are added so that all notations in a column are equivalent. The relationship between the columns is given by the scale factor.

Figure 20 summarizes the autocorrelation and spectral density functions of current interest in theoretical studies of hydrologic time series. Both pure deterministic and stochastic processes possess theoretical autocorrelation and spectral density functions. Cases (a) and (b) are examples of deterministic processes. The Dirac delta function $\delta(f)$ represents the unit spikes shown in cases (a) and (b); its properties are defined as follows:

$$\delta(f) = 0 \quad \text{for} \quad f \neq 0$$

$$\delta(0) = \infty$$

$$\int_{-\varepsilon}^{\varepsilon} \delta(f)\,df = 1 \quad \text{for any} \quad \varepsilon > 0 \tag{60}$$

$$\delta(-f) = \delta(f)$$

$$\int_{-\varepsilon}^{\varepsilon} G(f)\,\delta(f)\,df = G(0) \quad \text{for any} \quad G(f)$$

Its use facilitates mathematical analysis as, for example, in case (c) for determining $G_x(f)$ of fictitious white noise that has infinite power (variance). The autocorrelation function for *white noise* (uncorrelated random variables) is $R_x(\tau) = a\delta(\tau)$, which, according to the definitions of Eq. (60), indicates that the white noise is perfectly correlated at zero lag but is zero elsewhere. Case (d) shows that the autocorrelation function and power spectrum for the sine wave plus random noise are simply the sum of these respective functions for the sine wave and random noise. Case (e) shows the effect of filtering off the high-frequency components of white noise, whereas cases (f), (g), (h) are more representative of outputs of hydrologic measuring instruments. Natural attenuation and instrument smoothing of the original signal (for example, rainfall) tend to remove the high-frequency components, whereas a finite sampling interval tends to cut off the slower moving elements of the generating process. The remaining cases will be discussed in a later section on properties of generating processes.

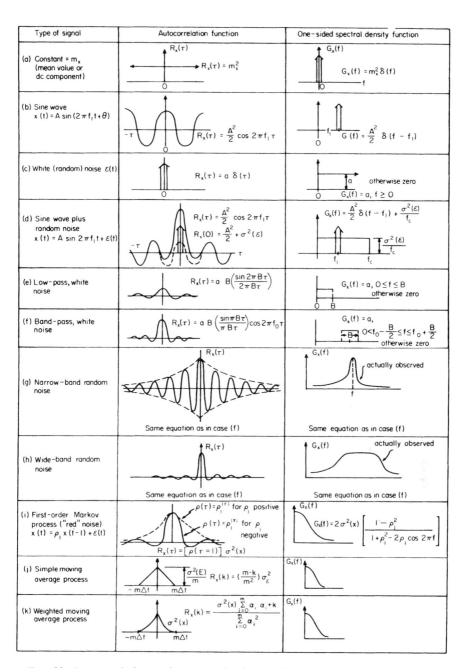

FIG. 20. Autocorrelation and spectral density functions of time series pertinent to hydrology (adapted in part from Bendat and Piersol [109], pp. 86, 87).

TABLE VI

THE CHAOS IN NOTATION FOR SIGNAL SPECTRA[a]

| | | Power density (random) Power (periodic) | | | | Energy density (aperiodic) |
| | | Two-sided | | One-sided | | Two-sided |
	Ave.[b]	ω	f	ω	f	ω
Scale factor	Ave.[b]	$(2\pi)^{-1}$	1	π^{-1}	2	$(2\pi)^{-1}$
Streets [120]	T	$\Phi(s)/2\pi$	—	—	—	$\Phi(s)/2\pi$
Korn and Korn [121]	T	$g(\omega)$	$\gamma(v)$	$G(\omega)$	$\Gamma(v)$	—
	E	$g(\omega)$	—	$G(\omega)$	—	—
Lee [10]	T, E	$\Phi(\omega)$	—	—	—	—
Newton et al. [122]	T, E	$\Phi(s)$	—	—	—	$I(s)/2\pi$
Middleton [123]	T	—	—	—	$W^{(J)}(f)$	$|W^{(J)}(f)|^2/2\pi$
	E	—	—	—	$W(f)$	—
Chang [124]	T, E	$\Phi(s)/2\pi$	—	—	—	$A(s)A(-s)/2\pi$
Bendat [125]	E	$G(\omega)/2$	$G(f)/2$	—	$G(f)$	—
Davenport and Root [14]	T	—	$G(f)$	—	—	—
	E	—	$S(f)$	—	—	—
Laning and Battin [126]	E	—	—	$G(\omega)$	—	$E(\omega, x)/2$
Solodovnikov [127]	T, E	$S(\omega)/2\pi$	—	—	—	—
Smith [128]	T	$\Phi(\lambda)/2\pi$	—	—	—	—
Blackman and Tukey [25]	T	—	$P(f)$	—	—	—
Aseltine [129], Truxal [130]	T	$\Phi(j\omega)/2\pi$	—	—	—	—
Horowitz [131]	E	—	—	—	—	$EM(s)N(-s)/2\pi$
Bendat and Piersol [109]	E	—	$G(f)/2$	—	$G(f)$	—

[a] Adapted from Streets [120].
[b] T = time, E = ensemble.

Time Series Analysis of Hydrologic Data

The main application for autocorrelation functions is to detect the effect of values of a sample record at any time on values of the same record at a future time. It is seen from Fig. 20d that, at large time displacements, the deterministic sine wave persists, whereas the effect of white noise is zero. On the other hand, the principal application of the power spectral density function is to establish the frequency composition of the data which, in turn, is related to the basic properties of the physical system. For example, let $h(t)$ represent the unit impulse response function of a linear time-invariant system (such as a linear reservoir) to a unit impulse (such as 1 in. of rainfall excess deposited instantaneously over a watershed) as shown in Fig. 21. The

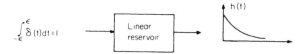

FIG. 21. Unit impulse response of a linear reservoir.

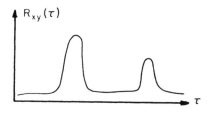

FIG. 22. Cross-correlogram.

Fourier transform of $h(t)$ is the system frequency response function $H(f)$ given as

$$H(f) = \int_{-\infty}^{\infty} h(t)e^{-j2\pi ft}\, dt \qquad (61)$$

and its inverse transform is

$$h(t) = \int_{-\infty}^{\infty} H(f)e^{j2\pi ft}\, df \qquad (62)$$

Note that $h(t)$ is the *instantaneous unit hydrograph*, that is, a unit hydrograph for a unit rainfall excess with zero duration. The frequency response functions of rain gages and stream gaging systems have been investigated by Eagleson and Shack [132]. The foregoing functions will be discussed further in a later section on responses of linear systems to single and multiple inputs.

The cross-correlation function $R_{xy}(\tau)$ of two sets of random data defines the linear dependence of the values of one set of data on the other. It is given by

$$R_{xy}(\tau) = \lim_{T \to \infty} T^{-1} \int_0^T x(t)y(t+\tau)\,dt \tag{63}$$

and is always a real-valued function that may be positive or negative. Its maximum is not necessarily at $\tau = 0$, and it is not an even function as is $R_{xx}(\tau)$, that is

$$R_{xy}(-\tau) = R_{yx}(\tau) \tag{64}$$

in which $R_{yx}(\tau)$ signifies the cross-correlation function of $y(t)$ with $x(t+\tau)$.

The Fourier transform of $R_{xy}(\tau)$ is the physically realizable one-sided cross-power spectral density function $G_{xy}(f)$:

$$G_{xy}(f) = 2S_{xy}(f), \quad 0 \leq f < \infty, \text{ otherwise zero} \tag{65}$$

$$G_{xy}(f) = 2 \int_{-\infty}^{\infty} R_{xy}(\tau) e^{-j2\pi f \tau}\,d\tau \tag{66}$$

Therefore,

$$R_{xy}(\tau) = \int_0^{\infty} G_{xy}(f) e^{j2\pi f \tau}\,df \tag{67}$$

Cross-power spectra permit the evaluation of the degree of correlation between low-frequency and high-frequency components of two time series.

Because a cross-correlation function is not an even function, $G_{xy}(f)$ is generally a complex number such that

$$G_{xy}(f) = C_{xy}(f) - jQ_{xy}(f) \tag{68}$$

where the real part, $C_{xy}(f)$, is called the *cospectral density function*, and the imaginary part, $Q_{xy}(f)$, is called the *quadrature spectral density function*.

In the frequency context, $C_{xy}(f)$ is the average product of $x(t)$ and $y(t)$ within a narrow frequency interval between f and $f + df$, divided by the frequency interval. This operation amounts to a measurement of the contribution of oscillations of different frequencies to the total cross-correlation at lag $\tau = 0$ between two time series. It is computed by averaging the cross-correlation at lag (τ) and at lag $(-\tau)$. The quadrature spectral density is identical except that either $x(t)$ or $y(t)$, not both, is shifted in time to give a 90° phase shift at frequency f. Both are defined as

$$C_{xy}(f) = \lim_{\Delta f \to 0} \lim_{T \to \infty} (\Delta f T)^{-1} \int_0^T x(t, f, \Delta f) y(t, f, \Delta f)\,dt \tag{69}$$

$$Q_{xy}(f) = \lim_{\Delta f \to 0} \lim_{T \to \infty} (\Delta f T)^{-1} \int_0^T x(t, f, \Delta f) y^{\circ}(t, f, \Delta f)\,dt \tag{70}$$

where $x(t, f, \Delta f)$ and $y(t, f, \Delta f)$ are the filtered portions of $x(t)$ and $y(t)$, respectively, and $y°(t, f, \Delta f)$ is the 90° phase shift from $y(t, f, \Delta f)$.

In complex polar notation, the cross-spectral density function is

$$G_{xy}(f) = |G_{xy}(f)| \exp[-j\theta_{xy}(f)] \tag{71}$$

where the modulus

$$|G_{xy}(f)| = [C_{xy}^2(f) + Q_{xy}^2(f)]^{1/2} \tag{72}$$

and the phase spectrum

$$\theta_{xy}(f) = \tan^{-1}\left[\frac{Q_{xy}(f)}{C_{xy}(f)}\right] \tag{73}$$

The phase spectrum is a measure of the relative phase of the harmonics of $x(t)$ and $y(t)$. In terms of $C_{xy}(f)$ and $Q_{xy}(f)$, the cross-correlation function

$$R_{xy}(\tau) = \int_0^\infty [C_{xy}(f) \cos 2\pi f\tau + Q_{xy}(f) \sin 2\pi f\tau] \, df \tag{74}$$

Furthermore, from Eqs. (64) and (68), one obtains

$$C_{xy}(f) = \int_0^\infty [R_{xy}(\tau) + R_{yx}(\tau)] \cos 2\pi f\tau \, d\tau = C_{xy}(-f) \tag{75}$$

$$Q_{xy}(f) = \int_0^\infty [R_{xy}(\tau) - R_{yx}(\tau)] \sin 2\pi f\tau \, d\tau = -Q_{xy}(-f) \tag{76}$$

Therefore, $C_{xy}(f)$ is a real-valued even function of f, whereas Q_{xy} is a real-valued odd function of f. In terms of $G_{xy}(f)$ and $G_{yx}(f)$

$$\begin{aligned} C_{xy}(f) &= \tfrac{1}{2}[G_{xy}(f) + G_{yx}(f)] \\ Q_{xy}(f) &= (j/2)[G_{xy}(f) - G_{yx}(f)] \end{aligned} \tag{77}$$

A rather important real-valued quantity, $\gamma_{xy}^2(f)$, the *coherence function*, is defined as

$$\gamma_{xy}^2(f) = \frac{|G_{xy}(f)|^2}{G_x(f)G_y(f)}, \qquad 0 \leq \gamma_{xy}^2(f) \leq 1 \tag{78}$$

It is analogous to the correlation coefficient squared and defines a normalized measure of coherence. It defines a linear measure of association as a function of frequency. There is some confusion in the literature on terminology concerning the coherence function [133]; for example, $\gamma_{xy}^2(f)$ is called the *squared coherence* by some. If $\gamma_{xy}^2(f) = 0$ at a particular frequency, $x(t)$ and $y(t)$ are *incoherent* at that frequency, that is, uncorrelated. If $x(t)$ and $y(t)$ are statistically independent, then $\gamma_{xy}^2(f) = 0$ at all frequencies. When $\gamma_{xy}^2(f) = 1$ for all f, then $x(t)$ and $y(t)$ are fully coherent, that is, completely correlated.

Wiener [134] in 1930 proposed the first adequate mathematical model for studying the phenomenon of coherence as first defined in optics. Amos and Koopmans [135] present the following instructive analogy on coherence:

> Coherence, literally the quality of sticking together or being in accord, was first used in the theory of optics to describe a property of a pair of time series. An experiment familiar to students of elementary physics demonstrates the qualitative aspect of coherence. One observes the interference of two rays of light from the same source when the rays travel paths of slightly different lengths. By varying the path lengths, one can superimpose the rays to produce darkness or a light more intense than that usually formed by the sum of two rays of their given intensities. Here the rays are called coherent. If one uses rays from different sources, no interference is observed. Such rays are called incoherent.
>
> The physicist's explanation of coherence is that the interference pattern produced by two sources of light depends on their relative displacement in phase. Since phase is a frequency-dependent concept, we may think of decomposing the light rays into their spectral components (say, by means of a prism) and isolating a nearly monochromatic beam of the same color from each. The resulting light beams theoretically would be composed of trains of sinusoidal oscillations each lasting only a minute fraction of a second. In light rays from different sources, the phase of one beam relative to that of the other would change rapidly in time in such a way that the phase difference would vary uniformly over its possible range of values in even very short intervals of time. Since the eye averages the light it receives over a small time interval, the instantaneous interference patterns established by the two beams of light would be blended into the uniform intensity that is characteristic of incoherent light rays. In light rays from the same source, although the phases of the monochromatic beams change rapidly in time, the relative phase difference remains constant and the interference patterns of perfectly coherent light are observed. Between these extremes the distribution of relative phase difference may range from uniform to concentrated at a single value thus generating a continuum of degrees of coherence ....

Measurements of cross-correlation functions as plotted on cross-correlograms (see Fig. 22) may be applied to measure time delays, to determine transmission paths, and to detect and recover signals in noise. In Fig. 22, the sharp peaks are evidence of a strong correlation between $x(t)$ and $y(t)$ for specific time displacements. The signal $x(t)$ may be a streamflow record on one of several tributaries into a natural or artificial reservoir. The time trace of the water levels $y(t)$ in the reservoir may be cross-correlated with

Time Series Analysis of Hydrologic Data

$x(t)$, as well as with the other tributaries, to determine travel times of various inflows and to detect the transmission path of the main effect on the reservoir. The cross-correlation function peaks at those time displacements of $x(t)$ or $y(t)$ relative to $y(t)$ or $x(t)$, respectively, required for the flood wave to pass through the linear river system. A separate peak will occur for each tributary that contributes significantly to the output. The peaks, as the average product of two linearly related pulses, will always be a maximum when the time displacement (or phase shift in the frequency domain) between the pulses is zero. The measurement may be fruitful in the study of the interrelationship between the input pulses to the reservoir and its outflow, between streamflow and ground water levels, between evaporation rates and atmospheric conditions, between a seismic disturbance and the oscillation of lakes, between lake levels at different locations [118], and between wind stresses and mean sea level [36]. The third application of cross-correlograms, to detect and recover deterministic signals buried in noise, is employed when the autocorrelogram does not uncover the desired signal. The procedure entails the cross-correlation of the input consisting of the signal plus noise with a stored version of the signals. The stored signal may be a base flow hydrograph derived from the drainage equations, known flood pulse, or the known input signal to a hydrologic instrument. The cross-correlation will produce a greater signal-to-noise ratio than will the autocorrelation function because the stored version of the postulated signal is free from noise, assuming that it represents a reasonable model of the natural phenomenon. The utility of cross-correlation measurements in hydrologic practice depends on availability, accuracy, and length of appropriate sample records.

Applications of cross-spectral density functions are similar to those for cross-correlation functions. This is to be expected, as each function is a transform of the other. The domain to be used for analysis is not entirely a matter of personal preference because of statistical inference problems associated with correlograms in contrast to power spectra and because the transmission velocity and/or path through the system may be frequency-sensitive, so that the cross-correlogram may not detect a distinct peak. The cross-spectral density functions may be used to measure frequency response functions, to measure time delays, and to obtain optimum linear predictors and filters. The time delay through the system at any frequency f is

$$\tau = \theta_{xy}(f)/2\pi f \qquad (79)$$

which permits the determination of the time delays as a function of frequency (not available otherwise in the lag-time domain). The determination of an optimum linear filter according to some established criterion such as the minimum mean square error, $R_e(0)$, allows one to transmit and predict desired signal information (as an instantaneous unit hydrograph) while rejecting undesired noise information. Consider the situation where one desires to estimate the runoff hydrograph $y(t)$ by a linear operation on the rainfall excess $x(t)$. Assume that the measurement of $y(t)$ is garbled by external noise $\xi(t)$ that is independent of $x(t)$. According to Bendat and Piersol [109], the optimum system frequency response function $H_{y/x}(f)$ according to a mean square error criterion has the form

$$H_{y/x}(f) = G_{xy}(f)/G_{xx}(f) \qquad (80)$$

in which y/x indicates that $y(t)$ is determined, given $x(t)$. Recall that $H(f)$ is the transform of $h(t)$, the instantaneous unit hydrograph or the optimmu weighting function. In the expression for the minimum mean square error [109, 135] for the optimum system

$$R_e(0) = \int_0^\infty G_y(f)[1 - \gamma_{xy}^2(f)]\, df \qquad (81)$$

it is seen that for $\gamma_{xy}^2(f) = 1$ a linear operation on $x(t)$ will estimate $y(t)$ with zero mean square error. The preceding result indicates in practice the goodness of the approximation between an actual system and an optimum system. The foregoing will be discussed in a later section on the Wiener-Hopf theorem as applied to linear systems in hydrology.

IV. Generating Processes

In addition to seeking out a clearer understanding of the physics of the hydrologic process, the aforesaid concepts are employed to postulate underlying generating mechanisms that may have produced the sample time series (or space series). Complete description, however, is not possible because of the complexities of the system. As a first approximation, proposed schemes assume a linear system, stationarity, and addition of a random component $\xi(t)$ that is independent of the main components [see Eqs. (11) and (15)]. Time series sequences generated by these models exhibit more regularity than do pure-random sequences. In fact, their general appearance is that of a disturbed periodic function in view of the active role played by the random component $\xi(t)$.

Properties of the univariate models given by Eqs. (11)–(13) and (15)–(17) will be emphasized in this section. Of the three schemes, the *linear cyclic process* (or *scheme of hidden periodicities*) has the most regularity, whereas the others, *moving average* [Eq. (16)] and *linear autoregressive process* [Eq. (17)], have weaker linkage with the past. According to Kendall [136], when Eq. (17) is solved recursively for successive values of the time series, $x(t)$ is found to be an infinite weighted sum of random terms. Therefore, the autoregression process is really a linear moving average process wherein the moving average is of infinite length ($m = \infty$).

Time Series Analysis of Hydrologic Data

If $\{X(t): t \in T\}$ is generated by a *linear cyclic process*, then

$$E\{X(t)\} = E\{A \cos 2\pi f_1 t + \xi(t)\} = 0 \tag{82}$$

on the assumption that data are standardized (mean value effect removed). The remaining properties are

$$\text{var}\{A \cos 2\pi f_1 t + \xi_t\} = (A^2/2) + \text{var}\{\xi(t)\} = R_x(0) \tag{83}$$

$$R_x(\tau) = E\{X(t)X(t+\tau)\} = (A^2/2) \cos 2\pi f_1 \tau \tag{84}$$

$$G_x(f) = \int_0^\infty (A^2/2) \cos 2\pi f_1 \tau \cos 2\pi f \tau \, d\tau = (A^2/2) + \text{var}\{\xi(t)\} \tag{85}$$

See Fig. 20d for a representation of this process. When more than one cyclic component is present [see Eq. (15)], the pertinent properties are

$$\text{var}\{X(t)\} = \tfrac{1}{2} \sum_{i=1}^{m} A_i^2 + \text{var}\{\xi(t)\} = R_x(0) \tag{86}$$

$$R_x(\tau) = \tfrac{1}{2} \sum_{i=1}^{m} A_i^2 \cos 2\pi f_i \tau \tag{87}$$

$$G_x(f) = \tfrac{1}{2} \sum_{i=1}^{m} A_i^2 + \text{var}\{\xi(t)\} \tag{88}$$

The linear cyclic model is of interest in other fields, such as economics [28], population and business trends [137–139], and industrial management [140]. Short-term forecasting on the order of days, weeks, or months is the objective. The parameters of the model (amplitudes, circular frequencies, trend coefficients) are adjusted as the phenomenon evolves in time. Digital computers facilitate such running adjustments of the model on a routine basis. In a recent spectral analysis of demand variations in water distribution systems of Champaign-Urbana, Illinois, Philadelphia, Pennsylvania, and Denver, Colorado, Gracie [141] suggests that inherent predictive values reside in such observed time series that reflect human habits. As the water resource system comes under greater human control, the statistics of releases from reservoir systems are conditioned not only by the stochastic inputs of runoff, but also by the more or less regular habits of the downstream users (municipal, industrial, power, agriculture, recreational, and pollution control). The situation is analogous to the problems attacked by automatic control theory. Such systems might be more efficiently operated if maximum use were made of the information that is available in the multiple time series. Recent efforts in hydrological forecasting are summarized elsewhere [142, 143]. Kohler [144], in a procedure on multicapacity basin accounting for predicting runoff from storm precipitation, suggests a model wherein the annual seasonal effect is included. The diurnal variations of evapotranspiration may be a useful addition to the synthesis of hourly or daily streamflow hydrographs. The importance of these variations in a hydrologic model is determined by the nature of the hydrologic system (for example, kinds and extent of vegetation, climate, shape of basin, subsurface conditions). In a spectral analysis of the effects of precipitation on the level of Lake Michigan-Huron [145], it is suggested that forecast value inheres in the historical time series. The linear cyclic models suggested are a first approximation; the pure random component $\xi(t)$ tends to distort the amplitude and period as $\xi(t)$ increases. In view of the uncertainty that inheres in natural phenomena and human habits, there is no reason to exclude the notion of a random sine wave [78, 91] in which either the amplitude or period or both are allowed to possess stochastic properties. This introduces a greater degree of complexity into the model.

If $\{X(t): t \in T\}$ is generated by a *moving average process*, its properties are

$$E\{X(t)\} = E\{\xi(t)\} \sum_{i=0}^{m} \alpha_i \tag{89}$$

$$\text{var}\{X(t)\} = \text{var}\{\xi(t)\} \sum_{i=0}^{m} \alpha_i^2 \tag{90}$$

$$R_x(\tau) = E\{X(t)X(t+\tau)\} = \frac{\text{var}\{X(t)\}\sum_{i=0}^{m}\alpha_i\alpha_{i+k}}{\sum_{i=0}^{m}\alpha_i^2}, \quad k \leq m \tag{91}$$

$$\rho_x(k) = \frac{R_x(k)}{\text{var}\{X(t)\}} = \frac{\sum_{i=0}^{m}\alpha_i\alpha_{i+k}}{\sum_{i=0}^{m}\alpha_i^2} \tag{92}$$

$$G_x(f) = 2\int_{-\infty}^{\infty} R_x(\tau)\cos 2\pi f\tau \, d\tau = 2\sum_{k=-\infty}^{\infty} R_x(k)\cos 2\pi fk$$

$$= \text{var}\{X(t)\}\left[1 + 2\sum_{k=1}^{\infty}\rho_x(k)\cos 2\pi fk\right] \tag{93}$$

wherein k is an integer representing the number of lag times Δt up to a maximum of m, m is an integer representing the extent of the moving average, and $\tau = k\,\Delta t$ to relate the continuous and discrete forms of the time series. For a simple symmetric moving average on random components $\xi(t)$, m is an odd integer, the weights α are equal to $1/m$, $\sum_{i=1}^{m}\alpha_i = 1$, $\sum_{i=1}^{m}\alpha_i^2 < 1$. The autocorrelation function

$$R_x(k) = \frac{m-k}{m^2}\text{var}\{\xi(t)\} \tag{94}$$

and

$$G_x(f) = \sum_{k=-m}^{m}\left(\frac{m-k}{m^2}\right)\text{var}\{\xi(t)\}\cos 2\pi fk$$

$$= \frac{\text{var}\{\xi(t)\}}{m}\left[1 + 2\sum_{k=1}^{m}\left(\frac{m-k}{m}\right)\cos 2\pi fk\right] \tag{95}$$

Time Series Analysis of Hydrologic Data

Both of these functions are illustrated in Fig. 20j. Note that at $f = \frac{1}{2}$ the one-sided spectral density function $G_x(f)$ reaches a minimum value, is a maximum at zero frequency, and between $f = 0$ and $f = \frac{1}{2}$ decreases monotonically.

From a practical standpoint, the properties of the moving average operator indicate the extent to which correlation is built into an original random time series by natural and artificial operations. Yevjevich [39, 40] and Matalas [42, 86] review the properties of the moving average process and its simpler version, the autoregressive process, in the context of hydrologic time series; whereas Kendall [136] demonstrates the effect of artificial moving average schemes on cyclic processes and pure random sequences.

A moving average process of extent m may be used to relate mean annual runoff, $X(t)$, and annual effective precipitation, $\xi(t - i)$. The latter is an independent random sequence. The *moving average coefficients*, α_i, given in

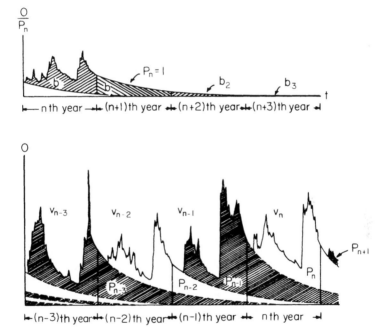

FIG. 23. A schematic representation of water carryover from year to year is found in the lower graph. The upper graph shows the proportions (b-coefficients) of an annual effective precipitation flowing out for successive years (from Yevjevich [39]). $V_n =$ annual flow (volume units); $Q =$ annual discharge rate; $P_n =$ annual effective precipitation; $P_{n-j} =$ annual effective precipitation for the year that precedes the nth year by j years; $b_j =$ coefficients that represent the proportions of P_n flowing out in successive years.

the model are defined as the fraction of the effective annual precipitation flowing out in the $(t-i)$th year, as illustrated in Fig. 23. Yevjevich [39] gives an empirical procedure for estimating these coefficients based on flow recession curves. In effect, the coefficients represent the strength of the effects of carryover from basin storage and of evapotranspiration from such storage from year to year. The carryover effect of prior precipitation extends back in time m years. This antecedent time varies from basin to basin and may not even be constant in the same basin. The coefficients $\alpha_i > 0$ for all i and $\alpha_0 > \alpha_1 > \cdots > \alpha_m$, whereby $\sum_{i=0}^{m} \alpha_i^2 < 1$. Since $\sum_{i=0}^{m} \alpha_i = 1$, then mean annual runoff equals annual precipitation. From Eq. (90), the variance of the annual runoff is *less* than the variance of the annual effective precipitation, thus giving evidence of the smoothing effect of a watershed on precipitation inputs. In Eq. (91) for $k=0$, $\rho_x(0)$ is a maximum, and, as k increases, $\rho_x(k)$ decreases monotonically to zero at $k = m$ as shown in Fig. 20k for a weighted moving average process. Similarly, the spectral function, $G_x(f)$, is a maximum at $f = 0$ and decreases monotonically to a minimum at $f = \frac{1}{2}$, which indicates that the variance of the runoff is predominantly distributed over the low frequencies (the so-called *red noise* spectrum).

The simple moving average with odd m has been used either to remove trend in order to study the residual stationary series or the oscillatory characteristics of a sample record, or to smooth out high-frequency components of the trace in order to obtain a clearer picture of the slower moving features of the phenomenon. Because one of the main purposes of time series analysis is to see what is inside the data, care must be taken not to induce in the time series artificial serial correlation or oscillatory movements through mathematical manipulation. The moving average may induce a cyclic movement in a sequence of pure random numbers. If the period is much greater than m, then the long-period oscillation is included as part of the trend. If the period is on the order of m or shorter, the oscillation tends to be dampened. The foregoing consequences of the moving average operation are referred to as the *Slutzky–Yule effect*. The trend determined by the moving average method is used to obtain a residual series about the trend line by subtracting the trend line (nonstationary) from the original series. It is generally presumed that the residual series is stationary. Because of the Slutzky–Yule effect, care must be taken in the interpretation of trends and cycles in hydrologic time series [146]. Figure 24 illustrates the effects of a moving average on a random sequence of annual precipitation totals.

The *linear autoregressive process* is a useful simplification of the moving average process as far fewer coefficients are involved. If one operates recursively on the moving average model, Eq. (16), the linear autoregressive model in Eq. (17) is obtained under the assumption that the effective annual precipitations are mutually independent random variables [40].

Time Series Analysis of Hydrologic Data

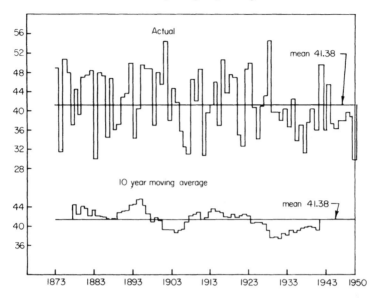

FIG. 24. Time series of "water year" precipitation values at Philadelphia, Pennsylvania, 1873–1950, after scrambling into random time sequence. The annual values are shown above, and the 10-year moving average of the annual values are shown below (from [38]).

Its properties, as summarized by Matalas [86], are

$$E\{X(t)\} = E\{\xi(t)\}\left[1 - \sum_{i=1}^{m} \beta_i\right]^{-1} \tag{96}$$

$$\text{var}\{X(t)\} = \text{var}\{\xi(t)\}\left[1 - \sum_{i=1}^{m} \beta_i \rho(i)\right]^{-1} \tag{97}$$

$$\rho_x(k) = \frac{R_x(k)}{\text{var}\{X(t)\}} = \sum_{s=1}^{m} \beta_s \rho(|s - |k||), \quad |k| = 0, 1, \ldots \tag{98}$$

$$S_x(f) = \frac{S_\xi(f)}{|1 - \beta_1 \exp(-j2\pi f) - \beta_2 \exp(-j4\pi f) + \cdots|^2} \tag{99}$$

$$S_x(f) = \text{var}\{X(t)\} \frac{[1 - \sum_{i=1}^{m} \beta_i \rho(i)]}{\left[\begin{array}{c} 1 + \sum_{i=1}^{m} \beta_i^2 - 2\sum_{i=1}^{m} \beta_i \cos i2\pi f \\ + 2\sum_{s=1}^{m-1}\sum_{i=s-1}^{m} \beta_s \beta_i \cos(s-i)2\pi f \end{array}\right]} \tag{100}$$

wherein $S_\xi(f) = \text{var}\{\xi(t)\}$ is the two-sided spectral density function for the random component. The normalized autocorrelation function, Eq. (98), may oscillate about zero, but $\rho_x(k) \to 0$ as $|k| \to \infty$ (see Fig. 20i). It is evident that $S_x(f)$ will have more than one maximum and minimum.

For the special case where $m \leq 2$, Matalas [86] outlines the solutions when the coefficients of the autoregressive process are expressed as $\beta_1 = p + q$ and $\beta_2 = -pq$, where $|p| < 1$ and $|q| < 1$, so that $|\beta_1| < 2$ and $|\beta_2| < 1$. When both p and q are real, then

$$\rho(|k|) = \frac{p(1 - q^2)}{(p - 2)(1 + pq)} p^{|k|} + \frac{q(1 - p^2)}{(q - p)(1 - pq)} q^{|k|}, \qquad |k| = 0, 1, \ldots \tag{101}$$

$$S_x(f) = \text{var}\{X(t)\} \frac{(1 - p^2)(1 - q^2)(1 - pq)}{(1 + pq)(1 + p^2 - 2p \cos 2\pi f)(1 + q^2 - 2q \cos 2\pi f)} \tag{102}$$

When both p and q are positive, $\rho(|k|)$ decreases monotonically from $\rho(|0|) = 1$ to $\rho(|\infty|) = 0$. If both p and q are negative, $\rho(|k|)$ is positive for even $|k|$ and negative for odd $|k|$ and the absolute value of $\rho(|k|)$ decreases in a monotonic manner from $|\rho(|0|)| = 1$ to $|\rho(|\infty|)| = 0$. For either p or q negative or for both negative, $\rho(|k|)$ alternates so that $\rho(|2k|)$ and $\rho(|2k + 1|)$ diminish monotonically as $|k| \to \infty$. $S_x(f)$ may have only one minimum but one or two maxima.

When p and q are complex conjugates of each other, $p = A + jB$ and $q = A - jB$ such that $\beta_1 = (p + q) = 2A$ and $\beta_2 = -pq = -(A^2 + B^2)$, where $|A| < 1$ and $|B| < 1$. Under these conditions, $-1 < \beta_2 < 0$ and $\beta_1^2 < 4|\beta_2|$, and the solution given by Matalas [86] is

$$\rho(|k|) = C^{|k|} \cos |k| 2\pi f + \frac{A(1 - C^2)}{B(1 + C^2)} C^{|k|} \sin |k| 2\pi f, \qquad k = 0, 1, \ldots \tag{103}$$

$$S_x(f) = \text{var}\{X(t)\} \frac{C^2(1 - C^2)(1 + C^4 - 2A^2 + 2B^2)}{(1 + C^2)\{[A(1 + C^2) - 2C^2 \cos 2\pi f]^2 + B^2(1 - C^2)\}} \tag{104}$$

in which $C = (A^2 + B^2)^{1/2}$. The autocorrelation coefficient $\rho(|k|)$ has the form of a damped cosine curve as $|k| \to \infty$. When $|A| < 2C^2/(1 + C^2)$, $S_x(f)$ is a maximum at $f = \cos^{-1} A(1 + C^2)/2C^2$ and has two minima at $f = 0$ and at $|f| = \frac{1}{2}$. For $|A| \geq 2C^2/(1 + C^2)$, $S_x(f)$ has a maximum at $f = 0$ and a minimum at $|f| = \frac{1}{2}$ if $A > 0$; its maximum and minimum occur at $|f| = \frac{1}{2}$ and $f = 0$, respectively, if $A < 0$. The *second-order Markov process* defined by these properties has been found by Algert [114] to represent the longitudinal bed profile of a sand channel wherein $X(l)$ instead of $X(t)$, wherein l replaces t and represents longitudinal distance along the channel. The random component $\xi(t)$, therefore, is seen to play an important role in determining the lengths and amplitudes of the sand dunes.

Consider the case when $q = 0$ so that $\beta_1 = p$ and $\beta_2 = 0$ and Eqs. (101) and (102) simplify to

$$\rho(|k|) = p^{|k|} = \beta_1^{|k|} = \rho^{|k|}(k=1) = \rho_1^{|k|}, \qquad |k| = 0, 1, \ldots \qquad (105)$$

$$S_x(f) = \text{var}\{X(t)\} \frac{(1 - \rho_1^2)}{(1 + \rho_1^2 - 2\rho_1 \cos 2\pi f)} \qquad (106)$$

which are the appropriate expressions for the *first-order Markov process* (so-called *persistence model* or *lag-one process*). When ρ_1 is positive, $\rho(|k|)$ decreases monotonically from $\rho(0) = 1$ to $\rho(|\infty|) = 0$, and $S_x(f)$ decreases in a similar manner from a maximum at $f = 0$ to a minimum at $|f| = \frac{1}{2}$. If ρ_1 is negative, then $\rho(|k|)$ oscillates about zero, since it is positive for even $|k|$ and negative for odd $|k|$. Absolute $\rho(|k|)$ falls away monotonically as $|k| \to \infty$. $S_x(f)$ has a minimum at $f = 0$ and increases monotonically as $|f| \to \frac{1}{2}$. Figure 20i graphically portrays $G_x(f) = 2S_x(f)$ for positive ρ_1.

In Fig. 25 are shown the normalized variance spectra of the first-order Markov process for varying degrees of autocorrelation. Note that, as the serial correlation decreases, the spectrum flattens out and approaches the flat spectrum typical of white noise. The area under each curve must equal one, indicative of the fact that the sum of the variance contribution at all frequency bands equals the total variance of the time series. The variance contribution of the low-frequency components tends to be strongest for a *red noise* spectrum. Figure 26 shows the fitting of a Markov spectrum to the power spectrum of daily precipitation amounts at Woodstock, Maryland. The Markov spectrum may be plotted on a special coordinate system to obtain a straight-line plot (Fig. 27). The sequential generation of hydrologic data from a linear lag-one model will be discussed in a later section.

V. Linear Systems

As a first approximation, it is common to assume linear transformation systems in hydrology. This is true for most of engineering and science. It is only in the past few years that nonlinear analysis and synthesis has begun to gain momentum [4, 13, 34, 147, 148].

The principles of superposition and proportionality characterize a linear system. These essential elements are incorporated in the *convolution integral* for a time-invariant (constant parameter) linear system:

$$f_0(t) = \int_0^\infty h(t - \tau) f_i(\tau)\, d\tau = \int_0^\infty f_i(t - \tau) h(\tau)\, d\tau \qquad (107)$$

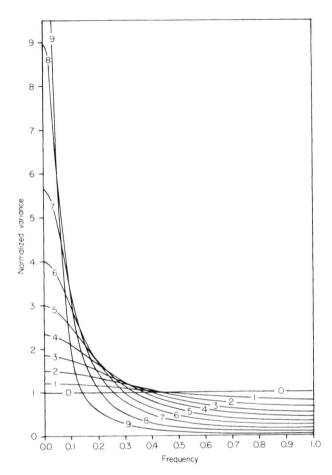

FIG. 25. Normalized variance spectra for a first-order Markov process (from Mitchell [33]). One-lag autocorrelation: $9 = 0.90$, $8 = 0.80$, $7 = 0.70$, $6 = 0.60$, $5 = 0.50$, $4 = 0.40$, $3 = 0.30$, $2 = 0.20$, $1 = 0.10$, $0 = 0.00$.

where $h(t - \tau)$ is a unit *impulse response function*[8] lagged in time by τ, $f_i(\tau)$ is the lagged input that is acted on by $h(t - \tau)$, and $f_0(t)$ is the output as a function of the argument t. The convolution integral embodies the important principle of antecedence which states that only past values can influence the future. The integral specifies the following sequential operations: displacement of $h(t)$ by τ, folding of $h(t)$ about the displacement axis, multiplication with the appropriate ordinate of $f_i(t)$, and summation of the contributions to

---

[8] It is also called a *system weighting function* or a *system time response function*.

Time Series Analysis of Hydrologic Data

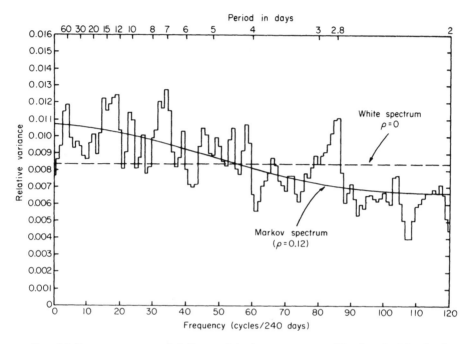

FIG. 26. Power spectrum of daily precipitation amounts at Woodstock, Maryland. Sample length, $N = 4383$ days. Maximum lag of analysis, $m = 120$ days. Fitted "red-noise" continuum is shown by solid curve (from Mitchell [38]).

$f_0(t)$ at t by all *past* pulses $f_i(t)$. As indicated previously, $h(t)$ is the instantaneous unit hydrograph in the context of short-term time series; for stochastic time series generated as linear moving average processes, the weighting coefficients are α and β. A useful first approximation of $h(t)$ for synthesis of hydrographs of storm runoff is Nash's deterministic model (93):

$$h(t) = [k\Gamma(n)]^{-1} e^{-t/k} (t/k)^{n-1} \qquad (108)$$

where k is the *storage delay time* (*routing constant*) for a single linear reservoir, n is the number of linear reservoirs postulated at the outlet of the watershed, and $\Gamma(n)$ is the gamma function. Nash's conceptual model represents the system time response to a unit impulse of rainfall excess routed sequentially through a series of n linear reservoirs with equal k. *The parameters n and k are assumed to be time-invariant. It should be emphasized to avoid confusion that the transformation system is deterministic for the case of the Nash model but the input and output signals may be deterministic or stochastic.* More realistically, the basin system is a random operator, which, in the case of Nash's model, means that n and k would vary according to some probability

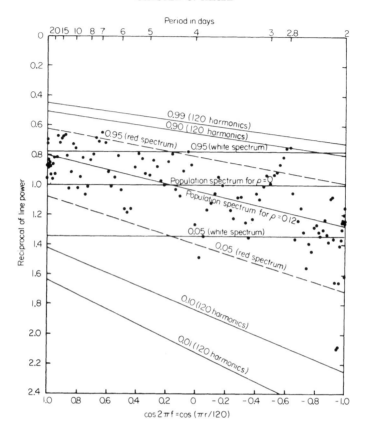

FIG. 27. Power spectrum of Fig. 26 transferred to special coordinate system in which "red-noise" continua and their confidence limits plot as straight lines. Note that $f = (rf_c/m)$, in which r is the lag number and m is the maximum number of days (from Mitchell [33]).

law. For a weakly stationary stochastic input $\xi(t)$ into a time-invariant linear system, the output $X(t)$ is a weighted function of present and past inputs:

$$X(t) = \sum_{i=0}^{\infty} \alpha_i \, \xi(t - i) \tag{109}$$

which is a discrete form of the convolution integral. The stochastic output is also weakly stationary.

If $f_i(t)$ in Eq. (107) is a periodic or transient function, then Fourier transforms of the convolution integral for each of these functions as given by Lee [10] are, respectively,

$$H(n) = F_0(n)/F_i(n) \tag{110}$$

$$H(f) = F_0(f)/F_i(f) \tag{111}$$

where $H(n)$ and $H(f)$ are variously referred to as the *system function* or the *frequency response function*; the subscripts refer to the output and the input, respectively. $F(n)$ and $F(f)$ or $F(\omega)$ are, respectively, the amplitude spectrum and the amplitude density spectrum. As indicated in Eqs. (61) and (62), $H(n)$ and $H(f)$ are the *Fourier transforms* of $h(t)$. Because the random process cannot be expressed in functional form, similar input-output relations are derived in terms of the variance spectra and cross-spectra of the input $\xi(t)$ and output $X(t)$:

$$|H_{x\xi}(f)|^2 = S_x(f)/S_\xi(f) \qquad (112)$$

$$H_{x\xi}(f) = S_{x\xi}(f)/S_\xi(f) \qquad (113)$$

in which $|H_{x\xi}(f)|$ is the modulus of the complex function $H(f)$ referred to as the *gain* factor. The system function is a ratio that indicates the amount of variation in the output which is associated with variation of similar period or frequency in the input. Note that Eq. (112) contains only the gain factor $|H_{x\xi}(f)|$, whereas Eq. (113) is actually a pair of equations containing both the gain factor and phase factor. Equation (112) allows the determination of the *output* spectral density function $S_x(f)$ from the input spectral density function $S_\xi(f)$ and the system gain factor $H_{x\xi}(f)$. It follows that the *variance of the output* (standardized) is given by

$$\text{var}\{X(t)\} = \int_0^\infty G_x(f)\,df = \int_0^\infty |H_{x\xi}(f)|^2 G_\xi(f)\,df \qquad (114)$$

Shen [149] suggests this expression as a basis for analytical study of the effects of various hydrologic transformation processes (watersheds, rivers, reservoirs, hydraulic structures, ground water systems) on the statistics of the output. As control of the water resource system increases, the effects of linear and nonlinear transformations on the probability distributions of outputs are a relevant consideration in the design and control of hydrologic systems downstream of the transformation region. The choice of combinations of input and output in hydrologic systems is myriad and can be made on physical grounds. For example, evapotranspiration rates may be related to soil moisture content, barometric pressure, and air temperature as inputs; or, for the analysis of a dissolved oxygen record and a temperature record, the temperature is the input and the dissolved oxygen the output.

Equation (111) for transient functions such as rainfall hyetographs and flood pulses has been recently applied to the study of the frequency response characteristics of rainfall and runoff measuring systems [132]. Figure 28 gives a definition sketch of system functions and a flow chart of signal filtering, both natural and instrumental. The measured rainfall, P', is the result of low-pass filtering of the actual rainfall, P, by the rain gage. According to Eq. (111),

$$F_P(\omega) = P'(\omega)/P(\omega) \qquad (115)$$

The actual rainfall is also transformed by the watershed to yield Q'; the entire process has respective system functions:

$$H(\omega) = Q(\omega)/P(\omega) \qquad (116)$$

$$F_Q(\omega) = Q'(\omega)/Q(\omega) \qquad (117)$$

whereupon

$$H(\omega) = \frac{Q'(\omega)}{P'(\omega)} \frac{F_P(\omega)}{F_Q(\omega)} \qquad (118)$$

If $F_P(\omega)/F_Q(\omega)$ is unity over the effective frequency band of $P(\omega)$, the system function $H(\omega)$ of the watershed for a unit impulse is obtainable from recorded discharge and rainfall. However, if the time constant or the high-frequency components of the signals are not adequately measured by the instruments, then $H(\omega)$ depends on the frequency response properties of these devices. On the assumption of cyclonic and uniform intensity storms for inputs and of the Nash model, Eq. (108), for the unit impulse response function, the required response characteristics of the hydrologic instruments are obtained. The frequency pass-band of natural and urban drainage basins is shown in Fig. 29. The main conclusion drawn from the analysis is that the adequacy of a 1-min sampling interval of the standard tipping bucket rainfall gage is open to question for urban basins with a time of concentration less than 21 min. The same 1-min sampling interval for runoff measurements was found to be adequate. The foregoing study suggests possibilities for evaluation of the response characteristics of other hydrologic instruments so as not to lose important information that inheres in the phenomenon. Useful methods for the dynamic calibration of transducers are available elsewhere [150]. Laboratory flow systems and surface and subsurface hydrologic flow systems may be explored by use of frequency response techniques to study the laminar and turbulent mixing responses of the system to a variety of pulses [151].

A constant-parameter linear system (for example, a linear ordinary differential equation) may be usefully defined in terms of the coherence function [see Eq. (78)]:

$$\gamma_{x\xi}^2(f) = \frac{|S_{x\xi}(f)|^2}{S_x(f)S_\xi(f)} = \frac{|H_{x\xi}(f)|^2 S_\xi^2(f)}{S_\xi(f)|H_{x\xi}(f)|^2 S_\xi(f)} = 1 \qquad (119)$$

which follows from Eqs. (78), (112), and (113). Hence, for the ideal case of a time-invariant linear system with a single input, $\xi(t)$, and output, $X(t)$, the

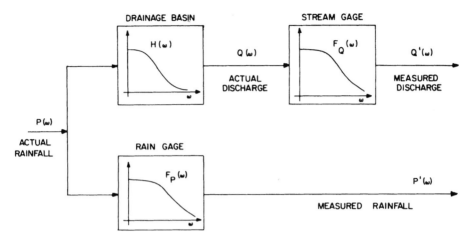

FIG. 28. Signal filtering in the study of drainage basin behavior (from Eagleson and Shack [132]).

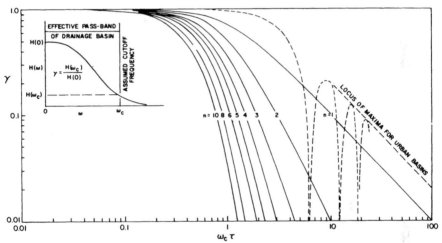

FIG. 29. Pass-band of natural and urban drainage basins (from Eagleson and Shack [132]). $H(\omega)$ = drainage basin system function; ω = frequency in (rad/hr); γ = normalized drainage basin system function; $\tau = t_c$ = basin concentration time; $n = 2.44 A^{0.1}$ = number of linear reservoirs in the Nash model (to the nearest integer); $\tau = k = 11.3 A^{0.3} S_0^{-0.3} \times L^{-0.1}$ = storage delay constant (hours) for a linear reservoir (both correlations are based on flood data on British catchments); L = main stream length (miles); A = drainage area (miles^2); S_0 = mean surface slope in vertical units per 10,000 horizontal units. The system function γ for natural basins, shown as a solid line, is based on the relation

$$\omega_c \tau = [\gamma^{-2/n} - 1]^{1/2}$$

The system function for urban basins, shown as a dashed line, is

$$\gamma = \left| \frac{\sin \omega_c \tau/2}{\omega_c \tau/2} \right|$$

coherence function will be unity at all frequencies. If $X(t)$ and $\xi(t)$ are completely unrelated, the coherence function will be zero. If the coherence function is greater than zero but less than unity, one or more of three possible situations may exist:

(a) Extraneous noise is present in the measurements.
(b) The system relating $X(t)$ and $\xi(t)$ is not linear.
(c) $X(t)$ is an output due to an input $\xi(t)$ as well as to other inputs.

Methods of evaluating case (a) are given by Bendat and Piersol [109] and Amos and Koopmans [135]. Methods for evaluating case (b) are discussed by Goodman [152], whereas case (c) is further discussed by Bendat and Piersol [109]. Dawdy and O'Donnell [52] have suggested the use of the coherency spectrum as a measure of the linearity of conceptual models of hydrologic systems.

Consider the case where $\{Y(t): t \in T\}$ defines the weakly stationary output of a second linear system in which the input $\eta(t)$ is weakly stationary. Let

$$Y(t) = \sum_{s=0}^{\infty} \beta_s \eta(t-s) \tag{120}$$

The properties of $\{Y(t): t \in T\}$ are defined as in Eqs. (89)–(93) for the output of the first linear system $\{X(t): t \in T\}$. The cross-covariance and cross-spectral functions for these two linear systems, as given by Lee [10] and amplified by Matalas [86] in the context of hydrology, are

$$R_{xy}(k) = \sum_{r=0}^{\infty} \sum_{s=0}^{\infty} \alpha_r \beta_s R_{\xi\eta}(-r-k+s) \tag{121}$$

$$S_{xy}(f) = \bar{H}_{x\xi}(f) H_{y\eta}(f) S_{\xi\eta}(f) \tag{122}$$

in which $H_{y\eta}(f)$ is the frequency response function of the linear system defined by $Y(t)$, and $\bar{H}_{x\xi}(f)$ is the complex conjugate of the frequency response function $H_{x\xi}(f)$ of the linear system defined by $X(t)$. The coherence $\gamma_{xy}^2(f)$ between $X(t)$ and $Y(t)$ is

$$\gamma_{xy}^2(f) = \frac{|S_{xy}(f)|^2}{S_x(f) S_y(f)} = \frac{|H_{x\xi}(f)|^2 |H_{y\eta}(f)|^2 S_{\xi\eta}(f) S_{\eta\xi}(f)}{S_\xi(f) |H_{x\xi}(f)|^2 S_\eta(f) |H_{y\eta}(f)|^2}$$

$$= \frac{S_{\xi\eta}(f) S_{\eta\xi}(f)}{S_\xi(f) S_\eta(f)} \tag{123}$$

which follows from Eqs. (112), (113), and (122). An interesting consequence of Eq. (123) arises if $R_{\xi\eta}(k)$ and $R_{\eta\xi}(k)$ are zero for all $|k| > 0$, in which case

Time Series Analysis of Hydrologic Data

$$\gamma_{xy}^2(f) = \frac{R_{\xi\eta}(0)R_{\eta\xi}(0)}{R_\xi(0)R_\eta(0)} = \frac{R_{\xi\eta}^2(0)}{R_\xi(0)R_\eta(0)}$$

$$\gamma_{xy}^2(f) = \cdots = \cdots = \frac{[\text{Cov}(\xi,\eta)]^2}{\sigma_\xi^2 \sigma_\eta^2}$$

(124)

as a result of Eqs. (56) and (66). Thus, γ_{xy}^2 is the square of the product moment correlation coefficient for the inputs and is independent of frequency. It is in this sense that the coherence spectrum may be regarded as a measure of the degree of linearity of systems. Matalas [86] suggests the application of the foregoing concepts for two inputs to the problem of determining the correlation between past climatic indices as tree ring sequences for two different areas in semiarid regions. Other climatic indices such as mud varves may be studied in a similar manner according to Amos and Koopmans [135]. Similarly, the coherency spectrum for annual water levels for large reservoirs may be used to determine the cross-correlation between variable sources of reservoir inflow. This problem of multiple inputs to a linear system (the reservoir) resulting in a single output has been considered in detail by Bendat and Piersol [109]. The concept of partial coherence between separate inputs is introduced and is analogous to the concept of partial correlation if serial correlation is not present within the input sequences [see Eq. (124)].

The notion of optimum discrete linear hydrologic systems with multiple inputs has recently been introduced to hydrologic analysis [153, 154]. These reports also constitute an excellent discussion of linear analysis and synthesis. In specifying a measured outflow hydrograph, $f_d(t)$, as the desired output of a basin filter to rainfall excess, $f_i(t)$, we know that an error, $e(t)$, exists between the desired output and the predicted output, $f_p(t)$:

$$e(t) = f_d(t) - f_p(t)$$

$$e(t) = f_d(t) - \int_0^\infty h(t-\tau)f_i(\tau)\,d\tau \qquad (125)$$

If the minimum mean square error $\overline{e^2(t)} = R_e(0)$ is used as the criterion for defining an optimum linear predictor, $h_{\text{opt}}(t)$, Lee [10] shows that the cross-correlation function, $R_{id}(\tau)$, relating $f_i(t)$ and $f_d(t)$ is the *Wiener-Hopf equation*:

$$R_{id}(\tau) = [R_{io}(\tau)]_{h(\text{opt})} \qquad \text{for} \quad \tau > 0$$

$$R_{id}(\tau) = \int_{-\infty}^{\infty} h_{\text{opt}}(\sigma)R_i(\tau-\sigma)\,d\sigma \qquad \text{for} \quad \tau \geq 0 \qquad (126)$$

in which $h_{\text{opt}}(\sigma)$ is the unit-impulse response of the optimum system, $R_i(\tau-\sigma)$ is the autocorrelation function of $f_i(t)$, and σ is a time variable. Note that Eq. (126) is a convolution type of equation. If the desired output hydrograph could be obtained without any error, the input-output cross-correlation of the optimum linear hydrologic system would hold not only for $\tau \geq 0$ but also for $\tau < 0$. Therefore, in terms of Eq. (113), the Wiener-Hopf equation in the frequency domain becomes

$$H_{opt}(\omega) = S_{id}(\omega)/S_{ii}(\omega) \tag{127}$$

wherein $H_{opt}(\omega)$ is the system function of the optimum linear system. Consequently, the coherency spectrum may now be defined according to Eq. (119) such that the minimum mean square error is

$$R_e(0) = \int_0^\infty G_d(f)[1 - \gamma_{id}^2(f)]\, df \tag{128}$$

as given earlier in Eq. (18). It is important to recognize that we do not know exactly what desired outflow hydrograph, appropriate to a given problem, can be obtained without error through the postulated linear system. Moreover, we demand more than the system can do so that minimization of mean-square error will give us the best approximation to the demand. Eagleson et al. [154] give the following discrete form of the Wiener-Hopf equation:

$$R_{id}(k-1) = \sum_{s=1}^{r} h_{opt}(s) R_{ii}(k-s) \tag{129}$$

in which

$$R_{id}(k-1) = \sum_{r=1}^{n} f_0(r+k-1) f_i(r) \tag{130}$$

$$R_{ii}(k-s) = \sum_{r=1}^{n} f_i(r+k-s) f_i(r) \tag{131}$$

Note that r and s are discrete time values, $\Delta r = \Delta s = 1$ are time increments, and n is the number of output points. Note the correlation functions are in the form defined earlier for transient functions. The convolution operations in Eq. (129) are shown graphically in Fig. 30, with $h(s)$ being the folded function rather than R_{ii}. The simplex method of linear programming is employed to solve Eq. (129). Figure 31 gives a typical result of runoff

---

FIG. 30. Definition sketch for discrete-time convolution using Wiener–Hopf equation (from Eagleson et al. [154]). The unit impulse response function $h(t)$, as shown in the middle figure, is folded about an argument value k of one time unit (shown in bottom figure). The folded $h(t)$ is in turn convoluted with R_{ii} according to Eq. (129) to obtain the cross-correlation function between the input forcing function and the desired output hydrograph.

FIG. 31. Prediction of runoff based upon m convolution equations via linear programming (from Eagleson et al., [154]). It was assumed that $l = m$, where m equals number of output values obtained by convolution with optimum linear predictor $h(t)$, and l equals number of unit hydrograph values, $h(t)$, consistent with the observed duration of the transient input and output functions. The program objective function is to minimize $[h(m+n-1) + \cdots + h(m)]$, with n being the number of input values and $f_d(t)$ forming the lower bound.

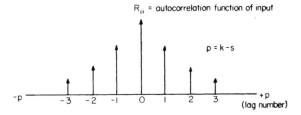

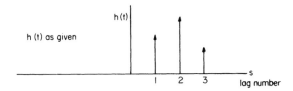

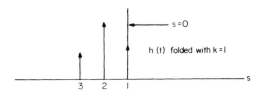

Fig. 30

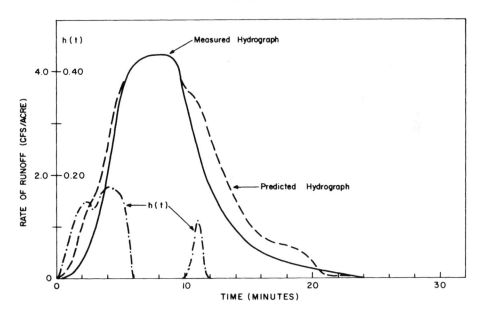

Fig. 31

prediction based upon m convolution equations solved by linear programming. The least-squares criterion optimizes the unit impulse responses of the system. Only when the system is reasonably linear and the conceptual model of the rainfall-runoff process is consistent with reality do we expect the model to have predictive value, because the optimum unit impulse responses are independent of the data used in the analysis.

O'Donnell [155] has made an interesting application of classical Fourier analysis to derive the harmonic coefficients of the instantaneous unit hydrograph, $h(t)$, as defined earlier, from records of rainfall excess, $f_i(t)$, and direct surface runoff, $f_0(t)$. The convolution integral, Eq. (105), embodies all three functions: $h(t)$, $f_i(t)$, $f_0(t)$. A sinusoidal recurrence pattern is postulated for both $f_i(t)$ and $f_0(t)$. Each of the three functions is expanded in an infinite Fourier series substituted into the convolution integral, which, in turn, is solved for the harmonic coefficients of $h(t)$ in terms of the harmonic coefficients of $f_i(t)$ and $f_0(t)$. Time invariance of the harmonic coefficients is assumed; otherwise, the coefficients must be determined for each combination of rainfall excess and storm runoff. O'Donnell's solution is well behaved but displays oscillations at the tail.

Bayazit [156] employed spectral analysis to derive the instantaneous unit hydrograph (IUH) from a continuous record of rainfall and runoff. The procedure is an extension of O'Donnell's method in that the harmonic coefficients of the IUH are expressed in terms of the cross-spectra between rainfall and runoff.

VI. Estimation of Correlation and Spectral Density Functions

The intelligent study of the information content of hydrologic data requires a reasonable understanding of both the physics of the hydrologic system and data analysis tools. As man increasingly manipulates his environment and exerts more control over the water resource system, it becomes more important to scrutinize hydrologic data in greater depth. An estimate of the statistics of controlled systems for more efficient control and design depends, in part, on a deeper understanding of the environmental behavior of the past. The increasing sophistication of analog and digital computers makes possible the economical use of more advanced tools of data analysis. The potential scope of hydrologic data analysis is outlined in Tables VII and VIII. In this section, emphasis is placed on the statistical estimation of covariance and spectral functions.

A. Autocorrelation and Variance Spectral Functions

In the calculation of autocorrelation and variance spectra, it is generally recommended that known periodicities (for example, seasonal oscillations)

Time Series Analysis of Hydrologic Data

and deterministic trends be removed and the analysis performed on the residual series [109]. Let us recall that, for a weakly stationary time series $R_x(\tau) = K_x(\tau) + m_x^2$, in which $R_x(\tau)$ is the autocorrelation function, $K_x(\tau)$ is the autocovariance function and m_x is the mean value. According to Matalas [86], among the proposed estimates of $K_x(\tau)$ are

$$\hat{K}_x(k)^{(1)} = (n-k)^{-1} \sum_{i=1}^{n-k} (x_i - \bar{x})(x_{i+k} - \bar{x}) \quad \text{for} \quad k = 0, 1, 2, \ldots \quad (132)$$

$$\hat{K}_x(k)^{(2)} = (n-k)^{-1} \sum_{i=l}^{n-k} (x_i - \bar{x})(x_{i+k} - \bar{x}_{i+k}) \quad \text{for} \quad k = 0, 1, 2, \ldots \quad (133)$$

in which the caret on K indicates an estimate of $K_x(\tau)$, the superscripts (1) and (2) are used to differentiate between the two estimators, k is the lag number, $k \Delta t = \tau$, $\Delta t = 1$, m is the maximum number of lags, $\bar{x} = (\sum_{i=1}^{n} x)/n$ is the sample mean irrespective of lag, and $\bar{x}_i = (\sum_{i=1}^{n-k} x)/(n-k)$ and $\bar{x}_{i+k} = (\sum_{i=1}^{n-k} x_{i+k})/(n-k)$ are sample means that depend on the lag. For $n \gg k$, both estimators in Eqs. (132) and (133) are unbiased, that is $E[\hat{K}_x] = K_x$. If n replaces $(n-k)$ in the *denominator* of Eq. (132), the resulting estimator of $R_x(k)$ is biased but has a smaller mean square error than $\hat{K}_x(k)^{(2)}$ given in Eq. (133). The estimates of $R_x(\tau)$ corresponding to Eq. (132) and (133) are, respectively,

$$\hat{R}_x(k)^{(1)} = \hat{K}_x(k)^{(2)} + (\bar{x})^2 \quad (134)$$

$$\hat{R}_x(k)^{(2)} = \hat{K}_x(k)^{(2)} + \bar{x}_{ii}\bar{x}_{+k} \quad (135)$$

Estimation of the correlation kernel, $\rho_x(\tau)$, commonly termed the serial correlation coefficient, depends on estimates of $\hat{R}_x(k)$ and the variance:

$$\hat{\sigma}_x^{2(1)} = (n-k)^{-1} \left[\sum_{i=1}^{n-k} (x_i - \bar{x})^2 \sum_{i=1}^{n-k} (x_{i+k} - \bar{x})^2 \right]^{1/2} \quad (136)$$

$$\hat{\sigma}_x^{2(2)} = (n-k)^{-1} \left\{ \left[\sum_{i=1}^{n-k} x_i^2 - (n-k)^{-1} \left(\sum_{i=1}^{n-k} x_i \right)^2 \right] \right.$$

$$\left. \times \left[\sum_{i=1}^{n-k} x_{i+k}^2 - (n-k)^{-1} \left(\sum_{i=1}^{n-k} x_{i+k} \right)^2 \right] \right\}^{1/2} \quad (137)$$

If $n \gg k$, then $\hat{\rho}_x(k) = \hat{K}_x(k)/\hat{\sigma}_x^2$ is an unbiased estimator of $\rho_x(k)$. Simpler equations arise if the $x(t)$ values are standardized and/or normalized (as is the habit of some authors) with respect to the estimators as defined in Eqs. (132), (133), (136), and (137); then

$$\hat{R}_k = \hat{R}_x(k) = (n-k)^{-1} \sum_{i=1}^{n-k} z_i z_{i+k} \quad (138)$$

$$\hat{\rho}_x(k) = \hat{R}_x(k)/\hat{\sigma}^2 \quad (139)$$

TABLE VII

Procedure for Analyzing Single Sample Records in Hydrology[a]

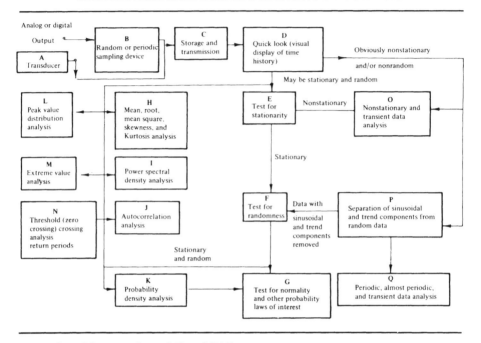

[a] Adapted from Bendat and Piersol [109].

wherein the z's signify the standardized values, that is, $z_i = x_i - \bar{x}_i$ and $z_{i+k} = x_{i+k} - \bar{x}_i$.

Current methods of spectral estimation employ a modification of the classical *Schuster periodogram* [157] in which the squared amplitude of each periodicity is plotted as a function of the Fourier frequency (see Figs. 14, 20). Strictly speaking, classical Fourier analysis is only applicable to true periodicities wherein *discrete line spectra* are the result. Bartlett [158] demonstrated that the periodogram is not an efficient estimator of the *continuous spectrum*; the variability of the raw spectral density function, $\tilde{G}(f)$, increases as $n \to \infty$.

For sampled data from a transformed record that is stationary with $\bar{x} = 0$, a *raw* estimate, $\tilde{G}_x(f)$, of a true one-sided spectral density function, $G_x(f)$, is

$$\tilde{G}_x(f) = \frac{2}{m}\left[\hat{R}_0 + 2\sum_{r=1}^{m-1}\hat{R}_r \cos\frac{(\pi r f)}{f_c} + \hat{R}_m \cos\frac{(\pi m f)}{f_c}\right] \quad (140)$$

which follows from Eq. (58), and wherein r is the lag number and $f_c = \frac{1}{2}/\Delta t$ is the cutoff (or Nyquist) frequency as shown in Fig. 32. Higher frequencies

Time Series Analysis of Hydrologic Data

TABLE VIII
PROCEDURE FOR ANALYZING A COLLECTION OF SAMPLE RECORDS IN HYDROLOGY[a]

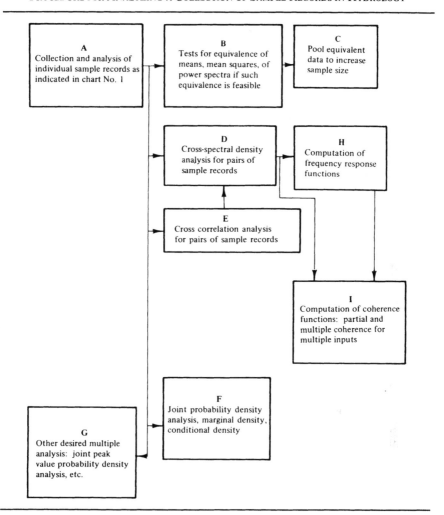

[a] Adapted from Bendat and Piersol [109].

with periods *less* than twice the sampling interval Δt are not detected by the sampling scheme; however, the variance contribution of the folding (higher frequency component) frequency is "seen" by the spectral analysis in that its variance reappears at a lower frequency ($f < f_c$). Their variance is folded into the variance spectrum over the "visible" frequency range, $0 < f \leq f_c$; Figs. 32 and 33 show the effect of *aliasing* or folding. In Eq. (140), we note

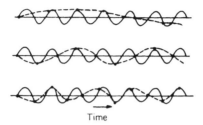

FIG. 32. Typical effects of observing periodic events at intervals of more than one-half the period (from Gunnerson [46]). Solid curve = true record; dashed curve = aliased record (observed); solid circles = sampling time.

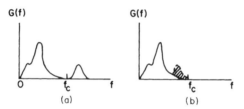

FIG. 33. Aliased power spectra due to folding (from Bendat and Piersol [109]): (a) true spectra; (b) aliased spectra.

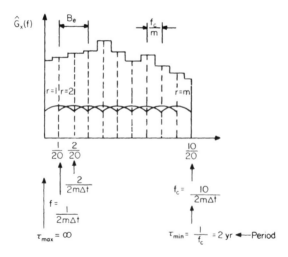

FIG. 34. Illustration of frequency bandwidths used in spectral analysis. k = harmonic number or lag; Δt = one year; $f_c = 1/(2\Delta t) = 0.5$ cycle/year; $m = 10$ = maximum number of lags; f_c/m = frequency bandwidth = $0.5/10 = 1/20$ cycle/year; $f = (kf_c)/m = k/20$ cycles/year; $m/2 = 5$ independent spectral estimates; B_e = equivalent bandwidth = $2f_c/m = 1/10$ cycle/year; τ = period.

that $(m + 1)$ estimates of R_k are required for each value of $G_x(f)$. The values of $G_x(f)$ are calculated only at $(m + 1)$ discrete frequencies, $f = kf_c/m$, where k/m equals the fraction of maximum lag, $\tau_{max} = m\, \Delta t$; thus,

$$\tilde{G}_x(f) = \tilde{G}_x\left(\frac{kf_c}{m}\right) = \frac{2}{m}\left[\hat{R}_0 + 2\sum_{r=1}^{m-1} \hat{R}_r \cos\frac{(\pi rk)}{m} + (-1)^k \hat{R}_m\right]^k \quad (141)$$

Corrections for end effects are not included, as is the case with some authors [109]. Equation (141) gives $m/2$ independent spectral estimates, since estimates at points less than $2f_c/m$ apart will be correlated as seen in Fig. 34. A useful check formula that requires all of the $(m + 1)$ estimates of $G_x(f)$ is

$$\hat{R}_x(0) = (2m\, \Delta t)^{-1}\left[\tfrac{1}{2}\tilde{G}_0 + \sum_{k=1}^{m-1} \tilde{G}_k + \tfrac{1}{2}\tilde{G}_m\right] \quad (142)$$

which follows from Eq. (55).

A *smooth* estimate of the variance spectral density function in each frequency band may be obtained by employing a *Hanning lag window weighting function*, D_r, as given by Blackman and Tukey [25]:

$$D_r = \tfrac{1}{2}[1 + \cos(\pi r/m)] \quad \text{for} \quad r = 0, 1, 2, \ldots, m$$
$$D_r = 0 \quad \text{for} \quad r = m \quad (143)$$

Note that $D_0 = 1$ and $D_m = 0$. Combining Eqs. (141) and (143), the *smooth* estimate of $\tilde{G}_k$ is

$$\hat{G}_k = \hat{G}_x\left(\frac{kf_c}{m}\right) = \frac{2}{m}\left[\hat{R}_0 + 2\sum_{r=1}^{m-1} D_r \hat{R}_r \cos\frac{(\pi rk)}{m}\right] \quad (144)$$

which gives, for $k = 0$,

$$\hat{G}_0 = 0.5\tilde{G}_0 + 0.5\tilde{G}_1 \quad (145)$$

Also, for $k = m$,

$$\hat{G}_m = 0.5\tilde{G}_{m-1} + 0.5\tilde{G}_m \quad (146)$$

Finally, for all $k = 1, 2, \ldots, (m-1)$,

$$\hat{G}_k = 0.25\tilde{G}_{k-1} + 0.5\tilde{G}_k + 0.25\tilde{G}_{k+1} \quad (147)$$

The foregoing equations give a variance spectrum to within a few percent of the total variance of the observed time series; moreover, the variance density estimate in a frequency band is a constant as seen in Fig. 34.

If the main goal is the computation of frequency response functions and the associated coherence functions, other lag weighting functions are appropriate to avoid negative estimates of $\hat{G}_k$ which are not physically realizable and to obtain values of the coherence function $\hat{\gamma}^2_{xy}(f)$ which are not greater than one. For these purposes, Parzen [159] has given

$$D'_r = 1 - 6(r/m)^2 + 6(r/m)^3 \quad \text{for} \quad r = 0, 1, 2, \ldots, m/2$$
$$= 2[1 - (r/m)]^3 \quad \text{for} \quad r = (m/2) + 1, \ldots, m$$
$$= 0 \quad \text{for} \quad r > m \quad (148)$$

Note that $D_0' = 1$ and $D_m' = 0$. Other lag weighting functions may be found elsewhere [25, 28, 159, 160].

The smoothed estimate, $\hat{G}_k$, of the "true" one-sided spectral density function represents a local average over the frequency band, f_c/m, surrounding the frequency, $f = k/2m \, \Delta t$, for which the variance is sought. As seen previously, the averaging process is carried out with the aid of a *window* or *kernel*. In general, the selection of a window is a compromise between two conflicting objectives: (a) to focus on a particular frequency and obtain an estimate of the variance (power) in the narrow band with as little variance of the estimate, $\hat{G}_k$, as possible, and (b) to estimate the variance (power) at adjacent frequencies as well. As the band narrows, or resolution increases, the greater is the variability in the estimates of G_k. As the frequency band widens, the more difficult it is to discern differences in power at adjacent frequencies, even though the variance of the estimate is much smaller. The foregoing discrimination problem is analogous to the difficulties encountered in the construction of histograms for hydrologic data; too few class intervals smudge the underlying probability law to an excessive degree, whereas too many class intervals of narrow width result in a very erratic histogram. Hence, the estimation of variance spectra is a problem in statistical inference.

Practical spectral analysis requires ample consideration of the interrelationship between sampling interval Δt, record length $T_r = N \Delta t$, sample size $N = m/\varepsilon^2$, maximum number of lags m, degrees of freedom $v = 2N/m$, equivalent bandwidth $B_e = 2(f_c/m)$, and normalized standard errors ε of the estimates of the spectral density function. Efficient use of spectral analysis as a data analysis tool in an environmental surveillance program requires an understanding of its rationale in relation to one's field objectives. This relationship between the data analysis tool and the design of experiments and field sampling schemes has been well illustrated in oceanography by Stommel [32] and in water quality control by Wastler [44] and Gunnerson [46].

The chosen sampling interval, Δt, must be small enough so that aliasing will not be a problem. Even though two points per cycle of the cutoff frequency is a theoretical requirement, more points are recommended in practice for improved results [109]. For accurate correlation function measurements where the correlation function has frequencies near f_c, one should choose $\Delta t = 1/(4f_c) = 0.25T_c$. However, if variance spectra measurements are

Time Series Analysis of Hydrologic Data

the prime consideration, then $\Delta t = 2/(5f_c) = 0.4T_c$ should be adequate. Figure 35 shows the effect of a change in sampling interval on the variance spectra for dissolved oxygen measurements in the Potomac River.

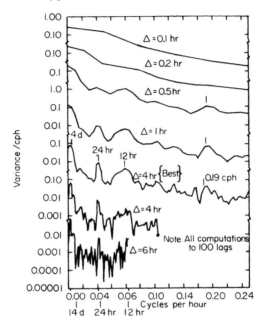

FIG. 35. Estimates of spectra density for dissolved oxygen in Potomac River, Washington, D.C., August 27–September 26, 1963. Note that Δ indicates sampling interval (from Gunnerson [46]).

Usually, the maximum number of lag values, m, is 10% of the sample size N. However, in some cases, as with census data, 20 to 30% may be used [137–139]. If the equivalent bandwidth B_e is fixed, then $m = 1/(B_e \Delta t)$. Thus, B_e will be small for a given Δt when m is large.

For a specified normalized standard error ε of the spectral density function, the required sample size N is m/ε^2. The associated minimum record length is $T_r = N \Delta t$. Note that the sample size requirements increase rapidly for a given m. Table IX shows this relationship for sampling intervals commonly employed in hydrology. As can be seen, error of the spectral estimate is an important consideration in the spectral analysis of *annual* hydrologic data. Bendat and Piersol [109] give the normalized mean square error of the estimate $\hat{G}_x(f)$ as

$$\varepsilon^2 = \frac{E\{[(\hat{G}_x(f) - G_x(f)]^2\}}{G_x^2(f)} \approx \frac{1}{B_e T_r} \tag{149}$$

They recommend $B_e T_r \geq 5$.

TABLE IX

Cutoff frequency f_c (cycle/unit time)	Normalized mean square error ε	Sampling interval $\Delta t = (2f_c)^{-1}$	Maximum number of lags m	Sample size $N = m\varepsilon^2$	Record length $T_r = N\Delta t$	Equivalent bandwidth $B_e = 2f_c/m$	Degrees of freedom $df \simeq 2N/m$
1 cycle/year		(years)			(years)	(cycles/year)	
	0.20	0.5	20	500	250	0.10	50
	0.10	0.5	20	2000	1000	0.10	200
	0.05	0.5	20	8000	4000	0.10	800
	0.01	0.5	20	200,000	100,000	0.10	20,000
1 cycle/month		(months)			(months)	(months)	
	0.20	0.5	20	500	250	0.10	50
	0.10	0.5	20	2000	1000	0.10	200
	0.05	0.5	20	8000	4000	0.10	800
	0.01	0.5	20	200,000	100,000	0.10	20,000
1 cycle/day		(days)			(days)	(days)	
	0.20	0.5	20	500	250	0.10	50
	0.10	0.5	20	2000	1000	0.10	200
	0.05	0.5	20	8000	4000	0.10	800
	0.01	0.5	20	200,000	100,000	0.10	20,000
½ cycle/day		(days)			(days)	(days)	
	0.20	1.00	20	500	500	0.05	50
	0.10	1.00	20	2000	2000	0.05	200
	0.05	1.00	20	8000	8000	0.05	800
	0.01	1.00	20	200,000	200,000	0.05	20,000

Time Series Analysis of Hydrologic Data

The number of degrees of freedom for spectral calculations is $v = 2N/m$. The choice of m in the analysis is critical. It must be kept small in proportion to N, say $\frac{1}{10}$ to $\frac{1}{20}$, so that the degrees of freedom might be as large as possible to maintain some degree of statistical reliability. Yet, m should be large enough to resolve significant features on the frequency scale. The greater m, the less is the statistical reliability of the estimates of the variance within a frequency band.

B. Cross-Correlation and Covariance Spectra Calculations

Estimates of the cross-covariance function for two time series evolving simultaneously in nature are similar to the estimates of the autocovariance function [see Eqs. (132) and (133)] and are given by

$$\hat{K}_{yx}(k)^{(1)} = \hat{K}_{xy}(-k)^{(1)} = (n-k)^{-1} \sum_{i=1}^{n-k} (x_i - \bar{x})(y_{i+k} - \bar{y}) \tag{150a}$$

$$\hat{K}_{xy}(k)^{(1)} = \hat{K}_{yx}(-k)^{(1)} = (n-k)^{-1} \sum_{i=1}^{n-k} (x_{i+k} - \bar{x})(y_i - \bar{y}) \tag{150b}$$

$$\hat{K}_{xy}(-k)^{(2)} = \hat{K}_{yx}(k)^{(2)} = (n-k)^{-1} \left[\sum_{i=1}^{n-k} x_i y_{i+k} - (n-k)^{-1} \sum_{i=1}^{n-k} x_i \sum_{i=1}^{n-k} y_{i+k} \right] \tag{151a}$$

$$\hat{K}_{yx}(-k)^{(2)} = \hat{K}_{xy}(k)^{(2)} = (n-k)^{-1} \left[\sum_{i=1}^{n-k} x_{i+k} y_i - (n-k)^{-1} \sum_{i=1}^{n-k} x_{i+k} \sum_{i=1}^{n-k} y_i \right] \tag{151b}$$

in which $\bar{x} = (\sum_{i=1}^{n} x_i)/n$ and $\bar{y} = (\sum_{i=1}^{m} y_i)/n$. Recall that the time series represented by the *second* subscripted letter is displaced by (k) or $(-k)$ lags with respect to the first time series. Simplified expressions are obtained if the original values are standardized with respect to the mean values used in Eqs. (150) or (151) and normalized by means of the variances defined in Eqs. (136) or (137); thus,

$$\hat{R}_{yx}(k) = \hat{R}_{xy}(-k) = (n-k)^{-1} \sum_{i=1}^{n-k} \dot{x}_i \dot{y}_{i+k} \tag{152a}$$

$$\hat{R}_{xy}(k) = \hat{R}_{yx}(-k) = (n-k)^{-1} \sum_{i=1}^{n-k} \dot{y}_i' \dot{x}_{i+k}' \tag{152b}$$

$$\hat{\rho}_{yx}(k) = \hat{\rho}_{xy}(-k) = \hat{R}_{xy}(-k)/[\hat{R}_x(0)]^{1/2}[\hat{R}_y(0)]^{1/2} \tag{153a}$$

$$\hat{\rho}_{xy}(k) = \hat{\rho}_{yx}(-k) = \hat{R}_{yx}(-k)/[\hat{R}_x(0)]^{1/2}[\hat{R}_y(0)]^{1/2} \tag{153b}$$

in which the dotted variables are the standardized values, and $\hat{R}(0) = \hat{\sigma}^2$.

Pertinent to the determination of the one-sided cross-spectral density function $\hat{G}_{xy}(f)$, between $x(t)$ and $y(t)$ are the *even* and *odd* parts of the cross-correlation function, respectively:

$$\hat{A}_k \equiv \hat{A}_{xy}(k) = 1/2[\hat{R}_{xy}(k) + \hat{R}_{yx}(k)] \quad \text{for} \quad k = 0, 1, \ldots, m \quad (154)$$

$$\hat{B}_k \equiv \hat{B}_{xy}(k) = 1/2[\hat{R}_{xy}(k) - \hat{R}_{yx}(k)] \quad \text{for} \quad k = 0, 1, \ldots, m \quad (155)$$

As before, $\tilde{G}_{xy}(f)$, in terms of $\tilde{C}_{xy}(f)$ and $\tilde{Q}_{xy}(f)$, the cospectrum and quadrature spectrum, respectively, is calculated at $(m+1)$ discrete frequencies, $f = kf_c/m$. At these frequencies,

$$\tilde{C}_k = \tilde{C}_{xy}(kf_c/m) = \frac{2}{m}\left[\hat{A}_0 + 2\sum_{r=1}^{m-1}\hat{A}_r \cos(\pi rk/m) + (-1)^k\hat{A}_m\right] \quad (156)$$

$$\tilde{Q}_k = \tilde{Q}_{xy}(kf_c/m) = \frac{4}{m}\sum_{r=1}^{m-1}\hat{B}_r \sin(\pi rk/m) \quad (157)$$

which follow from the earlier definitions of $C_{xy}(f)$ and $Q_{xy}(f)$.

"Smooth" estimates of $\tilde{C}_k$ and $\tilde{Q}_k$ are formed via the Hanning methods, which yield

$$\left.\begin{array}{l}\hat{C}_0 = 0.5\tilde{C}_0 + 0.5\tilde{C}_1 \\ \hat{Q}_0 = 0.5\tilde{Q}_0 + 0.5\tilde{Q}_1\end{array}\right\} \quad \text{for} \quad k = 0$$

$$\left.\begin{array}{l}\hat{C}_k = 0.25\tilde{C}_{k-1} + 0.5\tilde{C}_k + 0.25\tilde{C}_{k+1} \\ \hat{Q}_k = 0.25\tilde{Q}_{k-1} + 0.5\tilde{Q}_k + 0.25\tilde{Q}_{k+1}\end{array}\right\} \quad \text{for} \quad k = 1, 2, \ldots, (m-1) \quad (158)$$

$$\left.\begin{array}{l}\hat{C}_m = 0.5\tilde{C}_{m-1} + 0.5\tilde{C}_m \\ \hat{Q}_m = 0.5\tilde{Q}_{m-1} + 0.5\tilde{Q}_m\end{array}\right\} \quad \text{for} \quad k = m$$

Use of the Parzen window, described previously, would minimize the effects of leakage on estimates of the coherence spectrum.

Equations (158) are used to obtain smooth estimates of $G_{xy}(f)$ at the $(m+1)$ discrete frequency points:

$$\hat{G}_{xy}(f = kf_c/m) = \hat{C}_k - j\hat{Q}_k = |\hat{G}_{xy}(kf_c/m)| \exp[-j\hat{\theta}_{xy}(kf_c/m)] \quad (159)$$

in which

$$|\hat{G}_{xy}(kf_c/m)| = (\hat{C}_k^2 + \hat{Q}_k^2)^{1/2} \quad (160)$$

$$\hat{\theta}_{xy}(kf_c/m) = \tan^{-1}(\hat{Q}_k/\hat{C}_k) \quad (161)$$

The estimated coherence spectrum is

$$\hat{\gamma}_k^2 = \frac{\hat{C}_k^2 + \hat{Q}_k^2}{\hat{G}_{k,x}\hat{G}_{k,y}} \quad (162)$$

Time Series Analysis of Hydrologic Data

C. Tests of Significance

A test of significance for serial correlation coefficients, $\hat{\rho}_x(k)$, has been proposed by Anderson [161] based on a *circular* normal random series of length N. The last value of the sequence $x(t_n)$ is followed by the first value of the sequence $x(t_1)$. Although the circularity assumption is artificial, the test is one of the few available for testing the significance of correlograms. For the circular series, $\hat{\rho}_x(k)$ has a bounded normal distribution with $-1 \leq \hat{\rho}_x(k) \leq 1$ and with $E\{\hat{\rho}_x(k=1)\} = -1/(N-1)$ and $\text{var}\{\hat{\rho}_x(k=1)\} = (N-2)/(N-1)^2$. As $z_i = [\hat{\rho}_1 - E(\hat{\rho}_1)]/[\text{var}\{\hat{\rho}_1\}]^{1/2}$, the *confidence limits* for a computed value of $\hat{\rho}_1$ are given by

$$\text{C.L.}(\hat{\rho}_1) = \frac{-1 \pm z_\alpha (N-2)^{1/2}}{N-1} = \frac{-1 \pm 1.64(N-2)^{1/2}}{N-1} \tag{163}$$

in which $\hat{\rho}_1 = \hat{\rho}_x(k=1)$ and z_α is the *normal* variate corresponding to the α significance level. α is usually taken as 5%. If $\hat{\rho}_1$ falls outside the confidence limits, the null hypothesis that $\rho_1 = 0$ is rejected. For a one-tailed test, $z_\alpha = 1.64$. For small k/N, the foregoing test may be used for $k > 1$. In effect, we are testing how often a series of random numbers with $\hat{\rho}_1 = 0$ resembles the numbers making up our time series. Inasmuch as we cannot prove a hypothesis, we run the risk that, for $\alpha = 5\%$, five times out of a hundred, on the average, a false conclusion about the null hypothesis will be made, namely, that it is rejected when it is actually true (type I error). Figure 36 shows a correlogram of a pure random time series with the corresponding confidence limits. Note the fluctuations of the correlogram about $\hat{\rho}_k = 0$ even though from a theoretical standpoint the correlation at $k > 0$ should be zero. On the other hand, in Fig. 37 the $\hat{\rho}_1$ value is significant and suggests the applicability of a first-order linear autoregressive model to the time series. Additional examples of the applications of tests of significance to correlograms of hydrologic time series are given by Yevjevich [40], Matalas [41], and Mitchell [33, 38].

In hydrology, generally, the one-tailed significance test is appropriate because most alternatives to randomness in hydrologic series would tend to increase the value of $\hat{\rho}_1$. If *negative* values of $\hat{\rho}_1$ were encountered in nature, the interpretation may be that the time series contains a strong high-frequency oscillation. Thus, high values of streamflow follow low values of streamflow, and vice versa; these are difficult to explain in hydrologic terms. Yevjevich [40] found $\hat{\rho}_1 = -0.348$ for the Inn River near Reisach, Germany, based on 55 years of record, to be significant statistically at the 95% level. If this is the case, a two-tailed test may be preferable, in which case z_α in Eq. (163) becomes 1.96 if the time series has a normal distribution. Additional tests of significance on the randomness of hydrologic time series, based on runs, are given

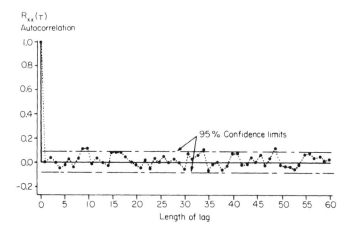

FIG. 36. The autocorrelation function of a simple random process (from Cunnyngham [137]). —, Expected values; ●·●·●, computed values based on sample; $N = 360$.

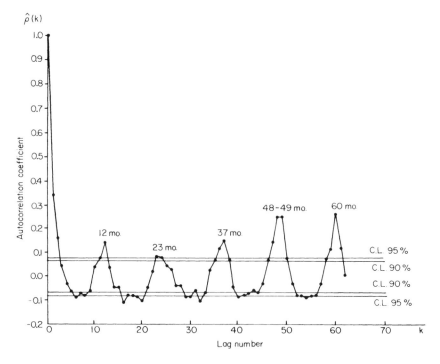

FIG. 37. Autocorrelation of runoff in Tonto Creek, Arizona, based on mean monthly values.

by Dawdy and Matalas [15] and Matalas [86]. Matalas [86] also presents a chi-square test to determine if a linear autoregression process is of order m.

The purpose of the *Tukey sampling theory* (see Tukey and Winsor [162]) is to obtain confidence limits on the *population spectrum* (or the *null continuum*) against which the estimated spectrum obtained from a truncated sample of data can be compared. The null continuum may be that for white noise (horizontal line) or that for a first-order Markov process. Assuming that the numbers of the time series are randomly drawn from a normal population, the sample spectrum estimates should be distributed about the population spectrum according to the chi-squared distribution divided by the number of degrees of freedom v, (χ^2/v). If it is assumed that the local value of the null continuum at a particular frequency is the true value of the population spectrum, then the local value corresponds to the 50% point of the χ^2/v distribution for sample size N and maximum lag m. The test statistic associated with each spectral estimate is the ratio of the magnitude of the spectral estimate, $\hat{G}_k$, to the local magnitude of the null continuum. This ratio is compared with the critical percentage-point levels of a χ^2/v distribution for the proper value of v. If only tables of χ^2 itself are available, rather than χ^2/v, then χ^2/v at a desired level of significance, α, is obtained by finding χ_α^2 that corresponds to v, and then dividing $\chi_{\alpha,v}^2$ by v. It is evident that the use of a smaller sampling interval for a given record length increases the number of measurements and, therefore, the degrees of freedom that result in a smaller confidence band. Thus, the reliability of the estimates of variance spectra is improved, but this luxury is costly and time-consuming.

The fitting of null-hypothesis continua to the computed spectrum depends on the statistical significance of $\hat{\rho}_1$. In general, if $\hat{\rho}_1$ is not statistically significant, the time series should be regarded as free from persistence even though $\hat{\rho}_1 = 0$ is not a sufficient condition for a pure random time series. For this case, the appropriate null continuum is that of a pure random process (horizontal line) whose spectral density in the frequency range $0 < f \leq f_c$ is constant and equal to the average of the values of all $(m + 1)$ raw spectral estimates in the computed spectrum. Figure 38 displays the white noise spectrum with its appropriate confidence bands. From tables of the χ^2/v distribution as given by Hald [163], one finds $(\chi_{0.05}^2/v) = 0.587$ and $(\chi_{0.95}^2/v) = 1.50$ at $v = 26$ degrees of freedom. The former is equivalent to the 95% significance point for a one-tailed test of a spectral *gap*, whereas the latter is equivalent to the 95% significance point for a one-tailed test of a spectral *peak*. As the test ratio corresponding to the 95% upper confidence limit is $1.50 = G_{0.95}(f)/\bar{G}_{\text{white}}(f)$, then $G_{0.95}(f) = 1.50 \times 0.45 = 0.675$ variance per unit circular frequency. Similarly, the 5% lower confidence limit is $0.587 = G_{0.05}(f)/\bar{G}_{\text{white}}(f)$, and so $G_{0.05}(f) = 0.587 \times 0.45 = 0.264$. These limits are accordingly shown in Fig. 38. It is evident that the low-frequency components are quite significant and give some basis for postulating a first-order Markov spectrum for annual flows at Lee Ferry on the Colorado River. Note that the area under *both* the null continuum and the computed spectrum must equal the total variance of the original time series.

If $\hat{\rho}_1$ differs significantly from zero, a check on a few greater lags is in order by assuming on *a priori* grounds a lag-one process, that is, $\hat{\rho}_2 = \hat{\rho}_1{}^2$ $\hat{\rho}_3 = \hat{\rho}_1{}^3$, etc. If these relations hold approximately, then the *red noise* Markov spectrum is assumed as the appropriate null continuum. The continuum is calculated by means of the following equation for the harmonic lag numbers $r = 0, 1, \ldots, m$:

$$G_r(f) = \bar{G}(f)\left[\frac{1 - \hat{\rho}_1{}^2}{1 + \hat{\rho}_1{}^2 - 2\hat{\rho}_1 \cos(\pi r/m)}\right] \quad (164)$$

in which $\bar{G}(f)$ is the average white noise value of all $(m + 1)$ raw spectrum estimates, $\tilde{G}(f)$, and $\hat{\rho}_1$ is an unbiased estimate of ρ_1. The values of $G_r(f)$ are superimposed on the sample spectrum as seen in Figs. 26 and 39. If it is found that higher-order serial correlations $\hat{\rho}_2$ and $\hat{\rho}_3$ do not bear a monotonic relation to $\hat{\rho}_1$, then a simple Markov persistence model is either not a dominant form of nonrandomness in the series or the persistence is coupled to another form of nonrandomness.

Once the foregoing null continuum is established, then the computed spectrum $\hat{G}(f)$ must be evaluated against $G_r(f)$ of the null continuum at the $(m + 1)$ discrete frequencies in the same manner as previously for the white noise spectrum. If *none* of the $(m + 1)$ spectrum estimates is significantly different from the local values of the null continuum, then we conclude that the null continuum is a reasonable model of the observed time series at the given significance level. If *one* or more of the $(m + 1)$ spectrum estimates deviate significantly from the continuum values, then we surmise that the postulated spectrum is not acceptable. We can either try another *red noise* continuum or discard the model entirely. In Fig. 39 are shown the confidence limits for a postulated Markov spectrum. Especially evident in this figure is the significant amount of variance contribution in the frequency bands between 3 and 4 years. However, Mitchell [33, 38] reasons that we should be asking: what is the joint probability of finding a peak and two of its harmonics in an arbitrary location in the spectrum? As a consequence, *a posteriori* tests of significance are introduced [33], according to which this particular result fails to exceed the 95% confidence level. It is concluded that the first-order Markov spectrum is a reasonable spectrum for summer total precipitation at St. Louis.

Cross-spectrum sampling theory is presented by Granger [28] and Jenkins [164], whereas Goodman [165] and Amos and Koopmans [135] report on the sampling distribution of the coherence function.

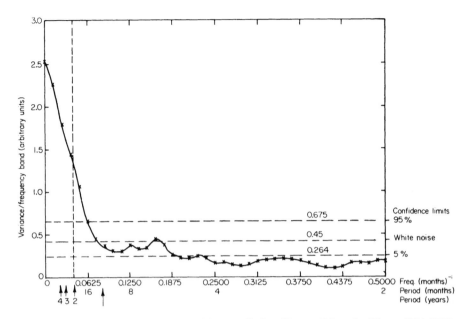

FIG. 38. Spectrum of gaged monthly runoff, Lee Ferry, Colorado River, 1914–1957 (from Julian [43]). $N = 528$, $m = 40$ lags, $\nu = 2N/m = 26$ degrees of freedom, $B_e = 0.25$ (months)$^{-1}$.

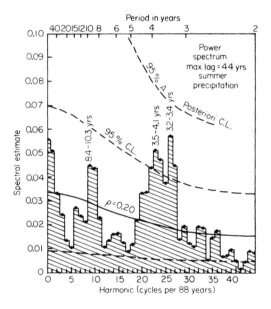

FIG. 39. Power spectrum of summer total precipitation at St. Louis, 1837–1963. Maximum lag of analysis, $m = 44$ years. Red noise continuum and associated 5 and 95% confidence limits are added. Part of the 95% *a posteriori* confidence limit for the same continuum is also indicated (from Mitchell [38]).

D. Limitations of Correlation and Power Spectrum Methods in Hydrology

Whether one analyzes a time series in the time domain or the frequency domain depends on several factors. First, as Tukey [166] points out, "Not everyone that's talking of frequencies needs a spectrum." One may extract considerable intelligence from a time series by skillful use of the low-pass and band-pass filters on the original time series [30, 145, 146, 167]. According to Jenkins [160], the use of the preceding filters may be a useful preliminary to spectrum analysis because there may be strong physical grounds for presuming that the observed time series consists of a mixture of possible stationary series. He argues that a single stationary series is an illusion.

In some cases, use of classical Fourier analysis is complementary to power spectra, as in the study of demand variations in water distribution systems [141] and the effect of precipitation on the level of the Great Lakes [145]. However, in general, most prefer power spectrum methods because it is felt that the structural components of the time series are more easily discerned in the frequency domain. Furthermore, in contrast to correlogram sampling theory, hypothesis testing is considerably simpler in power spectrum analysis. As indicated earlier, the test of hypothesis for serial correlation coefficients is based on a circular definition of the autocorrelation function, whose physical interpretation is obscure. Use of the methods is an art that develops as spectra and/or correlograms are interpreted in the light of an understanding of the physics of the system. Preliminary hypotheses are reinforced as similar spectra and/or correlograms are obtained in other locations of the environment. Spectrum analysis is not a substitute for a physical or engineering approach to a complex environmental situation. It is a means for obtaining valid environmental statistics to make possible an accurate scientific or engineering appraisal. In fact, spectrum analysis introduces new statistics, such as the coherence, and allows a merger of multivariate analysis and spectrum analysis [110, 133]. Stationarity is an important consideration in time series analysis. Usually, it is recommended that the mean value and "known" trends and periodicities be removed prior to analysis. There is no problem with the mean value as a dc component that supposedly contributes nothing to the total variance but tends to color the spectrum estimates [109]. However, removal of trends implies sufficient knowledge of the physical situation that gives rise to the trend. Usually, in view of short record lengths and inadequate knowledge of the physical system, stationarity is a valid assumption. Weak stationarity implies invariance of the first- and second-order moments of the generating process.

In practical terms, stationarity is relative. In an example given by Julian

Time Series Analysis of Hydrologic Data

[133], a sample of temperature fluctuations in the atmosphere permits study of small-scale turbulent fluctuations. A record length of a few hours will exhibit nonstationary behavior because of the known effects of diurnal variation on air temperatures. The nonstationarity may be either in the mean or the variance or both. Yet, if the time scale is increased to days, weeks, months, or years, other movements in the environment come to the fore, whereas the diurnal variations are smoothed out in the daily values, the daily variations in weekly averages, etc. As the time scale increases, it becomes increasingly more difficult to prove the existence of deterministic trends and oscillations. At all time scales, it is important to recognize that stationarity in mean is a necessary but not a sufficient condition for weak stationarity. Stationarity in variance and covariance is a complementary requirement.

Usually, the only realization of a time series $\{X(t): t \in T\}$ available to us is very much more disturbed in one stretch than another. It is natural to think the situation is nonstationary, and yet this conclusion may be wrong, as illustrated by the following example given by Tukey [166]:

> Consider a quiet woodland dell, sheltered from the breezes, where all precipitation is quiet snow. Suppose that exactly once during the calendar year 1967 a person is going to bring in a record player, a 200-watt amplifier, seven speakers and a recording of Tschaikowsky's 1812 Overture, and is going to play the record once at full volume. A four months recording of the sound in the dell is likely to be very patchy indeed. But the corresponding stochastic process may perfectly well be stationary. If the probability that the record will begin at any one time is the same as any other, shifting the time origin by a reasonable amount does not affect the process, so it must be stationary.
>
> Similarly, pistol shots occurring individually and collectively at random according to a Poisson process—with an average rate of seven shots per year are surely a stationary process. Yet every individual realization is very patchy. (This is little like the population with mean zero, each of whose members is not zero.)
>
> How do we get patchy realizations? Usually, as this example suggests, as a response to a Poisson process (in the dell, the Poisson event is starting the record). Never (that is, of course with zero probability) as it is easy to verify, do we get them from a Gaussian process, provided that the period of observation extends over many cycles of the lowest substantially-contributing frequency. (This is a heuristic statement, on balance quite correct, but requiring many unimportant conditions to be a theorem.)
>
> When we get patchiness, especially in geophysics, it is usually a sign of a Poisson source, not a symbol of nonstationarity.

In recent years, interest has evolved in spectrum analysis of nonstationary time series. Granger [28] introduces such methods for economic time series, and Bendat and Piersol [109] review the theory of nonstationary spectra in the context of problems in acoustics, vibration contol, and aerospace control.

VII. Examples of Correlograms and Power Spectra in Hydrology

In earlier sections, some examples of correlograms and spectra were presented not only to demonstrate the properties of theoretical stochastic and deterministic processes, but also to illustrate their use in the detection of nonrandom components and in the study of the frequency response characteristics of watersheds and of hydrologic instruments. Matalas [41, 86] and Yevjevich [40], in the main, have reported on the use of correlograms for rainfall and runoff sequences, whereas Julian [43, 133] has reported the earliest use of power spectrum methods on streamflow sequences.

Figures 40 and 41 present the estimated correlogram and variance spectrum,

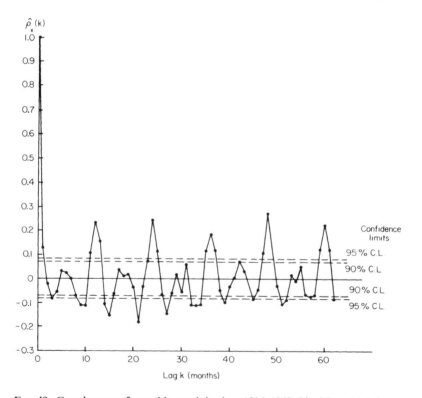

FIG. 40. Correlogram of monthly precipitation, 1896–1960, Pinal Ranch, Arizona.

Time Series Analysis of Hydrologic Data

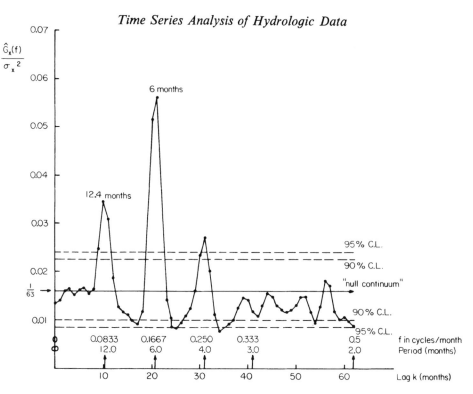

Fig. 41. Normalized power spectrum of monthly precipitation at Pinal Ranch, Arizona.

respectively, of monthly total precipitation at Pinal Ranch, Arizona. The correlogram tends to reach its maximum values at lags that are multiples of 12 months. A six-month component is also discernible. These same observations are shown as peaks at a period of one year and six months, respectively, on the variance spectrum. According to Sellers [87], these seasonal components are consistent with the meteorology of Arizona and New Mexico wherein winter and summer maxima are evident in the raw precipitation records. The presence of an additional peak at four months, more easily seen in the spectrum than in the correlogram, is indicative of the effects of oscillatory seasonal components in the record. As shown earlier in Fig. 39, interpretation of the spectrum is conditioned by the presence of such harmonics ($\frac{12}{3} = 4$ months) of phenomena that are not strictly periodic.

Also, Figs 40 and 41 indicate no statistically significant tendency toward persistence at Pinal Ranch; however, *persistence* is characteristic of many meteorological spectra such as daily precipitation (see earlier Figs. 26 and 39) and daily and monthly temperatures (see Fig. 42). In Fig. 42, the monthly temperature spectrum of Colville, Washington shows a steady decrease in spectrum density from the lower to the higher frequencies, whereas the

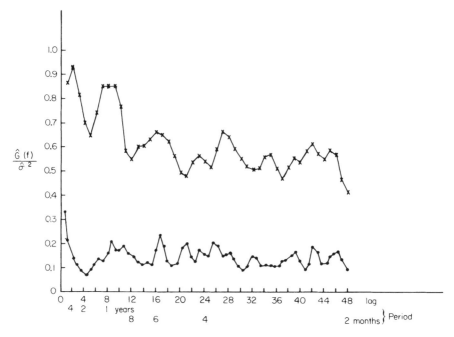

FIG. 42. Normalized power spectrum of monthly temperature and precipitation anomalies (from Julian [133]). × = Colville, Washington, 1900–1963, monthly temperature anomalies; ● = Tatoosh Is., Washington, 1884–1960, monthly precipitation anomalies.

monthly precipitation spectrum at Tatoosh Island, Washington has the characteristics of white noise except at the lowest frequencies. The foregoing pattern of persistence for monthly temperature and "white" precipitation sequences is characteristic of meteorological phenomena. The slight increase in spectrum density for the precipitation spectrum is the result of either of long-period climatic changes or of a nonstationary component in the monthly precipitation record. It should be noted that smoothing of daily precipitation totals to monthly precipitation totals tends virtually to eliminate the persistence pattern noted in daily precipitation totals (see Fig. 26). The persistent pattern in monthly temperature suggests the use of a persistence model of the evapotranspiration process for predictive purposes. Furthermore, the persistence in streamflow, to be discussed later, may be partly explained by the persistence of temperature anomalies and more so explained in terms of temperature persistence if the basin storage is negligible, as in the headwaters of a river basin. Cross-spectrum analysis of multiple time series may be of value in explaining the sources of persistence in streamflow for any given watershed.

Figures 43 and 44 give the correlogram and normalized variance spectrum of monthly runoff of the Salt River near Roosevelt Dam, Arizona, near the

Time Series Analysis of Hydrologic Data

precipitation gage at Pinal Ranch, Arizona. Significant variation is noted at periods of about one-half and one year and at low frequencies. Excluding the effects of the one- and two-year oscillatory components on the spectrum, a Markov persistence pattern of runoff is indicated for the watershed on both the correlogram and spectrum. The coherence spectrum of Fig. 45 indicates how much a given fluctuation at a given frequency in precipitation at Pinal Ranch manifests itself in a similar fluctuation in Salt River runoff. The coherence spectrum very clearly indicates a strong association, or coherence, between the monthly precipitation and runoff at periods of about 6 and 12 months.

Julian [133] has reported an interesting application of cross-spectra and coherence to evaluate the monthly standardized fluctuations of streamflow of the John Day and Umpqua Rivers in Oregon. Their streamflow spectra are given in Fig. 46, wherein the annual seasonal oscillation has been removed by the standardization procedure. Markov persistence tendencies are evident for both streams. Because streamflow fluctuations reflect the integrated effect of many hydrologic variables, the persistence may be due to time-dependence in monthly precipitation, evapotranspiration losses, carryover effect of basin storage, or to nonstationarity created by human consumptive use. It is reasoned that evapotranspiration and basin storage are primarily responsible for the persistence in streamflow because monthly precipitation has no persistence, as seen in Fig. 42, and because consumptive use cannot account for the magnitude of the effect. Better analytical models might be possible if the relative importance of evapotranspiration losses and basin storage were determined for each watershed of hydrologic interest. Adequate data is a key factor in such evaluations.

Figures 47 and 48 present, respectively, cross-power spectra and coherence between the John Day and Umpqua Rivers in Oregon. In general, the co-spectrum is consistently larger than the quadrature spectrum. Both spectra indicate that most of the covariance is concentrated in the lower frequencies (long-term movements on the order of a year or two. The fact that both the cospectrum and quad spectrum are positive at most of the frequencies indicates that the phase angle, $\theta_{xy}(f)$, as given by Eq. (73), is positive. Hence, the relative timing of fluctuations in the two basins is in the first quadrant (0° to 90°) and may be obtained from $\tau = \theta_{xy}(f)/2\pi f$.

Therefore, in view of the definitions of the cospectrum and quadspectrum, the positive displacement time τ indicates that the Umpqua River *leads* the John Day River at nearly all frequencies. From the theory, it is known that $\theta(f) = 0°$ indicates a *direct* relationship between the two streamflows, and $\theta(f) = 180°$ indicates an *inverse* relation between them. The flows are in phase, out of phase, or somewhere in between. Because $\theta(f)$ for the two streams is, in general, in the first quadrant, the interrelationship between the two stream-

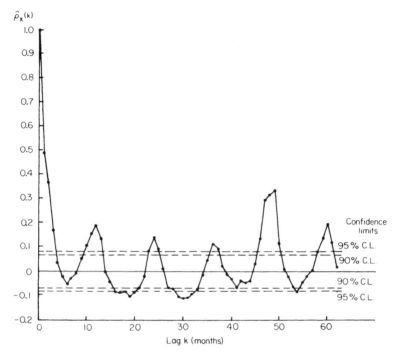

FIG. 43. Correlogram of monthly runoff of the Salt River above Roosevelt Dam, Arizona.

flows should, undoubtedly, be keyed to other causal factors, which, in common, influence the streamflow fluctuations. Both between-basin and within-basin effects must be evaluated. The size of the drainage areas is important inasmuch as larger basins, such as the John Day at its gaging point, tend to smooth out discharge fluctuations (and thus the variance) in contrast to smaller basins such as the Umpqua at its gaging point. The importance of drainage area as a consideration in the evaluation of variance spectra of individual streams is emphasized by Julian's 1961 report [43] in which the spectra of flows near the headwaters of a large basin were found to be typical "white" noise spectra and in which the spectra assumed a characteristic Markov spectrum for streamflows from larger and larger drainage areas on the same basin (upper Colorado River). It is evident that both basin storage and evapotranspiration losses tend to become more important further downstream.

The coherence spectrum for the John Day and Umpqua Rivers (Fig. 48) indicates a high degree of association, or correlation, of the streamflows for the two basins at most of the frequencies. The properties of the spectra for the Umpqua and John Day Rivers may reflect the extent of response of these

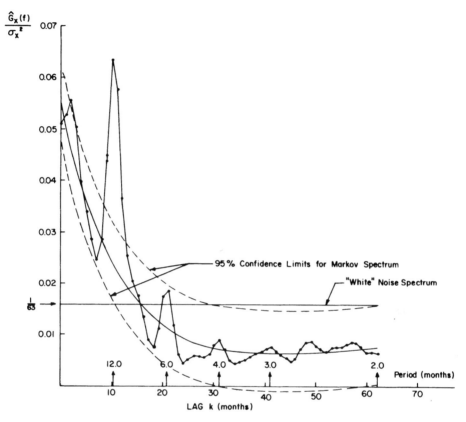

FIG. 44. Normalized power spectrum of monthly runoff of the Salt River above Roosevelt Dam, Arizona (Parzen filter used).

basins to a variety of storm systems and the degree of hydrologic "homogeneity" between the basins. These considerations are pertinent to the extension of streamflow records at one gaging station from a longer record at another station because current correlation methods [59, 88, 89] do not use the information from each frequency band in an optimal fashion to estimate the regression coefficient (see Hamon and Hannan [36]). Thus, the preceding representation of the time series in the frequency domain gives us more information than possible in the time domain. For example, if two time series appear to be correlated, it is pertinent to ask whether this correlation is due to a correlation between high- or low-frequency components. Alternatively, two time series may appear to be uncorrelated because the low-frequency components are negatively correlated and the high-frequency components positively correlated. The extension of spectrum theory into regression analysis is a relatively recent development [27, 36, 110, 133].

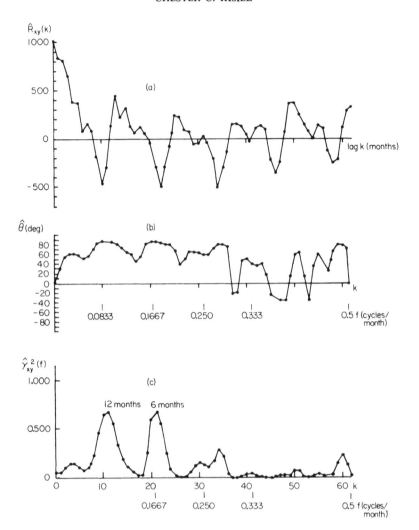

FIG. 45. (a) Cross-correlogram, (b) phase spectrum, and (c) coherence spectrum of monthly averages of precipitation totals at Pinal Ranch, Arizona, and monthly runoff of the Salt River above Roosevelt Dam, Arizona.

Roden [168], in an extensive study of the interrelationship between meteorological, oceanic, and estuarial variables along the eastern Pacific coast of the United States and Canada, reports on the influence of streamflow on the salinity of coastal waters. It is found that the coherence between salinity and river discharge is moderate to good for stations at the boundary between river and oceanic water and poor elsewhere, as shown in Fig. 49.

Time Series Analysis of Hydrologic Data

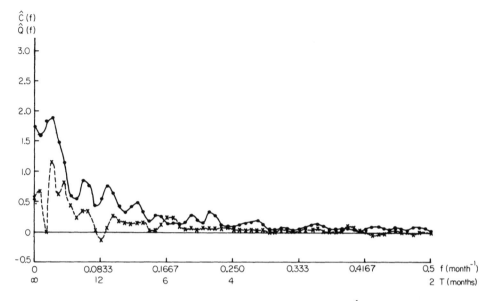

FIG. 46. Streamflow spectra (from Julian [133]). ●, cospectrum, $\hat{C}(f)$; × = quadrature spectrum, $\hat{Q}(f)$.

It is seen that about 70% of the San Francisco salinity variation is accountable by the river discharge. On the other hand, relatively low coherence is obtained between the salinity at Astoria and Columbia River flows at The Dalles, because the river at Astoria is seldom influenced by oceanic water. It is noted that the phase between salinity and runoff is 180° where the coherence in high; the result is consistent with theory and reality, inasmuch as $\theta = 180°$ indicates an inverse relation between salinity and runoff.

Wastler [44, 45] reported the use of spectrum analysis on the Potomac Estuary at Washington, D.C. to study the complex interrelationship between Potomac River discharges, semidiurnal tidal fluctuation, dissolved oxygen (D.O.) concentrations, biochemical oxygen demand (B.O.D.) of wastes diluted in the stream, solar radiation, and photosynthetic growth of algae. Samples were collected every four hours so as to evaluate the effects of the 12-hr tidal cycle. Each spectrum of Figs. 50a and 50b reflects the effects of long-period, diurnal, and semidiurnal fluctuations of dissolved oxygen (D.O.) and biochemical oxygen demand (B.O.D.). The long-period effects include all components whose periods are too long to be resolved in the computation. The diurnal (24-hr) components of the D.O. and B.O.D. spectra reflect the effects of diurnal variations in waste discharge and in photosynthetic activity of the planktonic population. Figures 50a and 50b illustrate how a major waste load at station 6 affects the D.O. and B.O.D. spectra over stations 5,

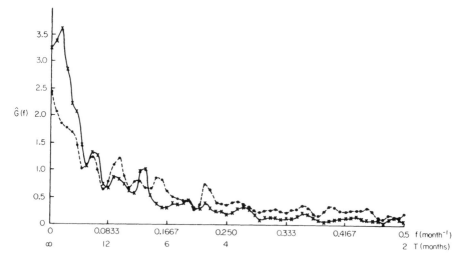

FIG. 47. Cross-power spectrum for the John Day (×) and Umpqua Rivers (●) in Oregon, 1906–1960 monthly anomalies (from Julian [133]).

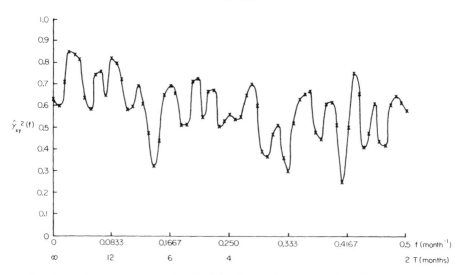

FIG. 48. Coherence spectrum for the John Day and Umpqua Rivers in Oregon, 1906–1960 monthly anomalies (from Julian [133]).

6, and 7, whereas stations 6, 7, and 8 are subject to considerable photosynthetic activity in addition to the diurnal waste load variations. Because the semidiurnal component reflects the advective motion of D.O. and stream waste load due to tidal action, one notes large effects at stations 3, 4, and 5

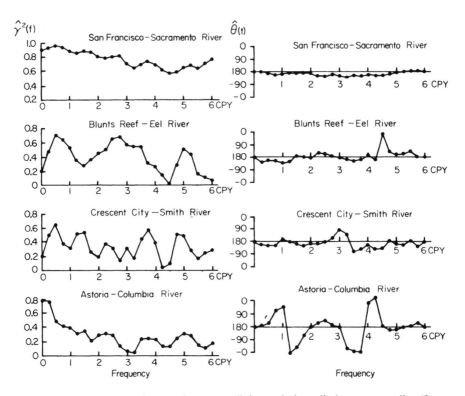

FIG. 49. Coherence and phase between salinity and river discharge anomalies (from Roden [168]). Note that cpy is cycles per year.

in the D.O. spectra, whereas the D.O. at station 7 is critically low, as shown by the small value of the spectral component. The latter result is reflected, in part, in the B.O.D. spectrum at stations 6, 7, and 8. Cross-spectra between dissolved oxygen and one of its forcing parameters, namely, solar radiation, were employed to show that solar radiation has a pronounced diurnal effect and a negligible semidiurnal effect, as seen in Fig. 51. The preceding example emphasizes once again that the final interpretation of a spectrum analysis must be based on an understanding of the natural process and not on magical manipulation of data.

A final example pertains to the effect of monthly basin precipitation on the monthly level of Lakes Michigan and Huron [145]. As shown in Fig. 52, 96.3% of the variance in lake levels is associated with a yearly cycle and longer period fluctuations. The plot was obtained by integrating the area under the spectrum curves of lake level and of precipitation, since these curves are expressions of variance per frequency interval. In contrast to lake

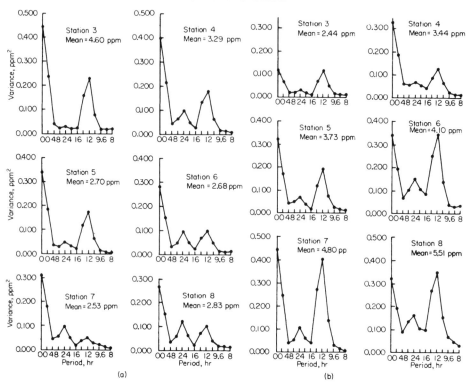

FIG. 50. (a) Dissolved oxygen spectra in the Potomac Estuary (from Wastler [44]); (b) biochemical oxygen demand spectra in the Potomac Estuary (from Wastler [44]).

levels, Fig. 52 shows that fluctuations of 12 months and longer account for only 32.6% of the variance in the precipitation, indicating the greater prominence of fluctuations shorter than 12 months and possibly the effects of sampling problems. Figures 53a and 53b indicate that there is a very high coherence squared (about 0.70) between basin precipitation and lake level in the frequency range of zero to one cycle per 12 months. For all cycles in this range, except for fluctuations longer than 20 years, the lake level lags the precipitation by about one-quarter cycle. It is concluded that most of the information concerning levels for the Lake Michigan–Huron system is contained in the record of basin precipitation. Thus, the quarter-cycle lag on future lake level changes is of predictive value, in that the end of a well-marked positive half-cycle in the range 20 to 40 months must be followed inevitably by an overall decline in lake levels lasting about 5 to 10 months. Even though the preceding findings possess forecast value, in addition to precipitation as a factor in long-term changes in lake level, one must consider in any predictive scheme the effects of diversion, changes in capacity of the outflow channels, and geological changes.

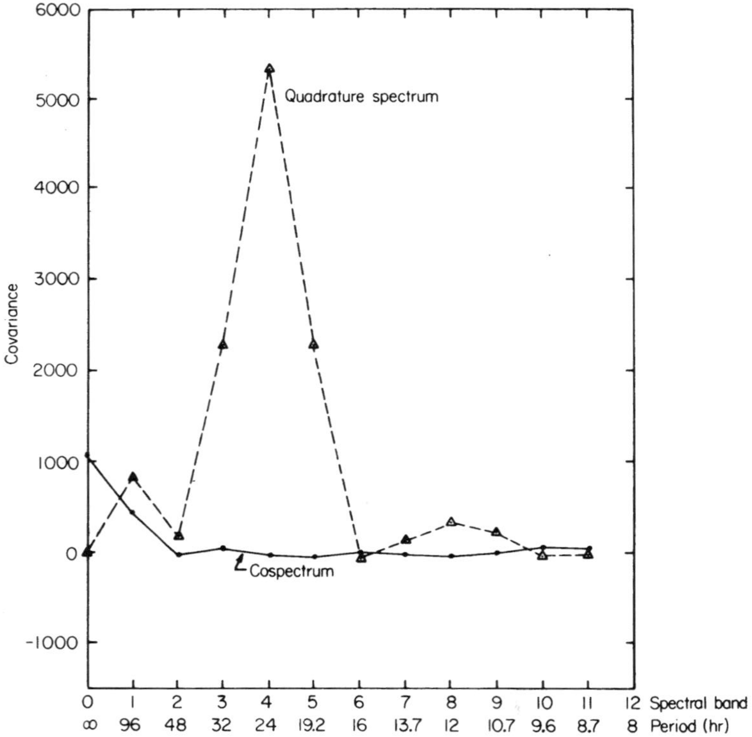

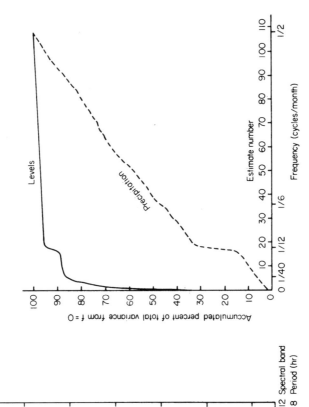

FIG. 51. Cross-power spectrum of solar radiation and dissolved oxygen at station 8 on the estuary of the Potomac River at Washington, D.C. (from Wastler [44]).

FIG. 52. The effect of monthly precipitation on the monthly level of Lakes Michigan and Huron (from Muller et al. [145]). The effect is shown in terms of the accumulated percent of total variance accounted for by the band from zero frequency to a given frequency.

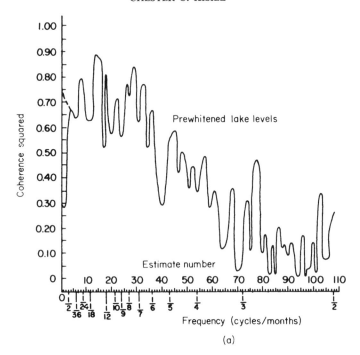

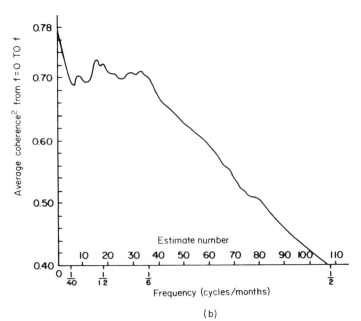

Time Series Analysis of Hydrologic Data

VIII. Importance of Time Dependence in Hydrologic Analysis and Synthesis

The presence of time dependence, or persistence, in a hydrologic time series implies repetition of information given by previous flows. A completely deterministic process supplies no new information, as it is completely defined for all time; whereas a pure random process has no "memory" of the past and supplies a new bit of information with each new event. Hydrologic phenomena possess a cluster tendency in that wet periods tend to follow wet periods, and dry periods follow dry periods. Recognition of this property of hydrologic phenomena requires an evaluation of the *reliability* of the statistical moments that characterize $\{X(t): t \in T\}$.

First, if one considers purely random phenomena with no persistence, it is well known that an unbiased estimate of $\sigma^2(x)$ is

$$s^2(x) = (N' - 1)^{-1} \sum_{i=1}^{N'} [X(t_i) - \bar{X}]^2 \qquad (165)$$

in which $\bar{X} = \sum_{i=1}^{N'} X(t_i)/N'$, and N' is the length of the uncorrelated random sequence, that is, $\hat{\rho}_1 = 0$. The variance of the mean $\bar{X}$, according to the central limit theorem, is given by

$$\sigma^2(\bar{X}) = \sigma_{N'}^2(x)/N' \qquad (166)$$

However, for correlated inputs into a basin or reservoir generated by a first-order Markov process, $\rho_k = \rho_1^{|k|}$, the expected value of $s^2(x)$ for N values is given by

$$E[s^2(x)] = \sigma^2(x)\left\{1 - \frac{2}{N(N-1)}\left[\frac{N\rho_1(1-\rho_1) - \rho_1(1-\rho_1^N)}{(1-\rho_1)^2}\right]\right\} \qquad (167)$$

in which $N > N'$ and is the length of the *correlated* record. If $\rho_x(k) = 0$ for all $|k| > 0$, Eq. (167) reduces to $E[s^2(x)] = \sigma^2(x)$. For $\rho > 0$, the term in the braces is less than one, and so $s_N^2(x)$ in Eq. (165) tends to underestimate $\sigma^2(x)$. If $s^2(x)$ in Eq. (165) is based on N rather than N', an unbiased estimate of $\sigma^2(x)$ is obtained by dividing the resulting $s^2(x)$ by the term in the braces in Eq. (167), where $\hat{\rho}_1$ estimates ρ_1. From another standpoint, N' must be larger than N if $s_{N'}^2(x)$ of the correlated sequence of length N' is to equal

---

FIG. 53(a). The effect of monthly precipitation on the monthly level of Lakes Michigan and Huron is expressed in terms of coherence-squared between prewhitened lake levels and basin precipitation at each frequency (from Muller *et al.* [145]); (b) the same effect as in Fig. 53(a) expressed as the average coherence-squared accounted for by the band from zero frequency to a given frequency.

$s_N^2(x)$ of the uncorrelated sequence of length N. In effect, N' is the *effective number of observations*, since the serial correlation reduces the net information of the time series.

The reliability of the mean of an uncorrelated random process is given by $\sigma^2(\bar{x}) = \sigma_{N'}^2(x)/N'$. If $\{X(t): t \in T\}$ is generated by a first-order linear autoregressive process, then, according to Brooks and Carruthers [169],

$$\sigma_N^2(\bar{x}) = \frac{\sigma^2(x)}{N}\left\{1 + \frac{2\rho_1}{N}\left[\frac{N(1-\rho_1)-(1-\rho_1^N)}{(1-\rho_1)^2}\right]\right\} = \sigma^2(x)f(N, \rho_1) \quad (168)$$

When $\rho_1 > 0$, the term in braces is greater than unity, and $\sigma_N^2(\bar{x}) > \sigma_{N'}^2(\bar{x})$ if $N' < N$ exceeds a certain critical value, N_c', referred to as the effective number of observations. Therefore,

$$N_c' = N\left\{1 + \frac{2\rho_1}{N}\left[\frac{N(1-\rho_1)-(1-\rho_1^N)}{(1-\rho_1)^2}\right]\right\} \quad (169)$$

which follows from $f(N, \rho_1) = 1/N_c'$. Hence, $\bar{X}$ based on N observations, correlated according to a first-order Markov process, is only as precise as $\bar{X}$ based on $N_c' < N$ uncorrelated observations.

Daily precipitation amounts, as a weakly stationary time series, $\{X(t): t \in T\}$, may be formed into a new time series by the addition of precipitation amounts for n consecutive, nonoverlapping days. Kotz and Neumann [170] have investigated the autocorrelation of the resulting time series $\{Y(\tau): \tau \in T\}$, where $Y(\tau_i) = X(t_{i+1}) + \cdots + X(t_{i+m})$, for the case where $\{X(t): T \in T\}$ is a first-order Markov process. It was found that the first-order autocorrelation coefficient is

$$\rho_{yy}(k=1) = \rho_{yy}(1) = \frac{\rho_1(1-\rho_1^N)^2}{N(1-\rho_1)^2 - 2\rho_1 + 2\rho_1^{N+1}} \quad (170)$$

where $\rho_1 = \rho_x(k=1)$. As N becomes large, $\rho_{yy}(1) \approx 1/N$; therefore, the autocorrelation is negligible for periods much greater than one day.

A stationary time series $\{X(t): t \in T\}$ may be replaced by $\{Z(t): t \in T\}$, where, for a reference value X', within the range of $X(t)$, $Z(t) = 1$ if $X(t) \geq X'$ and $Z(t) = 0$ if $X(t) < X'$. $\{Z(t): t \in T\}$ is called a chain with two states, 0 and 1. If the order of the chain, m, is zero, the outcome does not depend on occurrences at any previous time point.

For the case where $m = 1$, the outcome at time t depends upon the outcome at only one previous time point. Matalas [86] has shown that the expected value and variance of the return period, I, are, respectively,

$$E[I] = 1/p \quad (171)$$

$$\sigma^2(I) = \frac{1}{p}\left(\frac{1}{p}-1\right)\left(\frac{1+\rho}{1-\rho}\right) \quad (172)$$

in which $\rho = \rho_{zz}(k = 1)$, and p is the probability that the event will be equaled or exceeded in the future. It is seen that the mean value of the return period in unaffected by time dependence represented by $\rho_{zz}(k)$. But, for $\rho > 0$, the variance of the return period is larger for a chain of order $m = 1$ than for a chain of order $m = 0$, the usual case in hydrology. If $\rho = 0$,

$$\sigma^2(I) = p^{-1}(p^{-1} - 1),$$

which is recognized as the variance of the geometric probability law.

Let a normal stationary time series $\{X(t): t \in T\}$ represent the inflows to a reservoir. It is recalled from Eq. (3) that the partial sums of elements of a stochastic process define another stochastic process. The partial sums of cumulative departures of the inflows $X(t)$ from their sample mean $\bar{X}$ for n years of record are

$$W_j = \sum_{i=1}^{j} [X(t_i) - \bar{X}] \quad \text{for} \quad j = 1, \ldots, n \tag{173}$$

in which W_j is the cumulative departure from the mean at the *end* of the year j. Then the range R of the cumulative departures is

$$R = \max |W_j' - W_j''| \tag{174}$$

in which R is the overyear *storage* requirement in the reservoir, and W_j' and W_j'' denote the maximum and minimum values of W_j, respectively, which occur at the time points j' and j''. The range R is a stochastic variate and represents the storage needed in a reservoir of infinite capacity to assure an outflow $\{Y(t): t \in T\}$ equal to $\bar{X}$ at each time point.

Feller [68] in 1951 obtained the first two moments for the asymptotic distribution of the range as defined previously; for $\rho_x(k) = 0$ for all $|k| > 0$, the expected value of R, first given by Hurst [83], is

$$E(R) = \sigma(X)[N\pi/2]^{1/2} = 1.25\sigma(X)N^{1/2} \tag{175}$$

and the variance, given by Feller, is

$$\sigma^2(R) = [\pi^2/6 - \pi/2]\sigma^2(X)N = 0.074\sigma^2(X)N \tag{176}$$

In the context of the storage-yield relation, the results apply to situations where the variate is the annual flow of a stream, 100% regulation (annual draft equals mean annual inflow to reservoir) is prescribed, and the time period n is long. The annual flows of the stream are assumed to be normally distributed and to possess no serial correlation.

No analytical solutions have attempted to account for the effects of non-normal flows, serial correlation, and regulation less than 100%. However, in an empirical study of 690 time series, for 75 natural phenomena, Hurst [69] found that the ratio, $E(R)/\sigma(X) \propto (N/2)^r$, has a value of $r = 0.73$ rather than

$r = 0.5$ as given by Eq. (175). It was concluded by Hurst that the difference between 0.50 and 0.73 is due to serial correlation of the data. On the other hand, Moran [96] suggested that, because of serial correlation, the effective length of record N' for natural time series is not large enough for application of the asymptotic Eq. (175).

From Eq. (175), it is seen that the storage, R, depends only upon one statistical parameter, namely, the population standard deviation of the mean annual flows and varies as the square root of N. Fiering [88] and Matalas [89] have presented efficient methods for the estimation of $\sigma(X)$ by correlating streamflows at the design point with a larger body of information from adjacent gaging stations. Equation (175) also implies that the range R is quite unstable statistically, having a variability $\sigma(R)$ about $E(R)$. Hence, the value of the storage obtained by the mass diagram analysis of an historic record is likely to be erratic. The mass-diagram method presumes, for example, that a stream with a 50-year record will reproduce its past behavior almost exactly in the subsequent 50-year design life of a reservoir. This postulates a cyclic behavior and negates the inherent stochastic nature of natural phenomena.

Thomas and Fiering [71] undertook a simulation study of the reservoir storage-yield relation by postulating as the univariate stream model a first-order Markov process of the form

$$Q_{i+1} = \bar{Q}_{j+1} + \beta_j(Q_i - \bar{Q}_j) + \xi_{i+1} \tag{177}$$

in which Q_{i+1} and Q_i are the flows during the $(i+1)$st and ith time period respectively, initiated at the start of the *synthetic* sequence, $\bar{Q}_{j+1}$ and $\bar{Q}_j$ are the mean flows during the $(j+1)$st and jth period time over the entire length of historic record, respectively, within a repetitive annual cycle of the time period (e.g., 365 days, 52 weeks, 12 months, or 1 year), β_j is the least squares regression coefficient for estimating the flow in the $(j+1)$st time period from the jth time period, and ξ_{i+1} is a random component corresponding to the synthesized flow in the $(i+1)$st time period. When $\xi_i = 0$, Eq. (177) becomes

$$Q_{i+1} = \bar{Q}_{j+1} + \beta_j(Q_i - \bar{Q}_j) \tag{178}$$

which does not preserve the variance of the flows unless the serial correlation coefficient ρ_j between flows in the jth and $(j+1)$st season equals ± 1. Equation (178) is an *indeterministic model*[9] and not truly causal, in that the flow in the previous time period does *not* cause the flow in the present period. The flows possess a time-dependent structure because of other external driving forces.

---

[9] Indeterministic models would exclude those possessing deterministic and probabilistic properties (see Bunge [20]).

Time Series Analysis of Hydrologic Data

Because flows exhibit variability about the straight-line equation defined by Eq. (178), the uncorrelated random component ξ_{i+1} is added to form the *probabilistic model* given by Eq. (177). Whereas Eq. (178), if realistic, predicts the future with certainty, Eq. (177) predicts the future with uncertainty but with the added benefit of fluctuation about a mean value regression line. The probabilistic model of Eq. (177) is more akin to nature because it provides for the inherent natural variability about a mean value and preserves the time-dependent structure that inheres in many hydrologic time series.

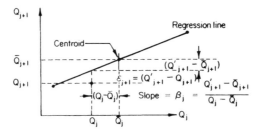

FIG. 54. Variability about regression line.

The variability $\sigma(\xi)$ about the linear lag-one model may be obtained by considering the departure ξ_i about the regression line as shown in Fig. 54. From that figure it is seen that

$$\xi_{j+1} = Q'_{j+1} - Q_{j+1} = (Q_{j+1} - \bar{Q}_{j+1}) - \beta_j(Q_j - \bar{Q}_j) \qquad (179)$$

in which Q'_{j+1} is the regression estimate. Its variance $\sigma^2(\xi)$ is

$$\sigma^2(\xi) = \mathrm{var}[(Q_{j+1} - \bar{Q}_{j+1}) - \beta_j(Q_j - \bar{Q}_j)] = \sigma^2(Q_{j+1})[1 - \rho_j^2] \qquad (180)$$

Recall that the standardized random deviate of a random variable is

$$t_i' = (Q_{j+1} - Q'_{j+1})/\sigma(\xi) = (\xi_{j+1})/\{\sigma(Q_{j+1})[1 - \rho_j^2]^{1/2}\}$$

which gives for the random component

$$\xi_{j+1} = t_i'\sigma_{j+1}(1 - \rho_j^2)^{1/2} \qquad (181)$$

The standardized random deviate t_i' is either chosen from a table of random numbers or generated as a pseudo-random number on a digital computer, preferably the latter. The deviate may be assigned a normal distribution with zero mean and unit variance, specified as $N(0, 1)$, a gamma distribution with different values of skewness, or a log-normal distribution. If the flows follow a normal distribution, only the first two moments must be preserved and Eq. (177) is an adequate recursion relation. If the flows are gamma distributed, the first three moments of the distribution must be preserved. Hence, the deviate t_i' must not only reflect the mean and the variance but also the coefficient of skewness γ. To preserve the skewness of the distribution of the historical flows, Thomas and Fiering [71] give the skewness of the random deviate t_i' as

$$\gamma_{t'} = \frac{\gamma_{j+1} - \rho_j^3 \gamma_j}{(1 - \rho_j^2)^{3/2}} \tag{182}$$

Note that t_i' is now gamma distributed with zero mean, unit variance, and skewness $\gamma_{t'}$ given by Eq. (182). If t_N' is a normal deviate, $N(0, 1)$, then t_N' can be transformed into a gamma deviate t_{γ}' whose distribution is almost like gamma by the transform

$$t_{\gamma'} = \frac{2}{\gamma_t}\left(1 + \frac{\gamma_{t'} t_N'}{6} - \frac{\gamma_{t'}^2}{36}\right)^3 - \frac{2}{\gamma_{t'}} \tag{183}$$

which is valid for both positive or negative skewness.

Therefore, to apply the first-order Markov process given by Eq. (177), the historical record must be analyzed to obtain $\bar{Q}_j$, β_j, ρ_j, σ_j, and γ_j. For example, if the time period of recorded values is one month, then 12 values of each of these constants are required for synthesis of monthly streamflows. In their simulation of mean annual streamflows, Thomas and Fiering [71] allowed ρ_j to assume the values 0, 0.1, and 0.2 and the underlying populations to be normal and gamma with skewness -0.5, 0.5, and 1.0. Furthermore, the degrees of reservoir regulation were 100, 90, and 80%, and the lengths of the time series were 10, 25, 50, and 100 years. The possible combinations of each specification are $3 \times 4 \times 3 \times 4 = 144$; each of these 144 combinations was replicated 100 times, thus resulting in 14,400 simulated time series on a computer, each of which never occurred in nature. Each generated sequence is indistinguishable from the original historical record as judged by statistical criteria. Each realization is just as likely as the next (see Fig. 9). It was found that the mean range (storage) is in close agreement with $E(R)$ for $N > 25$. The variance of the underlying distribution is far more important than the skewness. In a more recent paper, Fiering [73] emphasizes the operational importance of simulation of water resources systems in view of the fact that probability statements concerning cost-benefit studies naturally follow from the probabilistic character of the reservoir storage-yield relation.

More recently, Chow and Ramaseshan [102] have employed a relation of the type given by Eq. (177) to synthesize hourly rainfalls. From the historical data of the French Broad River Basin at Bent Creek, North Carolina, 1000 annual storms of 36-hr duration were generated by means of Monte Carlo methods. Annual storms are defined as those which produced the maximum peak discharge in a water year. Each of the synthetic storms was subsequently routed through n linear reservoirs with routing constant k according to Nash's conceptual model defined by Eq. (108). The synthetic floods were then summarized in the form of stochastic flow-duration curves. An important use of this simulation procedure, suggested by Chow and Ramaseshan, is in the evaluation of alternate system designs by routing of the stochastic floods through each design to determine an optimal system.

Fiering [107] postulated a Markov model as defined by Eq. (177) for generation of daily flows. Equation (177) may be simplified by use of $\beta_j = \rho_j \sigma_j / \sigma_{j-1}$, which in turn may be approximated by ρ_j:

$$Q_{i+1} = \rho_j Q_i + (1 - \rho_j)\bar{Q}_j + t'_{i+1}(1 - \rho_j^2)^{1/2} \qquad (184)$$

The preceding model is employed to obtain simulated relations between mean 7- and 15-day low flows and duration series. It is suggested that the procedure may be employed to estimate low flow values when gaging records are inadequate, to ascertain the extent to which a site should be gaged according to cost considerations, and to examine synthetic traces of peak flows, rather than low flows, so as to relate the mean n-day peak flow to a certain exceedance probability close to zero.

Two stationary time series $\{X(t): t \in T\}$ and $\{Y(t): t \in T\}$ may be related by the probabilistic model:

$$Y(t) = \alpha + \beta X(t) + \xi(t) \qquad (185)$$

in which α and β are the population parameters, independent of time, defining the regression relation, and $\xi(t)$ is the portion of $Y(t)$ unexplained by $X(t)$.[10] $X(t)$ and $Y(t)$ may be the streamflows of two proximate watersheds, concurrent rainfall and runoff time series, or other combinations of hydrologic variables. Given n pairs of observations of $X(t)$ and $Y(t)$, the method of least squares gives

$$Y(t) = a + bX(t) + \xi(t) \qquad (186)$$

such that the sum of squares of the error term $\xi(t)$ is a minimum about the regression line defined by the first three terms of Eq. (186). The constants a and b are consistent estimates of α and β, respectively, according to Wold [171].

The regression component, $a + bX(t)$, in Eq. (186) is not necessarily composed of elements whose variance is concentrated closely at one frequency f_0. If, however, such is the case, then no serial correlation exists in both $X(t)$ and $Y(t)$, and so the $\xi(t)$ is independently distributed in time. Hence, according to Hannan [27], the least squares method is efficient, and the

---

[10] $\xi(t)$ is defined as $t\sigma(y)[1 - \rho_{xy}^2]^{1/2}$ according to the earlier discussions concerning the univariate stream model.

variance of the least squares estimators of the regression coefficient b may be estimated by

$$\text{var } b = \frac{1}{n} \frac{\hat{G}_y(f_0) - b^2 \hat{G}_x(f_0)}{\hat{K}_x(0)} = \frac{\sigma^2(\xi)}{n\sigma^2(X)} \tag{187}$$

in which $\sigma^2(\xi) = \sigma^2(Y) - b^2\sigma^2(X)$.

For the model of Eq. (185) to be valid, β should be the same in all frequency bands. Both $X(t)$ and $Y(t)$ should be either in phase (positive β) or in antiphase (negative β) because there are no time lags inherent in the model. In either case, the quadrature spectra, $\hat{Q}_x(f)$ and $\hat{Q}_y(f)$, should be zero except for sampling fluctuations.

The model given by Eq. (185) for the special case where $\rho_x(k) = 0$ and $\rho_y(k) = 0$ for all $|k| > 0$ has been employed to extend streamflow records. Given a stream with time series $Y(t)$ of length N_1, it is desired to extend or fill in the gaps of $Y(t)$ from a longer time series $X(t)$ on a nearby stream with similar hydrologic properties. The concurrent portions of $X(t)$ are correlated with $Y(t)$. The method is particularly applicable to mean annual flows, or some suitable transform thereof (e.g., logarithms), which are normally distributed. Equation (185) represents a working hypothesis inasmuch as the flows $X(t)$ do not cause the flows $Y(t)$. The least squares method simply gives the best estimates of the parameters, a and b; the responsibility for the adequacy of the model rests with the user and not with the statistical procedure. The statistical precision of the extension of $Y(t)$ by N_2 time periods (years) may be evaluated by defining a relative information ratio I' for both the mean and the variance of $Y(t)$. For the mean, $I' = \text{var}(\bar{Y}_{N_1})/\text{var}(\bar{Y}_{N_1+N_2})$, and for the variance s^2, $I' = \text{var}(s^2_{N_1})/\text{var}(s^2_{N_1+N_2})$. If $I' > 1$, then we have a net gain in information, since the variance of Y based on a record length of $(N_1 + N_2)$ is then less than the variance of Y based on the original record length N_1. If $I' < 1$, then we have a loss in precision, and, therefore, the extension based on the original record alone should be more precise. The value of I' is determined by the strength of the cross-correlation $\hat{\rho}_{xy}$ between the two time series. When I' (mean) $= 1.0$, $\rho^2 = 1/(N_1 - 2)$ as given by Fiering [88]. For $N_1 = 20$ and 40 years, the critical values are, respectively, 0.25 and 0.16 whether or not $\xi(t)$ is zero in Eq. (185). Matalas [89], in reporting the latter result, has evaluated the effects of the noise term $\xi(t)$ on the critical ρ values for various N_2 years of record extension. His findings are presented in Table X for I' (variance) $= 1.0$. For typical values of N_1 and N_2 encountered in hydrology, ρ_{xy} is on the order of 0.8 without noise and 0.39 with noise for I' (variance) > 1.0. Consequently, conditions acceptable for improving estimates of the variance are much more restrictive than those for improving estimates of the mean. Furthermore, the noise term

Time Series Analysis of Hydrologic Data

TABLE X

CRITICAL MINIMUM VALUES OF ρ_{xy} FOR THE VARIANCE[a]

Critical minimum values of ρ_{xy} for the variance: noise $\varepsilon(t)$ added

N_2 \ N_1	10	15	20	25	30
10	0.65	0.54	0.52	0.42	0.38
15	0.65	0.54	0.51	0.42	0.39
20	0.65	0.54	0.51	0.42	0.39
25	0.65	0.54	0.50	0.42	0.39
30	0.65	0.54	0.50	0.42	0.39

Critical minimum values of ρ for the variance: no noise added

N_2 \ N_1	10	15	20	25	30
10	0.73	0.63	0.70	0.76	0.76
15	0.75	0.77	0.79	0.80	0.80
20	0.76	0.79	0.81	0.81	0.82
25	0.78	0.80	0.84	0.83	0.81
30	0.77	0.80	0.82	0.83	0.84

[a] From Matalas [89].

introduces more realism into the model and is less restrictive on extension of the record.

In contrast to the previous application in the extension of streamflow records, the more general case requires consideration of the effect of time-dependence in both $X(t)$ and $Y(t)$ on estimates of β. The problem is a realistic one in that it is desirable to predict hydrologic events on a daily, weekly, monthly, or annual basis in an optimal fashion. If the model of Eq. (185) is structurally valid from a hydrologic standpoint, the estimate of β from the band of frequencies around kf_c/m is given by

$$b_k = b(f = kf_c/m) = \hat{C}_{xy}(f)/\hat{G}_{xy}(f) \tag{188}$$

in which $\hat{C}_{xy}(f)$ and $\hat{G}_{xy}(f)$ are, respectively, the smoothed estimates of the cospectrum and cross-power spectrum. The constant α is estimated by $a = Y - bX$. In the foregoing estimate of β, it is assumed that $\xi(t)$ is independent of $X(t)$. The assumption may be open to question if other factors in the form of a third or fourth time series are related to $X(t)$ and if $X(t)$ contains measurement errors. In Eq. (188), each $b(f)$ is not equally useful as an estimate of β because high frequencies may be affected more by measurement noise in $X(t)$, which will tend to reduce $b(f)$ in absolute value. The $b(f)$ values from bands where the *signal-to-noise ratio* $N(f)$ [that is, the variance in $\beta X(t)$ in that frequency band divided by the variance in $\xi(t)$ for the band] is high will be the most useful. An estimate of $N(f)$ which suggests the formation of the weighted average of $b(f)$ is given by Hamon and Hannan [36] as

$$\hat{N}(f) = \hat{G}_x(f)/\{\hat{G}_{yy}(f)[1 - \hat{\gamma}_{xy}^2(f)]\} \tag{189}$$

The weighted average of $b(f)$ is

$$b = \frac{\sum_{f=0}^{f_c} \delta(f)\hat{N}(f)\hat{C}_{xy}(f)/\hat{G}_x(f)}{\sum_{f=0}^{f_c} \delta(f)\hat{N}(f)} \tag{190}$$

which may be calculated directly from

$$b = \frac{\sum_{f=0}^{f_c} \{[\delta(f)\hat{C}_{xy}(f)]/\hat{G}_{xy}(f)[1 - \hat{\gamma}_{xy}^2(f)]\}}{\sum_{f=0}^{f_c} \{[\delta(f)\hat{G}_{xx}(f)]/\hat{G}_{yy}(f)[1 - \hat{\gamma}_{xy}^2(f)]\}} \tag{191}$$

Above $\delta(f) = \frac{1}{2}$ when $f = 0$ or f_c and is unity otherwise. $\delta(f)$ represents a correction for end effects. The variance of b is

$$\text{var } b = n^{-1}\left\{m^{-1}\sum_{f=0}^{f_c} \{[\delta(f)\hat{G}_x(f)]/\hat{G}_y(f)[1 - \hat{\gamma}_{xy}^2(f)]\}\right\} \tag{192}$$

in which m is the maximum number of lags and n is the sample size. A plot of $b(f)$ as a function of f establishes the extent to which $b(f)$ is independent of f. The validity of the simple regression model, Eq. (185), may be evaluated in this manner. Hamon and Hannan [36] present methods for estimating the β's for multiple regression and lagged regression models when more than two time series are involved. Each model was evaluated by them as a predictor of daily mean sea level from atmospheric pressure and components of wind stress. For this example, it is found that the differences between the coefficients obtained by normal linear or multiple regression methods and those obtained by spectrum methods are not particularly large. This is explained by the unusually large coherence between atmospheric pressure and sea level in most of the frequency range. The spectral methods would be more useful for time series in which there is a greater variability of the signal-to-noise ratio with frequency.

Hannan [27] indicates that the foregoing procedures are *robust* against considerable departures from stationarity. Two advantages of spectral methods over traditional regression methods include (a) explicit accounting of the variation of $N(f)$ with f, and (b) the adjustment for the effects of instrumental frequency response and the frequency response of any numerical filters applied to the data [36].

Time Series Analysis of Hydrologic Data

Jones [110] has applied spectral analysis in the linear prediction of multivariate meteorological time series. Julian [133] suggests the extension of the notions of partial regression to define partial coherence functions as a measure of the conditional dependence of two hydrologic time series in a complex system with multiple inputs. Bendat and Piersol [109] present an excellent discussion on the theory of multiple-input linear systems in both the time and frequecy domain. In the context of hydrology, the following vector process may be postulated as a first approximation:

$$\underline{Y}(t) = \underline{\beta}\,\underline{X}(t) + \underline{\xi}(t) \qquad (193)$$

wherein each term represents a vector. For example, $\underline{Y}(t)$ may represent outflow from the hydrologic system in the form of streamflow, groundwater discharge from an open watershed,[11] and evapotranspiration (vapor discharge). $\underline{X}(t)$ may represent precipitation, air temperature, relative humidity, and groundwater movement. The model of Eq. (193) may also be made more realistic by the inclusion of lagged effects.

IX. Summary

In conclusion, the complex interrelationships in the hydrologic system can be better understood through wise design of data collection programs in the field. Spectrum analysis as a data analysis tool conditions the organization and implementation of these programs. The extensive gains by geophysicists in the application of numerical spectrum analysis to data on the earth and its atmosphere have been summarized by Tukey [166]; hydrologists, likewise, can reap similar benefits, particularly through analysis of *short-term* time series on the order of hours, days, and months.

It has been shown that time series analysis in hydrology is of value in defining the properties of the hydrologic cycle. The techniques recognize the inherent variability of the environment and the presence of measurement noise. Better definitions of the interactions between various geophysical processes of the hydrologic cycle should enable the hydrologist of the future to synthesize more realistic parametric and stochastic models of universal utility. The challenge is in the detection of the nature of the nonlinear and nonstationary components and in the subsequent formulation of tractable models that reasonably incorporate the inherent complexity of nature. However, many hydrologic problems are amenable to analysis and synthesis by assuming *stationary stochastic* inputs into *linear time-invariant* (or *time-variable*) systems with a *stationary stochastic* output. The parametric or

---

[11] An open watershed is one wherein the input does not equal the streamflow, vapor discharge, and storage. Leakage into adjacent watersheds occurs because of geologic properties of the system (see Amorocho and Hart [3]).

deterministic models are a special case of this general conception of the hydrologic system; undoubtedly, such models will continue to be useful in many applications.

Symbols

A_i Amplitude of the ith harmonic in a linear cyclic process

a_{xn} Cosine harmonic coefficient (amplitude) for nth harmonic of the time series $x(t)$

A_k Even part of the cross-correlation function

a Estimate of α in the linear regression model

b_{xn} Sine harmonic coefficient (amplitude) for the nth harmonic of the time series $x(t)$

B_e Equivalent bandwidth

b Estimate of the population regression β, independent of frequency

$b(f) = b_k$ Frequency spectrum of the regression coefficients b

c_{xn} Amplitude of the nth harmonic of the time series $x(t)$

$C_{xy}(f)$ Cospectral density function or cospectrum

d Horizontal distance from point of firing to point of impact for projectile

D_r Hanning lag window weighting function wherein r is the lag number

D_r' Parzen lag window weighting function

E Evaporation volume

$E[X(t)]$ Expected value of $X(t)$

e 2.718

f_c Cutoff frequency (circular)

$F[X(t_i)]$ Distribution function associated with the observed sample value $x(t_i)$ at the end of the ith time interval

f Circular frequency: discrete or continuous, e.g., f_1 indicates the fundamental circular frequency

$F_x(n)$ Complex amplitude spectrum of the periodic function $x(t)$; also the Fourier transform of $x(t)$

$\bar{F}(n)$ Complex conjugate of $F(n)$

$F(f)$ or $F(\omega)$ Complex spectrum for either transient or random time series

g Gravitational acceleration

$G_{xx}(f) = G_x(f)$ One-sided spectral density function for the single time series $x(t)$

$G_{xy}(f)$ One-sided cross-spectral density function for the two time series, $x(t)$ and $y(t)$

h Height of projectile

$H(f)$ or $H(\omega)$ System frequency response function

$h(t)$ Unit impulse response function of a linear time-invariant system

I Return period

I' Net gain of information

$j = \sqrt{-1}$ Imaginary number

k Sample number, i.e., $x_k(t_i)$ indicates the observed value in the kth sample at the end of the ith time interval from the stochastic process, $\{X(t): t \in T\}$

$K(s, t)$ Covariance kernel

k Integer representing the

Time Series Analysis of Hydrologic Data

	number of lag times Δt up to a maximum of m; note that $\tau = k\,\Delta t$
k	Storage delay time in Nash's model
$m(t)$	Mean value function of a stochastic process evolving in time
m	An integer representing the extent in time units of Δt of the moving average process
m	An integer representing maximum number of lags for use of the weighting function D_r
m	Order of the Markov chain
n	Sample size in terms of time units, Δt, from the starting point of the stochastic process
N	Total number of data intervals in the time interval $T = N\,\Delta t$
N_c'	Effective number of observations
$N(f)$	Signal-to-noise ratio
$N(0, 1)$	Normal probability law with zero mean and unit variance
P	Precipitation volume
$Q_{xy}(f)$	Quadrature spectral density function or quadspectrum
Q_i and Q_j	Synthetic streamflow and historic streamflow, respectively
R	Runoff volume
R_{xy}	Cross-correlation function
R_{xx}	Autocorrelation function
$R_e(0)$	Mean square error for the optimum system, Eq. (81)
R	Range of the cumulative departures of a stochastic variable with respect to the mean value function; the overyear storage requirement in a reservoir
S	Storage
$s = t - \tau$	Time shift of original record
S_n	Sum of n random variables
$S_{xy}(f)$	Cross-power spectrum for periodic functions, $x(t)$ and $y(t)$; also the cross-power spectral density function for random functions, $x(t)$ and $y(t)$
$S_{xx}(f) = S_x(f)$	Power spectrum or power density spectrum depending on nature of time series
$s^2(x)$	Unbiased estimate of $\sigma^2(x)$
t	Time
T	Index set of all possible values of time t or space
T_i	Fundamental period
t_i'	Standardized random deviate
v_0	Initial velocity of projectile
W_j	Partial sum of cumulative departures of the inflows $X(t)$ from their sample mean $\bar{X}$ for n years of record at the *end* of the year j
$X(t_i)$	Stochastic population at time t_i from which the sample records $x(t_i), \ldots, x(t_n)$ are obtained
$\bar{x}$	Sample mean of $x(t_i)$
$\bar{y}$	Sample mean of $y(t_i)$
Z	Index set of all possible points in space or of all possible sediment particles

Greek Letters

α	Weighting factor in the moving average process; level of significance for the chi-squared test

β Coefficients in the linear autoregressive process: a measure of the strength of the past value in relation to the present value in a series; also it is the least squares regression coefficient for estimating the flow in the $(j+1)$st time period from the jth time period

$\Gamma(n)$ Gamma function wherein the n is the number of linear reservoirs as defined in Nash's model

$\gamma_{xy}(f)$ Coherence function

ν Integer number of degrees of freedom for spectral calculations

γ_t Skewness of the random deviate

δ Dirac delta function

$\in$ Belongs to, e.g., $t \in T$

ε Normalized standard error of the estimates of the spectral density function

ζ Sample point in space or a single sediment particle, e.g., $\zeta \in Z$ for a stochastic process evolving in space

θ Angle at which a projectile is fired with respect to a horizontal plane

θ_i Phase angle of the ith harmonic for a linear cyclic process

θ_{xn} Phase angle for the nth harmonic of the time series $x(t)$

$\theta_{xy}(f)$ Phase spectrum: measure of the relative phase of the harmonics of $x(t)$ and $y(t)$

$\varepsilon(t) = \varepsilon_t$ Pure random component (error or noise term in a time series)

$\rho(s, t)$ Correlation kernel for a single time series

$\sigma^2(t)$ Variance function of a stochastic process evolving in time

τ Delay or lag time

χ^2 Chi-squared

ω_i Angular frequency of the ith harmonic; also ω represents continuous frequency

References

1. Committee on Surface-Water Hydrology, Research needs in surface-water hydrology. *J. Hydraulics Div. Am. Soc. Civil Engrs.* **91**, No. HY1, 75–83 (1965).
2. Chow, V. T., Laboratory study of watershed hydrology. *Proc. Intern. Hydrol. Symp., Fort Collins, Colorado* **1**, 194–202 (1967).
3. Amorocho, J., and Hart, W. E., The use of laboratory catchments in the study of hydrologic systems. *J. Hydrol.* **3**, 106–123 (1965).
4. Amorocho, J., and Hart, W. E., A critique of current methods in hydrologic systems investigation. *Trans. Am. Geophys. Union* **25**, 307–321 (1964).
4a. von Schelling, H., Most frequent random walks. Rept. No. 64GL92, p. 19. Advanced Technol. Labs., Gen. Elec. Co., Schenectady, New York, June 1964.
4b. Korn, G. A., Random Process and Noise. Lecture notes. Dept. Elec. Eng., Univ. of Arizona, Tucson, Arizona, 1966.
5. Crawford, N. H., and Linsley, R. K., The synthesis of continuous streamflow hydrographs on a digital computer. Tech. Rept. No. 12. Dept. Civil Engr., Stanford Univ., Stanford, California, 1962.
6. Bagley, J. H., An application of stochastic process theory to the rainfall-runoff process. Tech. Rept. No. 35. Dept. Civil Engr., Stanford Univ., Stanford, California, 1964.

7. Ishihara, T., and Takasao, T., A study of runoff pattern and its characteristics. *Disaster Prevent. Res. Inst. Kyoto Univ. Bull.* **19**, 1–23 (1964).
8. Cheng, D. K., "Analysis of Linear Systems." Addison-Wesley, Reading, Massachusetts, 1959.
9. Chow, V. T., Runoff. *In* "Handbook of Applied Hydrology" (V. T. Chow, ed.), pp. 14-13–14-35. McGraw-Hill, New York, 1964.
10. Lee, Y. W., "Statistical Theory of Communication." Wiley, New York, 1960.
11. Stubberud, A. R., "Analysis and Synthesis of Linear Time-Variable Systems." Univ. of California Press, Berkeley, California, 1964.
12. Amorocho, J., and Orlob, G. T., Nonlinear analysis of hydrologic systems. *Calif. Univ. Water Resources Center, Contrib.* **40** (1961).
13. Marshall, J. L., "Introduction to Signal Theory." Intern. Textbook Co., Scranton, Pennsylvania, 1965.
14. Davenport, W. B., and Root, W. L., "An Introduction to the Theory of Random Signals and Noise." McGraw-Hill, New York, 1958.
15. Dawdy, D. R., and Matalas, N. C., Analysis of variance, covariance and time series. *In* "Handbook of Applied Hydrology" (V. T. Chow, ed.), Sect. 8-III. McGraw-Hill, New York, 1964.
16. Minshall, N. E., Predicting storm runoff on small experimental watersheds. *J. Hydraulics Div. Am. Soc. Civil Engrs.* **86**, No. HY8, 17–38 (1960).
17. Chow, V. T., Frequency analysis. *In* "Handbook of Applied Hydrology" (V. T. Chow, ed.), Sect. 8-I. McGraw-Hill, New York, 1964.
18. Laurenson, E. M., Storage routing methods of flood estimation. *Trans. Inst. Engrs. Australia* **CE7**, 39–47 (1965).
19. Rockwood, D. M., Columbia basin streamflow routing by computer. *J. Waterways Harbors Div. Am. Soc. Civil Engrs.* **84**, No. WW5, 1–15, (1958).
20. Bunge, M., Causality, chance and law. *Am. Scientist* **49**, 432–448 (1961).
21. Cramer, H., Model building with the aid of stochastic processes. *Technometrics* **6**, 133–160 (1964).
22. Fair, G., and Geyer, J., "Water Supply and Waste-Water Disposal." Wiley, New York, 1954.
23. Wallis, J. R., Multivariate statistical methods in hydrology—a comparison using data of known functional relationship. *Water Resources Res.* **1**, 447–461 (1965).
24. Benson, M. A., Spurious correlation in hydraulics and hydrology. *J. Hydraulics Div. Am. Soc. Civil Engrs.* **91**, No. HY4, 35–42 (1965).
25. Blackman, R. B., and Tukey, J. W., "The Measurement of Power Spectra." Dover, New York, 1958.
26. Parzen, E., "Empirical Time Series Analysis." Holden-Day, San Francisco, California, 1967.
27. Hannan, E. J., The statistical analysis of hydrological time series. *Proc. Natl. Symp. Water Resources, Use, Management, Australian Acad. Science, Camberra, Australia, September, 1963*.
28. Granger, C. W. J., "Spectral Analysis of Economic Time Series." Princeton Univ. Press, Princeton, New Jersey, 1964.
29. Panofsky, H. A., and Brier, G. W., "Some Applications of Statistics to Meteorology." Pennsylvania State Univ., University Park, Pennsylvania, 1958.
30. Holloway, J. Leith, Jr., Smoothing and filtering of time series and space fields. *Advan. Geophys.* **4**, 1–43 (1957).
31. Van Isacker, J., Generalized harmonic analysis. *Advan. Geophys.* **7**, 189–214 (1961).
32. Stommel, H., Varieties of oceanographic experience. *Science* **139**, 572–576 (1963).

33. Mitchell, J. M., Jr., Some practical considerations in the analysis of geophysical time series. *Paper Am. Geophys. Union* (1963).
34. Tukey, J. W., Data analysis and the frontiers of geophysics. *Science* **148**, 1283–1289 (1965).
35. Frenkiel, F. N., and Schwartzchild, M., Additional data for the turbulence spectrum of the solar photosphere at long wave lengths. *Astrophys. J.* **121**, 216–223 (1955).
36. Hamon, B. V., and Hannan, E. J., Estimating relations between time series. *J. Geophys. Res.* **68**, 6033–6041 (1963).
37. Munk, W. H., Snodgrass, F. E., and Tucker, M. J., Spectra of low-frequency ocean waves. *Bull. Scripps Inst. Oceanog. Univ. Calif.* **7**, 283–362 (1959).
38. Mitchell, J. M., Jr., A critical appraisal of periodicities in climate. Weather and Our Food. CAED Rept. 20, pp. 189–227. Iowa State Univ., Ames, Iowa, 1964.
39. Yevjevich, V. M., Fluctuations of wet and dry years, Part I: Research data assembly and mathematical models. Hydrology Papers No. 1. Colorado State Univ., Fort Collins, Colorado, 1963.
40. Yevjevich, V. M., Fluctuations of wet and dry years, Part II: Analysis by serial correlations. Hydrology Papers No. 4. Colorado State Univ., Fort Collins, Colorado, 1963.
41. Matalas, N. C., Autocorrelation of rainfall and streamflow minimums. *U.S. Geol. Surv. Profess. Papers* **434-B** (1963).
42. Matalas, N. C., Statistics of a runoff-precipitation relation. *U.S. Geol. Surv. Profess. Papers* **434-D** (1963).
43. Julian, P. R., A study of the statistical predictability of stream runoff in the upper Colorado River Basin. Bur. Econ. Res., Univ. of Colorado, Boulder, Colorado, 1961.
44. Wastler, T. A., Application of spectral analysis to stream and estuary field surveys, Part I: Individual power spectra. *U.S. Public Health Serv. Publ.* **99-WP-7** (1963).
45. U.S. Public Health Service, Technical appendix on cross-spectral analysis to Pt. VII of the Rept. on the Potomac River Basin Studies: Report on needs for water supply and flow regulation for quality control in the Washington Standard Metropolitan Area. Public Health Service, Washington, D.C., 1962.
46. Gunnerson, C. G., Optimizing sample intervals in tidal estuaries. *J. Sanit. Eng. Div. Am. Soc. Civil Engrs.* **92**, No. SA2, 103–125 (1966).
47. Whittle, P., "Hypothesis Testing in Time Series Analysis." Almqvist & Wiksell, Uppsala, 1951.
48. Cornell, C. A., Stochastic process models in structural engineering. Tech. Rept. No. 24. Dept. Civil Engr., Stanford Univ., Stanford, California, 1964.
49. Snyder, W. M., Men, models, methods and machines in hydrologic analysis. *J. Hydraulics Div. Am. Soc. Civil Engrs.* **91**, No. HY2, 85–99 (1965).
50. Doob, J. L., Some problems concerning the consistency of mathematical models. *In* "Information and Decision Processes" (R. E. Machol, ed.), pp. 27–33. McGraw-Hill, New York, 1960.
51. Kinsman, B., Use and misuse of statistics in geophysics. *Tellus* **9**, 408–418 (1957).
52. Dawdy, D. R., and O'Donnell, T., Mathematical model of catchment behavior. *J. Hydraulics Div. Am. Soc. Civil Engrs.* **91**, 123–137, No. HY4 (1965).
53. Loucks, D. P., A probabilistic analysis of wastewater treatment systems. Ph.D. Thesis, Cornell Univ., Ithaca, New York, 1965.
54. Shapiro, R., and Ward, F., The time-space spectrum of the geostrophic meridional kinetic energy. *J. Meteorol.* **17**, 621–626 (1960).
55. Wu, I. P., Design hydrographs for small watersheds in Indiana. *J. Hydraulics Div. Am. Soc. Civil Engrs.* **89**, No. HY6, 35–66 (1963).

56. Linsley, R. K., Kohler, M., and Paulhus, J. L., "Hydrology for Engineers." McGraw-Hill, New York, 1958.
57. Linsley, R. K., and Franzini, J., "Water Resources Engineering." McGraw-Hill, New York, 1964.
58. Thomas, H. A., Jr., Frequency of minor floods. *J. Boston Soc. Civil Engrs.* **35**, 425–442 (1948).
59. Searcy, J. K., Graphical correlation of gaging-station records. *U.S. Geol. Surv. Water Supply Papers* **1541**-C, 67–100 (1960).
60. Gilman, C. S., Rainfall. *In* "Handbook of Applied Hydrology" (V. T. Chow, ed.), pp. 26–49. McGraw-Hill, New York, 1964.
61. Paulhus, J. L. H., and Gilman, C. S., Evaluation of probable maximum precipitation. *Trans. Am. Geophys. Union* **34**, 701–708 (1953).
62. Hershfield, D. M., Rainfall frequency atlas of the United States. Tech. Rept. No. 40. U.S. Weather Bureau, Wahington, D.C., 1961.
63. Dalrymple, T., Flood frequency analyses. *U.S. Geol. Surv. Water Supply Papers* **1543**-A, (1960).
64. Busch, W. F., and Shaw, L. C., Floods in Pennsylvania, frequency and magnitude. *U.S. Geol. Surv. Open-File Rept.* (1960).
65. Hely, A. G., Areal variations of mean annual runoff. *J. Hydraulics Div. Am. Soc. Civil Engrs.* **90**, No. HY5, 61–68 (1964).
66. Caffey, J. E., Interstation correlations in annual precipitation and in annual effective precipitation. Hydrology Papers No. 6. Colorado State Univ., Fort Collins, Colorado, 1965.
67. International Association of Scientific Hydrology, "Design of Hydrometeorological Networks." Intern. Assoc. Sci. Hydrol., Quebec, Canada, 1965.
68. Feller, W., The asymptotic distribution of the range of sums of independent random variables. *Ann. Math. Statist.* **22**, 427–432 (1951).
69. Hurst, H. E., Methods of using long-term storage in reservoirs. *Proc. Inst. Civil Engrs.* (*London*), **5**, 519–590 (1965).
70. Gould, B. W., Statistical methods for reservoir yield estimation. *Water Res. Foundation, Australia, 1964*, Rept. 8.
71. Thomas, H. A., Jr., and Fiering, M. B., Statistical analysis of the reservoir storage-yield relation. *In* "Operations Research in Water Quality Management" R. P. Burden, ed.) Harvard Univ. Cambridge, Massachusetts, 1963. Final Rept. to U.S. Public Health Service, Washington, D.C.
72. Yevjevich, V. M., Some general aspects of fluctuations of annual runoff in the upper Colorado River Basin. Rept. No. CER 61 VMY 54. Colorado State Univ., Fort Collins, Colorado, 1961.
73. Fiering, M. B., Whither synthetic hydrology. *Western Resources Conf., 1965*. Colorado State Univ., Fort Collins, Colorado, 1965.
74. Yevjevich, V. M., The application of surplus, deficit, and range in hydrology. Hydrology Papers No. 10. Colorado State Univ., Fort Collins, Colorado, 1965.
75. Hurst, H. E., Black, R. P., and Simaika, Y. M., "Long-Term Storage: An Experimental Study." Constable Press, London, 1965.
76. Kotz, S., and Neumann, J., On the distribution of precipitation amounts for periods of increasing length, *J. Geophys. Res.* **68**, 3635–3640 (1963).
77. Alexander, G. N., Karoly, A., and Susts, A. B., Divertible flow of streams. *Proc. Hydrol. Symp., 1965*, Paper No. 2012, pp. 37–47. The Institution of Engineers, Melbourne, Australia.
78. Parzen, E., "Stochastic Processes." Holden-Day, San Francisco, California, 1962.

79. Prabhu, N. U., "Stochastic Processes." Macmillan, New York, 1965.
80. Johnson, N. L., and Leone, F. C., "Statistics and Experimental Design." Wiley, New York, 1964.
81. Wilks, S. S., "Mathematical Statistics." Wiley, New York, 1962.
82. Kazmann, R., New problems in hydrology. *J. Hydrol.* **2**, No. 2, 92–100 (1964).
83. Hurst, H. E., Long-term storage capacity of reservoirs. *Trans. Am. Soc. Civil Engrs.* **116**, 770–799 (1951).
84. Alexander, G. N., Some aspects of time series in hydrology. *J. Inst. Engrs., Australia* **26**, No. 9, 16–26 (1954).
85. Leopold, L. B., Probability analysis applied to a water-supply problem. *U.S. Geol. Surv. Circ.* **410** (1959).
86. Matalas, N. C., Some aspects of time series analysis in hydrologic studies. *Proc. 5th Hydrol. Symp. McGill Univ. Montreal, Canada, 1966.*
87. Sellers, W. D., Precipitation trends in Arizona and western New Mexico. Tech. Rept. No. 43. Inst. Atmospheric Phys., Univ. of Arizona, Tucson, Arizona, 1960.
88. Fiering, M. B., Use of correlation to improve estimates of the mean and variance. *U.S. Geol. Surv. Profess. Papers* **434-C**, (1963).
89. Matalas, N. C., A correlation procedure for augmenting hydrologic data. *U.S. Geol. Surv. Profess. Papers* **434-E**, (1964).
90. Fiering, M. B., Discussion of, Men, models, methods, and machines in hydrologic analysis. *J. Hydraulics Div. Am. Soc. Civil Engrs.* **91**, No. HYS, 311–313 (1965).
91. Papoulis, A., "Probability, Random Variables and Stochastic Processes." McGraw-Hill, New York, 1965.
92. Schneck, H., Jr., "Theories of Engineering Experimentation." McGraw-Hill, New York, 1961.
93. Nash, J. E., The form of the instantaneous unit hydrograph. *Intern. Assoc. Sci. Hydrol. Publ.* **45** 3, 114–121 (1957).
94. Cox, D. R., "Renewal Theory." Methuen, London, 1962.
95. Cox, D. R., and Smith, W. L., "Queues." Methuen, London, 1961.
96. Moran, P. A. P., "Theory of Storage." Methuen, London, 1959.
97. Langbein, W. B., Queuing theory and water storage. *J. Hydraulics Div. Am. Soc. Civil Engrs.* **84**, No. HY5, 1–24 (1958).
98. Fiering, M. B., Queuing theory and simulation in reservoir design. *J. Hydraulics Div. Am. Soc. Civil Engrs.* **87**, No. HY6, 39–69 (1961).
99. Lloyd, E. H., A probability theory of reservoirs with serially correlated inputs. *J. Hydrol.* **1**, 99–128 (1963).
100. Pattison, A., Synthesis of hourly rainfall data. *Water Resources Res.* **1**, 489–498 (1965).
101. Weiss, L. L., Sequences of wet or dry days described by a Markov chain probability model. *Monthly Weather Rev.* **92**, 169–176 (1964).
102. Chow, V. T., and Ramaseshan, S., Sequential generation of rainfall and runoff data. *J. Hydraulics Div. Am. Soc. Civil Engrs.* **91**, No. HY4, 205–223 (1965).
103. Brittan, M. R., Probability analysis applied to the development of synthetic hydrology for the Colorado River. Bur. Econ. Res., Univ. of Colorado, Boulder, Colorado, 1961.
104. Thomas, H. A., Jr., and Fiering, M. B., Mathematical synthesis of streamflow sequences for the analysis of river basins by simulation. *In* "Design of Water-Resource Systems" (Arthur Maass *et al.*, eds.). Harvard Univ. Press, Cambridge, Massachusetts, 1962.
105. Yagil, S., Generation of input data for simulations. *IBM Syst. J.* **2**, 288–296 (1963).

106. Beard, L. R., Use of interrelated records to simulate streamflow. *J. Hydraulics Div. Am. Soc. Civil Engrs.* **91**, 13–22 (1965).
107. Fiering, M. B., A Markov model for low-flow analysis. *Bull. Intern. Assoc. Sci. Hydrol.* **9**, 37–47 (1964).
108. Fiering, M. B., Multivariate techniques for synthetic hydrology. *Proc. Am. Soc. Civil Engrs., J. Hydraulics Div.* **90**, 43–60 (1964).
109. Bendat, J. S., and Piersol, A. G., "Measurement and Analysis of Random Data." Wiley, New York, 1966.
110. Jones, R. H., Prediction of multivariate time series. *J. Appl. Meteorol.* **3**, 285–289 (1964).
111. Speight, J. G., Meander spectra of the Angabunga River. *J. Hydrol.* **3**, 1–15 (1965).
112. Melton, M. A., Methods for measuring the effect of environmental factors on channel properties. *J. Geophys. Res.* **67**, 1485–1490 (1962).
113. Kozin, F., Cote, L. G., and Bogdanoff, J. L., Statistical studies of stable ground roughness. Rept. No. 8391 LL 95. U.S. Army Tank-Automotive Center, Warren, Michigan, 1963.
114. Algert, J. A., A statistical study of bed forms in alluvial channels. Rept. No. CER 65 JHA 26. Colorado State Univ., Fort Collins, Colorado, 1965.
115. Brier, G. W., Diurnal and semidiurnal atmospheric tides in relation to precipitation variations. *Monthly Weather Rev.* **93**, 93–100 (1965).
116. Brier, G. W., and Bradley, D. A., The lunar synodical period and precipitation in the United States. *J. Atmospheric Sci.* **21**, 386–395 (1964).
117. Angell, J. K., and Korshover, J., Harmonic analysis of the biennial zonal-wind and temperature regimes. *Monthly Weather Rev.* **91**, 537–548 (1963).
118. Platzman, G. W., and Rao, D. B., Spectra of Lake Erie water levels. *J. Geophys. Res.* **69**, 2525–2535 (1964).
119. Fritts, H. C., Tree-ring evidence for climate changes in Western North America. *Monthly Weather Rev.* **93**, 421–443 (1965).
120. Streets, R. B., Jr., Integration of analytical and conventional design techniques for optimum control systems. Ph.D. Thesis, Univ. of Arizona, Tucson, Arizona, 1964.
121. Korn, G. A., and Korn, T. M., "Mathematical Handbook for Scientists and Engineers." McGraw-Hill, New York, 1961.
122. Newton, G. C., Gould, L. A., and Kaiser, J. F., "Analytical Design of Linear Feedback Controls." Wiley, New York, 1957.
123. Middleton, D., "An Introduction to Statistical Communication Theory." McGraw-Hill, New York, 1960.
124. Chang, S. S. L., "Synthesis of Optimum Control Systems." McGraw-Hill, New York, 1961.
125. Bendat, J. S., "Principles and Applications of Random Noise Theory." Wiley, New York, 1958.
126. Laning, J. H., and Battin, R. H., "Random Processes in Automatic Control." McGraw-Hill, New York, 1956.
127. Solodovnikov, V. V., "Introduction to the Statistical Dynamics of Automatic Control Systems." Dover, New York, 1960.
128. Smith, O. J. M., "Feedback Control Systems." McGraw-Hill, New York, 1958.
129. Aseltine, J. A., "Transform Method in Linear System Analysis." McGraw-Hill, New York, 1958.
130. Truxal, J. G., "Automatic Feedback Control System Synthesis." McGraw-Hill, New York, 1955.
131. Horowitz, I. M., "Synthesis of Feedback Systems." Academic Press, New York, 1963.

132. Eagleson, P. S., and Shack, W. J., Some criteria for the measurement of rainfall and runoff. *Water Resources Res.* **2**, 427–436 (1966).
133. Julian, P. R., Variance spectrum analysis. *Proc. Ann. Meeting Am. Geophys. Union. Washington, D.C. 1966.*
134. Wiener, N., Generalized harmonic analysis. *Acta Math.* **55**, 117–258 (1930).
135. Amos, D. E., and Koopmans, L. H., Tables of the distribution of the coefficient of coherence for stationary bivariate Gaussian processes. Monograph SCR-483. Sandia Corp., Albuguerque, New Mexico, 1963.
136. Kendall, M. B., "The Advanced Theory of Statistics." Hafner Press, New York, 1951.
137. Cunnyngham, J., The spectral analysis of economic time series. *U.S. Bur. Census Working Paper* **14**, (1963).
138. Nettheim, N. F., A spectral study of "overadjustment" for seasonability. *U.S. Bur. Census Working Paper* **21**, (1965).
139. Rosenblatt, H. M., Spectral analysis and parametric methods for seasonal adjustment of economic time series. *Bur. Census Working Paper* **23** (1965).
140. Brown, R. G., "Smoothing, forecasting and prediction of discrete time series." Prentice-Hall, Englewood Cliffs, New Jersey, 1963.
141. Gracie, G., Analysis of distribution demand variations. *J. Am. Water Works Assoc.* **58**, 51–66 (1966).
142. United Nations, Methods of hydrological forecasting for the utilization of water resources. *Trans. Inter-Regional Seminar, Bangkok, Thailand, Water Resources Seri.* No. 27 (1964).
143. Bratranek, A., Long term forecasts of river flows and their importance for an economic operation of reservoirs. *Hydraulic Res. Inst., Prague-Podbaba, Czechoslovakia, 1962.*
144. Kohler, M. A., Multicapacity basin accounting for predicting runoff from storm precipitation. *J. Geophys. Res.* **67**, 5196 (1962).
145. Muller, F. B., Gervais, J. G., and Shaw, R. W., The effect of precipitation on the level of Lake Michigan/Huron. *Can. Dept. Transport Circ.* 4264 (1965).
146. Brier, G. W., Interpretation of trends and cycles. *Proc. Symp. Environ. Measurements, 1964, Cincinnati, Ohio*, pp. 265–272. U.S. Public Health Serv., Washington, D.C., 1964.
147. Saaty, T. L., and Bram, J., "Nonlinear Mathematics." McGraw-Hill, New York, 1964.
148. Kuznetsov, P. I., Stratonovich, R. L., and Tikhonov, V. I., "Nonlinear Transformation of Stochastic Processes." Pergamon Press, Oxford, 1965.
149. Shen, J., Use of analog models in the analysis of flood runoff. *U.S. Geol. Surv. Profess. Papers* **506-A**, (1965).
150. Schweppe, J. L., Eichberger, L. C., Muster, E. L., and Paskusz, G. F., Methods for the dynamic calibration of pressure transducers. *Natl. Bur. Std. U.S. Monograph* **67** (1963).
151. Hays, J. R., Schnelle, K. B., Jr., and Krenkel, P. A., Application of frequency response technique to the analysis of turbulent diffusion phenomena. *Proc. 19th Ann. Ind. Waste Conf., Purdue, Lafayette, Indiana, 1964.*
152. Goodman, N. R., Measurement of matrix frequency response functions and multiple coherence functions. Res. and Technol. Div. Rept. No. AFFDL TR 65–66. Wright-Patterson AFB, Dayton, Ohio, 1965.
153. Restrepo, J. C. O., and Eagleson, P. S., Optimum discrete linear hydrologic systems with multiple inputs. Rept. 80. Hydrodynamics Lab., M.I.T., Cambridge, Massachusetts, 1965.

Time Series Analysis of Hydrologic Data

154. Eagleson, P. S., Mejia, R., and March, F., The computation of optimum realizable unit hydrographs from rainfall and runoff data. Rept. 40. Hydrodynamics Lab., M.I.T., Cambridge, Massachusetts, 1965.
155. O'Donnell, T., Instantaneous unit hydrograph derivation by harmonic analyses. *Intern. Assoc. Scientific Hydrol. Publ.* **51**, 546–557 (1960).
156. Bayazit, M., Instantaneous unit hydrograph derivation by spectral analysis and its numerical application. *Proc. CENTO Symp. Hydrol. Water Resource Develop., Ankara, Turkey, 1966.*
157. Schuster, A., The investigation of hidden periodicities. *J. Terrestrial Magnetism* **3**, 13 (1898).
158. Bartlett, M. S., Periodogram analysis and continuous spectra. *Biometrika* **37**, 1–16 (1950).
159. Parzen, E., Mathematical considerations in the estimation of spectra. *Technometrics* **3**, 167–190 (1961).
160. Jenkins, E., General considerations in the analysis of spectra. *Technometrics* **3**, 133–166 (1961).
161. Anderson, R. L., Distribution of the serial correlation coefficient. *Ann. Math. Statist.* **8**, 1–13 (1941).
162. Tukey, J. W., and Winsor, C. P., Note on some chi-square normalizations. Rept. 29. Statist. Res. Group, Princeton Univ., Princeton, New Jersey, 1949.
163. Hald, A., "Statistical Tables and Formulas." Wiley, New York, 1952.
164. Jenkins, G. M., Cross-spectral analysis and the estimation of linear open loop transfer functions. *In* "Time Series Analysis" (M. Rosenblatt, ed.), Chapter 18. Wiley, New York, 1963.
165. Goodman, N. R., On the joint estimation of the spectra, cospectrum and quadrature spectrum of a two-dimensional stationary Gaussian process. Paper No. 10. Engr. Statist. Lab., New York Univ., New York, 1957.
166. Tukey, J. W., Uses of numerical spectrum analysis in geophysics. *Intern. Statist. Inst., Beograd, Yugoslavia, 1965.*
167. Gargett, A., Long term fluctuations in the Toronto temperature and precipitation record. *Can. Dept. Transport Circ.* 4199 (1965).
168. Roden, G. I., On nonseasonal temperature and salinity variations along the West Coast of the United States and Canada. *Bull. Scripps Inst. Oceanogr. Univ. Calif.* **3**, 95–119 (1959–60).
169. Brooks, C. E. P., and Carruthers, N., "Handbook of Statistical Methods in Meteorology." H.M. Stationery Office, London, 1953.
170. Kotz, S., and Neumann, J., Autocorrelation in precipitation amounts. *J. Meteorol.* **16**, 683–685 (1959).
171. Wold, H., On least square regression with autocorrelated variables and residuals. *Bull. Intern. Statist. Inst.* **32**, No. 2 (1950).

THEORY OF SEEPAGE FROM OPEN CHANNELS

HERMAN BOUWER

U.S. Water Conservation Laboratory
U.S. Department of Agriculture, Phoenix, Arizona

I. Introduction 121
II. Hydrodynamics of Seepage 122
    A. Basic Flow Conditions 122
    B. Channels in Uniform Soil with an Impermeable Floor . . 124
    C. Channels in Uniform Soil with a Permeable Floor . . . 131
    D. Channels with Clogged Soil at Their Perimeter 139
    E. Channels in Nonuniform Soil 145
    F. Effect of Channel Geometry on Seepage 148
    G. Unsaturated Flow 151
    H. Transient Flow Systems 157
III. Measurement of Soil Hydraulic Conductivity 163
    A. Hydraulic Conductivity Measurements of Soil below the Water Table 164
    B. Hydraulic Conductivity Measurements of Soil above the Water Table 164
    C. Hydraulic Conductivity Measurements below Surface Inundations 168
IV. Summary and Conclusions 169
Symbols 169
References 170

I. Introduction

Seepage in the context of this chapter refers to the process of water movement into and through the soil from a body of surface water as may occur in canals, streams, or impoundments. Quantitative knowledge of seepage rates may be desirable in determining seepage losses from canals or streams, in evaluating surface-subsurface water relationships, etc. Such knowledge can be obtained by prediction or by direct measurement. Prediction of seepage is based on knowledge of the relevant hydraulic properties of the soil and of the boundary conditions, subjecting the flow system in question to a hydrodynamic analysis. Direct evaluation of seepage from flowing channels is based on obtaining discharge measurements at various points (inflow-outflow technique), or by measuring the rate of water movement into the bottom or bank (seepage meter or tracer techniques). The material presented in this chapter will be concerned with the prediction of seepage, i.e., the hydrodynamics of seepage flow systems and the measurement of hydraulic properties of soil.

This article does not represent an exhaustive review of past and present literature. Of the "classic" seepage analyses found in the various textbooks, only the resulting equations or graphs are included. The chief aim of this article is to bring together some of the advances in various scientific disciplines and their application to the analysis and prediction of seepage losses from open channels.

II. Hydrodynamics of Seepage

A. Basic Flow Conditions

1. Field Conditions

In actuality, seepage flow systems are characterized by channels of irregular cross section, nonuniformity of soil in horizontal as well as vertical extent, changing elevations of the water surface in the channel and of the water table in the soil, and other complications. The inherent heterogeneity of natural soils is augmented by processes in the channel (erosion, sedimentation, biological action, etc.) and by the effect of the chemical constituents of the water on the hydraulic properties of the soil. This causes *hydraulic conductivity* to change in time as well as in space. The hydraulic conductivity is also affected by the air content of the soil. Complete saturation, even below the water table, will be the exception rather than the rule. The volume of entrapped air below the water table can vary with time depending on the air content and the temperature of the seepage water, the temperature of the soil, and the barometric pressure. Unsaturated conditions also occur where the soil-water pressures are less than atmospheric.

The free boundary of seepage flow systems is normally taken as the *water table*, which is the zero (gage) pressure isopiestic line. For steady flow systems and relatively deep water tables, this boundary can be taken as a solid boundary. If the water table is sufficiently close to the surface of the soil to support evaporation, an upward flux across the water table will be maintained. The magnitude of this flux depends on the evaporative demand and on the depth of the water table below field surface or the root zone of the vegetation. Excess rainfall or deep percolation from irrigation causes a downward flux across the water table. Where the position of the free boundary is not constant (transient systems), fluxes across the boundary occur because of draining or filling of pore space in the soil. The seepage flow system following entry of water in dry channels is a problem of two-dimensional infiltration.

In view of the multitude of complexities with which seepage flow systems are afflicted in nature, theoretical treatment must begin with simplification of the soil and boundary conditions, particularly for mathematical treatment. With numerical techniques using resistance network analogs or digital computers, nonuniform soil conditions and other complexities can be included with relative ease. The accuracy of the seepage predicted for a given channel in that way then depends largely on how well the pertinent soil, water table, and boundary conditions can be characterized. Although solution techniques may yield exact or accurate answers, the fact that the model for which the solution is obtained is always a simplified version of the field situation renders the solution an estimate at best.

2. Theoretical Models

There are three basic conditions to which the multitude of natural profiles of soil hydraulic conductivity can be reduced for theoretical treatment of seepage flow systems. These conditions, labeled A, B, and C in accordance with Bouwer [1], are as follows:

Condition A. The soil in which the channel is imbedded is uniform and underlain by more permeable (considered as infinitely permeable) material.

Condition B. The soil in which the channel is imbedded is uniform and underlain by less permeable (considered as impermeable) material.

Condition C. The soil in which the channel is imbedded is of much lower hydraulic conductivity than the original soil for a relatively short distance normal to the channel perimeter (clogged soil, semipermeable linings).

To obtain general solutions, the depth of the infinitely permeable or impermeable material for conditions A and B should be treated as a variable. If the depth of this material is infinity, the case of seepage into uniform soil of infinite (great) depth is obtained. The position of the water table at some distance from the channel should also be treated as a variable. For steady-state systems, the simplest function of the water table is that of a solid boundary. For transient systems, filling or draining of pore space causes fluxes across the free boundary. The case of seepage to a free-draining permeable layer in the subsoil is a special case of condition A, and it is obtained by letting the water table be at or below the top of the permeable material. This condition will be labeled A'.

Solutions for seepage flow systems under conditions A, B, and C will be presented in the following sections. These solutions all apply to the steady-state condition. Some typically transient systems are discussed in a separate section. Other sections refer to including the flow at negative soil-water pressures in the analysis, to the treatment of flow systems in layered or nonuniform soil, and to the field measurement of soil hydraulic conductivity.

B. CHANNELS IN UNIFORM SOIL WITH AN IMPERMEABLE FLOOR

Geometry and symbols for the case of seepage from channels in uniform soil underlain by impermeable material (condition B) are shown in Fig. 1. For a discussion of how low the hydraulic conductivity of the underlying material must be before it can be considered zero, reference is made to Sect. II.E.3.

The water table is considered a solid boundary. The point where the water table is characterized by D_w should be at sufficient distance from the channel for complete development of *Dupuit-Forchheimer flow*. The horizontal distance between the channel center and this point is called L.

The seepage rate is expressed in terms of volume rate per unit length of channel and per unit width of the water surface in the channel. The resulting parameter, I_s, has the dimension of a velocity, and it can be visualized as the rate of fall of the water surface in the channel due to seepage as if the channel were ponded. The term I_s is expressed per unit hydraulic conductivity K of the soil in which the channel is imbedded to yield the dimensionless parameter I_s/K. Some of the approaches that have been presented to relate I_s/K to the geometry of the flow system will be discussed in the following sections.

1. Dupuit-Forchheimer Method

The simplest solution for seepage under condition B is obtained with the Dupuit-Forchheimer assumption (D-F assumption) of horizontal flow. If the channel is rectangular and the bottom of the channel extends to (or into) the impermeable layer, the D-F assumption yields an exact expression for the seepage ([2] and references therein). For that case, the distance between the channel bank and the control point is $L - \tfrac{1}{2}W_b$, and the average height of the flow system is $H_w - \tfrac{1}{2}D_w$. Thus, the following equation can be written:

$$\frac{I_s}{K} = \frac{2D_w}{W_s} \frac{H_w - \tfrac{1}{2}D_w}{L - \tfrac{1}{2}W_b} \qquad (1)$$

Equation (1) can be extended to a channel with a sloping bank and with the impermeable layer at some distance D_i below the channel bottom as follows:

$$\frac{I_s}{K} = \frac{2D_w}{W_s} \frac{H_w + D_i - \tfrac{1}{2}D_w}{L - \tfrac{1}{4}(W_b + W_s)} \qquad (2)$$

This equation is no longer exact, and the error in I_s/K will increase with increasing D_i. This is illustrated in Fig. 2, where Eq. (2) is compared with

Theory of Seepage from Open Channels

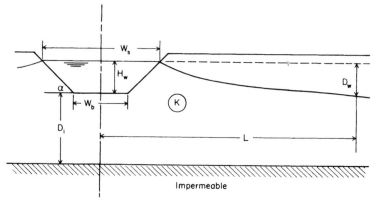

FIG. 1. Geometry and symbols for channels in soil underlain by impermeable material.

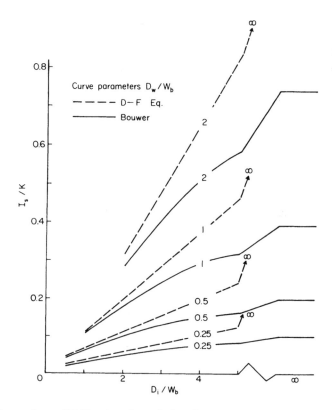

FIG. 2. Comparison of D-F and analog solutions for seepage from a trapezoidal channel for condition B ($\alpha = 45°$, $H_w/W_b = 0.75$).

solutions obtained with an electrical resistance network analog for a trapezoidal canal with $\alpha = 45°$ and $H_w = 0.75 W_b$. The analog solutions, taken from Bouwer [1], are discussed in detail in Sect. II.B.3. The continued linearity between I_s/K and D_i implied by the D-F assumption leads to complete failure of Eq. (2) when D_i approaches infinity. For relatively small D_i, for example, $D_i < 3W_b$, reasonably accurate solutions can be obtained with Eq. (2).

The D-F solution for condition B has been followed by more refined analyses, which are not restricted to relatively small values of D_i, as discussed in the following two sections. The present utility of the D-F solution is mainly that it forms a convenient basis for the analysis of transient systems (Sect. II.H.2).

2. Combination Methods

The error in Eq. (2) is due to the curvature and divergence of the streamlines in the vicinity of the channel. To account for the extra head losses in this zone, the flow in the region near the channel can be analyzed as a separate flow system, which is then joined to the D-F system assumed to occur at some distance from the channel. The developments by Dachler [3] and Ernst [4] are based on this approach.

Dachler ([3], see also Muskat [5]) divided the flow system on the basis of model studies into a region with curvilinear flow (region I) and one with D-F flow (region II), with the dividing line between the two systems at a distance

$$L_1 = \frac{W_s + H_w + D_i}{2}$$

from the center of the channel (Fig. 3). The flow in region I was analyzed with an approximate equation for the potential and streamline distribution under a plane source of finite width. The pattern of streamlines and equipotentials calculated with the equation was used to develop factors F that enabled calculation of the flow in region I as

$$\frac{I_s}{K} = \frac{2F \Delta H}{W_s} \tag{3}$$

where ΔH is the vertical distance between the water surface in the canal and the ground water table at the dividing line between the two flow regions. Values of F were presented in relation to $W_s/(D_i + H_w)$ and $(H_w + D_i - \Delta H)/(H_w + D_i)$ for relatively deep and for relatively shallow channels, using $W_s/$w.p. as criterion (Fig. 4). Because the effect of channel shape on seepage

Theory of Seepage from Open Channels

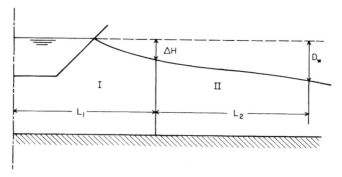

Fig. 3. Division of flow system in regions I and II for Dachler's analysis.

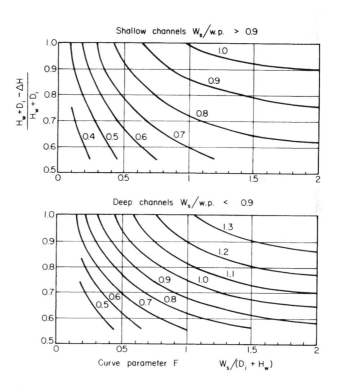

Fig. 4. Dachler's values of F for shallow and for deep channels.

is minor for condition B, this procedure of ignoring the actual shape and depth of the channel is not objectionable (see Sect. II.F.1).

The flow in region II can be expressed with the D-F theory as

$$\frac{I_s}{K} = \frac{2(D_w - \Delta H)}{W_s L_2} [D_i + H_w - \tfrac{1}{2}(\Delta H + D_w)] \tag{4}$$

In general, the problem will be to calculate the seepage for a given value of D_w at a distance $L_1 + L_2$ from the channel center, and ΔH will not be known initially. In that case, a trial-and-error procedure is employed, assuming different values of ΔH and calculating I_s/K with Eqs. (3) and (4). The correct magnitude of ΔH is then found as the value yielding equal values of I_s/K.

To compare the results from Dachler's method with those obtained by analog (see Sect. II.B.3), I_s/K was calculated for a trapezoidal channel with $\alpha = 45°$ and $H_w/W_b = 0.75$, taking $L_1 + L_2 = 10W_b$ and using different values for D_w and D_i. The results (Fig. 5) show excellent agreement. Because the effect on I_s/K of a given increase in D_i decreases with increasing D_i (see Sect. II.B.3), the I_s/K values obtained with Dachler's method for the largest D_i values for which F factors are available appear to be already relatively close to those for the theoretical case of $D_i = \infty$.

Ernst ([4], see also Van Beers [6]) analyzed the flow toward parallel drainage canals by resolving the total flow system into three component systems, i.e., vertical flow, horizontal flow, and "radial" flow. The latter refers to the convergence of streamlines as they approach the drainage channel. Each component system is described by a separate equation, and the flow rate is found by equating the total head loss in the flow system to the sum of the head losses of the individual component systems. Using this approach, Ernst developed approximate solutions for the flow toward parallel drainage channels for various profiles of layered soil.

Applying Ernst's approach to the problem of channel seepage under condition B, the head loss h_r due to radial flow can be described as

$$h_r = \frac{I_s W_s}{\pi K} \ln \frac{D_i + H_w}{\text{w.p.}} \tag{5}$$

This is an approximate expression based on an analysis of the flow system to a line sink. The head loss h_h due to horizontal flow is calculated with the D-F theory as

$$h_h = \frac{I_s W_s}{2K} \frac{L}{D_i + H_w - \tfrac{1}{2}D_w} \tag{6}$$

The head loss due to vertical flow is only significant where the flow has to pass through a local layer of low hydraulic conductivity. Thus, for seepage

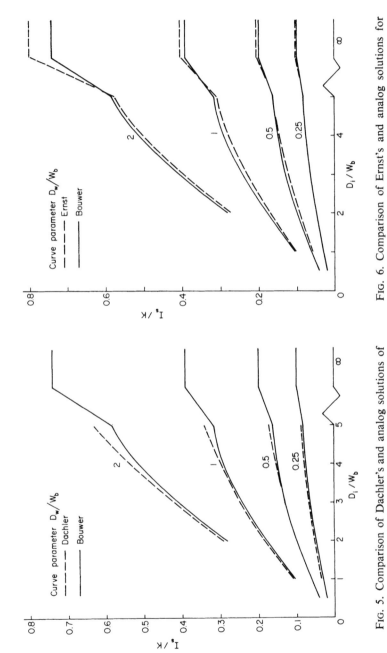

FIG. 5. Comparison of Dachler's and analog solutions of seepage from a trapezoidal channel ($\alpha = 45°$, $H_w/W_b = 0.75$).

FIG. 6. Comparison of Ernst's and analog solutions for same channel as in Fig. 5 ($H_w/W_b = 0.75$).

flow systems under condition B, $D_w = h_r + h_h$, which, after substitution of Eqs. (5) and (6) and rearrangement of terms, yields

$$\frac{I_s}{K} = \frac{D_w/W_s}{(1/\pi)\ln[(D_i + H_w)/\text{w.p.}] + [\tfrac{1}{2}L/(D_i + H_w - \tfrac{1}{2}D_w)]} \quad (7)$$

Equation (5) was developed for semicircular channels of radius r, where the wetted perimeter is πr. The equation can be used for channels of other shapes by substituting the actual wetted perimeter as shown in the equation. For shallow channels ($W_s \gg H_w$), h_r is more accurately estimated by the following expression:

$$h_r = \frac{I_s W_s}{\pi K} \ln \frac{4(D_i + H_w)}{\pi W_s} \quad (8)$$

Values of I_s/K calculated with Eq. (7), modified by using Eq. (8) for h_r instead of Eq. (5), are compared with results obtained by resistance network analog (see Sect. II.B.3) in Fig. 6 for a trapezoidal canal with $H_w = 0.75 W_b$ and $\alpha = 45°$. Equation (7) was developed for relatively small values of D_i and D_w, and, as shown in Fig. 6, excellent agreement is obtained with the analog results in this region.

The restraint on D_i of Ernst's analysis can be illustrated by calculating I_s/K for increasing D_i at constant D_w. In theory, I_s/K increases with D_i at decreasing rates to reach a maximum at $D_i = \infty$ (see Sect. II.B.3). However, Eq. (7) yields a maximum I_s/K at finite D_i, after which I_s/K decreases to 0 as D_i approaches ∞. For relatively small D_w, this maximum appears to agree with I_s/K at infinite D_i as obtained by analog. This is illustrated in Fig. 6, where Ernst's values of I_s/K for $D_i = \infty$ are taken as the maximum values yielded by Eq. (7).

3. Analog Solutions

The solutions for condition B discussed in the previous section apply to relatively small values of D_i and D_w. Solutions for the complete range of D_i and D_w, including the theoretical limits of infinite values, were obtained by Bouwer using a resistance network analog [1, 7]. The results of the analog analyses were expressed in dimensionless graphs, showing I_s/K as a function of D_w/W_b for different values of D_i/W_b (Fig. 7). The analyses were performed for trapezoidal canals with $\alpha = 45°$ and three different water depths (expressed as H_w/W_b), yielding one graph for each water depth. In addition to condition B, solutions were obtained for conditions A and A', as will be discussed in Sect. II.C.3.

The distance L for characterizing the water table position by D_w (Fig. 1) was selected as $10W_b$ for the analyses of Fig. 7. Inspection of flow nets of completed systems indicated that this distance was adequate for development of Dupuit-Forchheimer flow. An example of streamlines and equipotentials for a seepage system for condition B is shown in Fig. 8.

To apply the graphs of Fig. 7 to channels of other shapes, W_b can be computed from the actual values of W_s and H_w as if the channel were trapezoidal with $\alpha = 45°$, or the cross section can be replaced by the best-fitting trapezoidal cross section with $\alpha = 45°$. Values of I_s/K for water depths other than those in Fig. 7 can be evaluated by interpolation.

As regards the effect of D_i on seepage for a given water table position, Fig. 7 shows that I_s/K initially increases linearly with increasing D_i, but that the rate of increase decreases as D_i becomes relatively large. For $D_i > 5W_b$, I_s/K is already relatively close to the values for $D_i = \infty$. Thus, for the trapezoidal channels of Fig. 7, impermeable floors have a significant effect on seepage only if their distance below the channel is less than $5W_b$.

A similar trend is observed regarding the effect of D_w on seepage. Taking the curve for $D_i = \infty$, for example, I_s/K increases initially almost linearly with D_w, but at decreasing rates as D_w becomes relatively large. For all three graphs of Fig. 7, the value of I_s/K is already close to that for $D_w = \infty$ when D_w has reached a value of about three times the width of the water surface in the channel. Thus, a general lowering of the ground water table due, for example, to pumping, would result in significant increases in seepage only if the initial depth of the ground water level were considerably less than $3W_s$ below the water surface in the channel.

C. Channels in Uniform Soil with a Permeable Floor

Geometry and symbols for the case of seepage from channels in uniform soil underlain by very permeable material (condition A) are shown in Fig. 9. For theoretical treatment, the underlying material is considered to be of infinite K. For a discussion of the relative value of K of the permeable material whereby this condition is approached, reference is made to Sect. II.E.3.

The water table is again considered a solid boundary. Since the underlying material is taken as infinitely permeable, the water table approaches a horizontal line with increasing distance from the channel. The point where the water table is characterized by D_w should be at sufficient distance from the channel so that the water table is essentially horizontal.

If the water table is below the top of the permeable material, i.e., $D_w > (D_p + H_w)$, the flow system is one of seepage to a free-draining permeable layer (Fig. 14). In that case, the top of the permeable material can be taken as a fixed boundary of atmospheric pressure, provided that this material is sufficiently coarse (see Sect. II.D.1). Thus, the top of the permeable layer

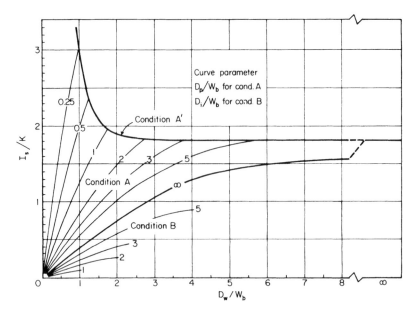

Fig. 7a. Results of seepage analyses with electric analog for trapezoidal channel with $\alpha = 45°$ and $H_w/W_b = 0.75$.

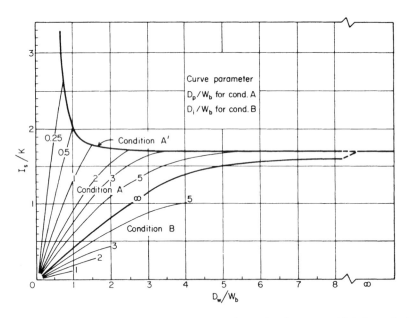

Fig. 7b. Results of seepage analyses with electric analog for trapezoidal channel with $\alpha = 45°$ and $H_w/W_b = 0.5$.

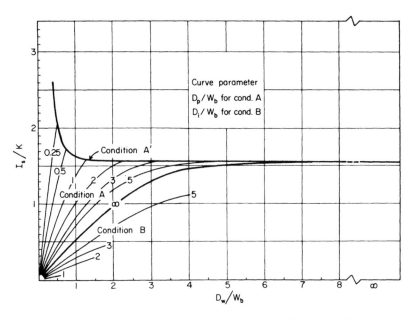

Fig. 7c. Results of seepage analyses with electric analog for trapezoidal channel with $\alpha = 45°$ and $H_w/W_b = 0.25$.

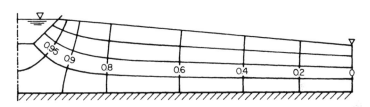

Fig. 8. Flow system obtained by analog for condition B.

serves as a lower limit for the effective water table position for condition A. The condition of seepage to a free-draining, permeable layer is a special case of condition A, and it will be labeled condition A'. It is characterized by an effective D_w value of $H_w + D_p$, although the actual depth of the water table may be greater than $H_w + D_p$.

Analytical solution of seepage for condition A' has received considerable attention in the literature, and it will be discussed first. The section will then continue with solutions of seepage for condition A, and conclude with the presentation of analog solutions for both conditions A and A'.

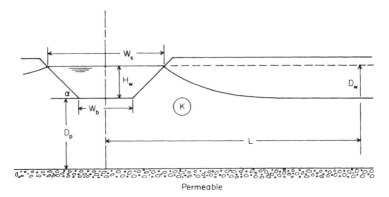

FIG. 9. Geometry and symbols for channels in soil underlain by permeable material.

1. Condition A'

If D_p is allowed to become infinity, condition A' represents the case of seepage into uniform soil of infinite depth with an infinitely deep water table. This case was first solved by Kozeny [8], using Zhukovsky's function, and later by Pavlovsky [8a] (see also Muskat [5], Harr [9], Polubarinova-Kochina [10]). The shape of the channel in this analysis was not selected *a priori*, but was taken as trochoidal so that it would fit the equation describing the equipotentials in the flow system. The resulting expression for the seepage rate was

$$\frac{I_s}{K} = 1 + 2\frac{H_w}{W_s} \qquad (9)$$

Since the flow at great depth becomes vertically downward under unit hydraulic gradient, the right-hand part of Eq. (9) expresses the ratio of the width of the flow system at great depth to the width of the water surface in the channel. Vertically downward flow and maximum width of the flow system are essentially reached at a distance of $1.5(W_s + 2H_w)$ below the water level in the channel.

Exact solutions for trapezoidal canals and finite as well as infinite values of D_p were obtained by Vedernikov ([11], see also Muskat [5], Harr [9], Polubarinova-Kochina [10]), using hodograph and conformal mapping techniques. The solution by Vedernikov for $D_p = \infty$ takes the form

$$\frac{I_s}{K} = 1 + A\frac{H_w}{W_s} \qquad (10)$$

where A is a function of the geometry of the system. In addition to an analytical expression for A, Vedernikov presented a graph relating A to W_s/H_w for different values of α (Fig. 10).

To calculate I_s/K for triangular channels with Eq. (10), Vedernikov presented a graph of A versus α for channels with $W_b = 0$ (Fig. 11). If $\alpha = 45°$, $A = 2$ and Eq. (10) is the same as Eq. (9) for the trochoidal channel. The

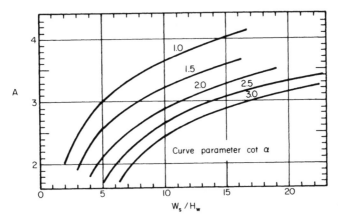

Fig. 10. Graph showing A as a function of W_s/H_w for different values of α.

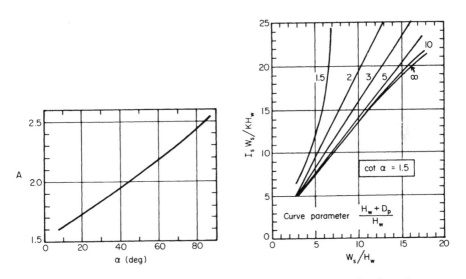

Fig. 11 (left). Graph showing A as a function of α for triangular channels.

Fig. 12 (right). Graph showing $I_s W_s/KH_w$ as a function of W_s/H_w for different values of $(H_w + D_p)/H_w$ for trapezoidal canal under condition A'.

following approximate equation for A in Eq. (10) was also presented by Vedernikov for calculating I_s/K from triangular channels for condition A' with $D_p = \infty$:

$$A = 2.12(\alpha/180) + 1.47 \tag{11}$$

Vedernikov's analysis of seepage under condition A' with finite values of D_p does not yield an explicit expression for I_s, and the resulting solution is difficult to handle ([11], see also Muskat [5], Harr [9], Polubarinova-Kochina [10]). A graph relating $I_s W_s/KH_w$ to W_s/H_w for different values of $(H_w + D_p)/H_w$ for a trapezoidal canal with cot $\alpha = 1.5$, based on Vedernikov's equations, was presented by Muskat [5]. This graph, with the additional curve $(H_w + D_p)/H_w = 1.5$ from Harr [9], is shown in Fig. 12.

2. Condition A with Relatively Small Values of D_w

Seepage flow systems for condition A with $D_w < (H_w + D_p)$ have been analyzed by Hammad [11a], using two steps of conformal mapping. The shape of the channel was considered curvilinear and characterized by H_w and W_s. The resulting equation for the seepage was

$$I_s = K \frac{D_w}{W_s} \frac{2K_1}{K_1' - c} \tag{12}$$

where K_1 and K_1' are the complete elliptic integrals of the first kind corresponding to the modulus k_1 and the complementary modulus k_1', respectively. The moduli are defined as

$$k_1 = \frac{1}{2}\left[\frac{W_s'}{2} + \left(\frac{W_s'^2}{4} - 2H_w'^2\right)^{1/2}\right] \tag{13}$$

and

$$k_1' = (1 - k_1^2)^{1/2} \tag{14}$$

The other terms are defined as

$$c = H_w'/k_1 \tag{15}$$

$$H_w' = \tan\left[\frac{\pi H_w}{2(H_w + D_p)}\right] \quad \text{for} \quad H_w < D_p \tag{16}$$

and

$$\tfrac{1}{2}W_s' = \tanh\left[\frac{\pi W_s}{4(H_w + D_p)}\right] \quad \text{for} \quad H_w < D_p \tag{17}$$

Theory of Seepage from Open Channels

The solution by Hammad implies linearity between I_s and D_w, which is valid only if D_w is small compared to $H_w + D_p$ (see Fig. 15).

Channel seepage for condition A with relatively small values of D_w can also be calculated with Ernst's solution for the two-layered soil profile (Sect. II.E.3). Condition A for the layered profile is obtained by taking K_2 in Eq. (26) infinitely large, which yields

$$\frac{I_s}{K} = \frac{\pi D_w}{W_s \ln[a(D_p + H_w)/\text{w.p.}]} \tag{18}$$

The most applicable value of a in this equation is 4.3, which is evaluated from Fig. 20 for the largest ratios of K_2/K_1 and $D_2/(D_1 + H_w)$. As with Hammad's solution, the equation based on Ernst's approach also implies linearity between I_s and D_w, which restricts the validity of Eq. (18) to relatively small values of D_w (see Fig. 15).

3. Analog Solutions for Conditions A and A'

The analysis of channel seepage with a resistance network analog [1] also included seepage under conditions A and A'. The resulting curves are shown in Fig. 7. As was done for condition B, the distance where the water table was characterized by D_w was selected as $10W_b$. Inspection of completed flow systems showed that this distance was sufficient for the water table to become essentially horizontal. The curves for condition A' in Fig. 7 are the loci of the endpoints of the curves for condition A. At these points, $D_w = D_p + H_w$, and further lowering of the water table does not increase the effective value of D_w. Thus, for condition A', the D_w values at the abscissa should be interpreted as $D_p + H_w$.

The curves for condition A show that the effect on seepage of a permeable layer in the subsoil becomes rather small if $D_p > 5W_b$. High values of I_s/K are obtained for relatively small values of D_p, particularly if D_w approaches $D_p + H_w$. For condition A', I_s/K decreases rapidly with increasing value of the effective D_w, i.e., $D_p + H_w$, and, when D_p has become equal to approximately $3H_w$, I_s/K is already close to I_s/K for $D_p = \infty$.

Examples of flow systems for conditions A and A', obtained by resistance network analog, are shown in Figs. 13 and 14, respectively.

A comparison of some of the results obtained with the resistance network analog and the exact and approximate analytical solutions by Vedernikov, Hammad, and Ernst, respectively, is shown in Fig. 15 for a trapezoidal canal with $\alpha = 45°$ and $H_w = 0.5W_b$. The points based on Vedernikov's analysis were obtained using Fig. 10 for the case of infinite D_p and Fig. 12 for the case of finite D_p. Because Fig. 12 applies to $\cot \alpha = 1.5$, the A values were

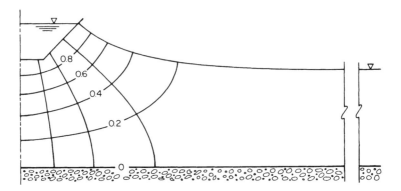

FIG. 13. Flow system for seepage under condition A obtained by analog.

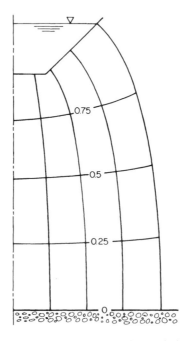

FIG. 14. Flow system for seepage under condition A' obtained by analog.

evaluated for a channel with the same depth H_w and the same width at half-depth as the actual channel with $\alpha = 45°$. The points based on Hammad's solution were taken from an earlier comparison presented by Hammad [12]. Ernst's points were calculated with Eq. (18). The agreement between the analog results and Vedernikov's analysis is excellent. Ernst's and Hammad's

Theory of Seepage from Open Channels

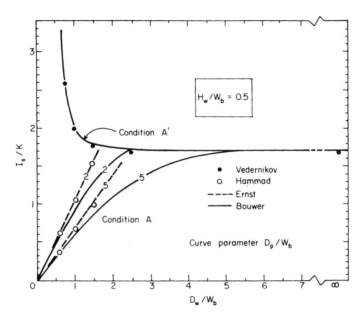

FIG. 15. Comparison of several solutions for I_s/K for a trapezoidal channel with $\alpha = 45°$ under conditions A and A'.

solutions give reasonable estimates of I_s/K at relatively low values of D_w/W_b, but increasingly erroneous values as D_w/W_b increases.

D. Channels with Clogged Soil at Their Perimeter

1. Complete Impeding Layers

Geometry and symbols for the case of seepage from channels with a relatively thin layer of low hydraulic conductivity along their wetted perimeter (condition C) are shown in Fig. 16. Such a seepage-restricting layer may be due to sedimentation of fine particles, biological action, or other natural causes. The layer could also be of artificial origin, such as sealing layers formed by waterborne chemicals or earth linings for seepage control.

If the hydraulic conductivity, K_a, of the restricting layer is sufficiently small to cause the downward flow rate in the underlying soil to be numerically less than K of this soil, the soil beneath the restricting layer will be unsaturated (provided that the water table is sufficiently deep for the channel bottom to be well above the capillary fringe and that air has access to the underlying

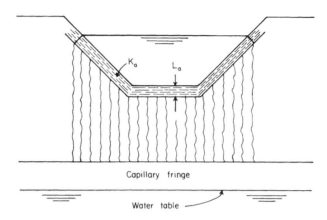

FIG. 16. Geometry and symbols for channel with a thin layer of low hydraulic conductivity at its perimeter.

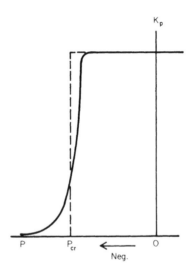

FIG. 17. Schematic relationship between K_p and P.

soil). In that case, the flow beneath the restricting layer will be due to gravity alone and thus at unit hydraulic gradient.

The (negative) soil-water pressure head, P, and the unsaturated hydraulic conductivity, K_p, in the soil beneath the restricting layer, are functionally related, as in any unsaturated porous material. The relationship between K_p and P is a characteristic of the soil, and its determination for a given soil is done by experimental procedure ([13] and references therein). Curves relating

K_p to P are usually sigmoid (Fig. 17), and they can, according to Gardner (see Bouwer [13]) be approximately described by the equation

$$K_p = \frac{a}{(-P)^n + b} \tag{19}$$

where a/b is the K_p value at $P = 0$, referred to as K in this article. Orders of magnitude of the factors a, b, and n for different soils are shown in Table I.

TABLE I

ORDERS OF MAGNITUDE FOR a, b, AND n IN EQ. (19) FOR STRUCTURELESS SOILS

	K at $P = 0$ (cm/day)	a	b	n
Medium sands	500	5×10^9	10^7	5
Fine sands, sandy loams	50	5×10^6	10^5	3
Loams and clays	1	5×10^3	5×10^3	1.5

Because the unsaturated flow beneath the restricting layer is due to gravity alone, P will be uniform in the zone between the restricting layer and the top of the capillary fringe. Thus, the infiltration rate, i, at any point of the bottom can be described with Darcy's equation as

$$i = K_a \frac{H_w + L_a - P}{L_a} \tag{20}$$

Because of the unit hydraulic gradient in the material underlying the restricting layer, K_p and i are numerically equal. Therefore, Eq. (19) can be written as

$$i = \frac{a}{(-P)^n + b} \tag{21}$$

Solving this equation for P and substituting the resulting expression into Eq. (20) yields

$$i = K_a \frac{H_w + L_a + [(a/i) - b]^{1/n}}{L_a} \tag{22}$$

The expression for the seepage through the channel banks can be derived in similar fashion. The resulting equations then enable the calculation of I_s.

Since i is not explicit in Eq. (22), a procedure of successive approximations should be used. However, as explained in the following paragraphs, a simpler expression for I_s can be developed.

Because of the sigmoid nature of K_p versus P curves, most of the reduction in K_p takes place over a relatively narrow range of P. This is particularly true for uniform, granular materials where the K_p-P curve already approaches a step function. The center, P_{cr}, of this P range (Fig. 17) was defined by Bouwer [13] as

$$P_{cr} = \frac{\int K_p \, dP}{K} \tag{23}$$

where K is K_p at $P = 0$. The integration should be carried out between $P = 0$ and a value of P where K_p has become insignificantly small compared to K [13]. The actual value of P below the restricting layer can be expected to be relatively close to the value of P_{cr} of the underlying soil for a considerable range of values of K_p, and thus of i. Therefore, P_{cr} of the soil beneath the restricting layer can be substituted for P in Eq. (20), instead of the P value calculated with Eq. (21).

Another simplification can be made by considering that restricting layers are usually rather thin (clogged surfaces, sediment layers, etc.). In that case, L_a will be relatively small compared to $H_w - P_{cr}$ as appearing in Eq. (20) after substitution of P_{cr}, and can be neglected. Furthermore, where L_a is relatively small, the actual value of L_a may be difficult to determine. This may also be true for K_a of the restricting layer. In that case, the hydraulic property of the restricting layer is more conveniently expressed and measured (see Sect. III.C.2) in terms of its hydraulic impedance, R_a, defined as L_a/K_a.

Substituting P_{cr} and R_a in Eq. (20) and neglecting L_a in the numerator yields

$$i = \frac{H_w - P_{cr}}{R_a} \tag{24}$$

Applying this equation to the trapezoidal channel of Fig. 15 yields the following expression for I_s:

$$I_s = (W_s R_a)^{-1}[(H_w - P_{cr})W_b + (H_w - 2P_{cr})(H_w/\sin \alpha)] \tag{25}$$

For triangular channels, this equation is used with $W_b = 0$. For rectangular channels, $\sin \alpha = 1$.

The replacement of the actual P value below the slowly permeable layer by P_{cr} is valid for granular soils where K_p versus P already approaches a step function. For cohesive soils, the reduction in K_p is more gradual, and, if i is quite small compared to K (for example, $i < 0.01K$), the actual value of P below the impeding layer may be considerably less (more negative) than P_{cr}. Numerical values of P_{cr} range from approximately -20 cm water for coarse and medium sands, to -50 to -100 cm water for fine sands and sandy loams, and to -150 cm and less for structureless loams and clays. Techniques for measuring the air-entry value of soils *in situ* as estimates of P_{cr} and use of the seepage meter technique for direct measurement of R_a of relatively thin, restricting layers are discussed in Sects. III.B.1 and III.C.2, respectively.

If the channel extends into the capillary fringe, Eq. (24) can be applied to calculate the seepage for that part of the channel perimeter which is above the capillary fringe (using the appropriate value of the water depth). The same equation can be used for calculating the seepage for the rest of the perimeter. The P value to be used in this equation is then found from the vertical distance between the restricting layer and the water table. (P is negative if the water table is below the point in question of the restricting layer, and positive if the water table is above that point.) This procedure is valid if K in the capillary fringe and below the water table is large compared to i, so that the build-up of hydraulic head in the capillary fringe and below the water table can be ignored.

Applying the discussion in the preceding paragraphs to the case of seepage to a permeable drainage layer (condition A'), it will be clear that P at the top of this layer, which forms the lower boundary value in the analysis of the flow system, is governed by the K_p-P relationship of the permeable material. In the same manner as for condition C, the value of P at the top of the drainage layer can be taken as P_{cr} of the material in this layer. The usual assumption that $P = 0$ at the top of the permeable layer (see Sect. II.C.1) is valid only if P_{cr} of the permeable material is numerically small compared to $H_w + D_p$, which is the total head loss in the flow system if P at the top of the permeable layer is taken as zero.

2. Partial Impeding Layers or Linings

Natural as well as artificial restricting layers are not always present over the entire wetted perimeter. Sediment may be deposited only on the channel bottom, or on only part of the bottom where the channel is curved. Artificial linings are sometimes applied only to the bottom or to one or both banks, or portions of the lining may have badly deteriorated and lost their effectiveness.

The seepage-reducing effect of natural or artificial layers of low hydraulic conductivity along the channel perimeter can be largely lost if such layers do not cover the entire wetted perimeter. Lining of the channel banks alone did not have a significant effect on the water table position, and thus not on seepage, in a viscous-flow model study of seepage under condition B with relatively large D_i [14]. For smaller values of D_i, more significant seepage reductions can be expected, and, if $D_i = 0$, lining the channel banks alone will essentially be as effective as lining the entire channel perimeter (depending on K of the "impermeable" material!).

Studies with a resistance network analog [1] showed that lining of the channel bottom alone does not result in appreciable seepage reduction, except for small values of H_w and/or of D_p, when most of the original seepage already takes place through the bottom. This is illustrated in Fig. 18, where the

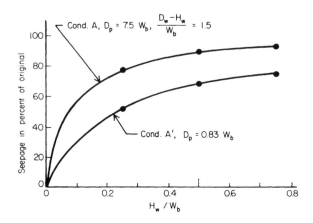

FIG. 18. Seepage in percent of original seepage by covering channel bottom with an impermeable layer.

reduction in seepage by covering the channel bottom with an impermeable layer is plotted in relation to H_w/W_b for condition A with a relatively large value of D_p, and for condition A' with a relatively small value of D_p. The curve for condition A, which can be considered to be generally indicative of channels in soils that are uniform to great depth, shows that significant reductions in seepage are obtained only if $H_w \ll 0.1 W_b$. For condition A', the seepage reductions are somewhat larger because of the high proportion of bottom seepage prior to sealing the bottom [1]. Curves for condition A' with D_p values larger than $0.83 W_b$ will be above the A' curve in Fig. 18, and vice versa. Thus, unless the water depth in the channel is very small compared to the bottom width, or unless permeable, free-draining layers are present at relatively small depth below the channel bottom, the presence of a seepage-restricting layer on the bottom will have only a minor effect on the seepage rate from the channel.

Theory of Seepage from Open Channels

E. CHANNELS IN NONUNIFORM SOIL

1. Heterogeneous Soil

Seepage flow systems in nonuniform soils can be analyzed using a numerical technique based on Laplace's equation in finite-difference form. The tediousness of earlier manual approaches, such as the relaxation technique [15], has been eliminated with the application of high-speed digital computers [16–18] and electrical analogs [7, 19, 20]. Of the electrical analogs, the resistance network analog is particularly suitable for solution of flow system in heterogeneous or anisotropic soil [7]. With computer or analog techniques, an almost endless variety of soil conditions can be represented, including irregular nonuniformities or anisotropies. The major problem in applying these techniques to obtain a prediction of seepage for a given situation will be the collection of the necessary input information.

Physical models, although not as versatile as computer or analog systems, can also be used to obtain solutions for nonuniform media. The Hele-Shaw viscous-flow model [21, 22] has been applied in various seepage studies [14, 22]. Sand models were used by Dachler [3].

Solutions obtained with numerical techniques or with electrical or physical models apply only to a specific system. Generalization of the results requires repetitive solutions in which the principal parameters are treated as variables [1]. The tank-analog study by Todd and Bear [20] was designed for the Sacramento River valley. Seepage solutions were obtained for a two-layer profile, and different values for the thickness and K value of each layer were employed. Anisotropy was included by transforming layers with anisotropic soil into equivalent layers of isotropic soil (see next section). Other variables were the depth and width of the river and the width of the levee base. Sloping and horizontal water tables at fixed position were simulated. Part of the objective of the study was to determine the upward flux at the water table as a function of distance from the river.

2. Anisotropic Soil

Because seepage flow systems are generally treated as two-dimensional systems, consideration of soil anisotropy is usually limited to the case where the hydraulic conductivity in vertical direction, K_z, differs from that in the horizontal direction, K_x. Thus, the soil is assumed to be horizontally isotropic. With digital computers or resistance network analogs, anisotropy is included by letting the values of K in the horizontal lines between the network nodes differ from those in the vertical lines between the nodes in accordance with the K_x/K_z ratio of the medium in question.

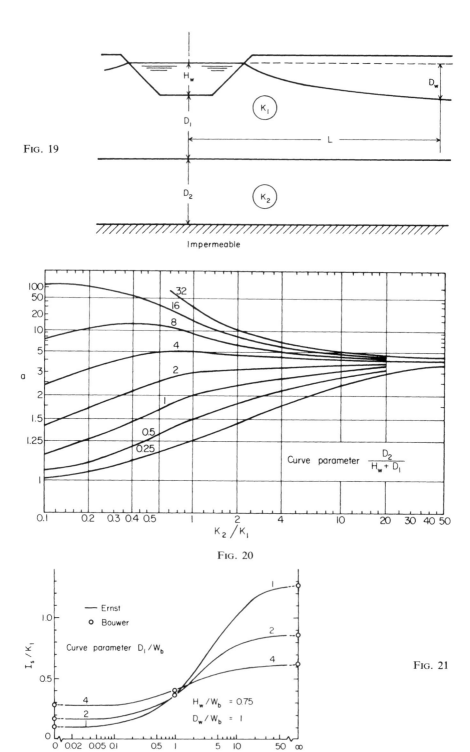

Fig. 19

Fig. 20

Fig. 21

With analytical solutions or electrolytic-tank analogs, anisotropy is not handled as readily as with the numerical techniques, and transformation to an equivalent isotropic medium will be necessary. To transform an anisotropic medium to an equivalent isotropic medium [23, 24], the vertical dimensions of the anisotropic medium are multiplied by $(K_x/K_z)^{1/2}$. The flow system is then treated as one in a fictitious isotropic medium with a K value of $(K_x K_z)^{1/2}$.

3. Layered Soil

The case of seepage in a two-layered soil underlain by an impermeable layer with the channel in the upper layer (Fig. 19) was analyzed by Ernst [4]. To calculate the radial resistance, a factor a was introduced behind the logarithm symbol in Eq. (5). Based on solutions obtained with the relaxation technique, a graph was constructed relating a to K_2/K_1 for different values of $D_2/(D_1 + H_w)$. This graph, taken from Van Beers [6], is shown in Fig. 20. Calculating the horizontal resistance on the basis of the product of hydraulic conductivity and average saturated height of each layer and adjusting Ernst's equation from the case of drainage with uniform flux across the water table to that of seepage with the water table taken as a solid boundary yields the following expression:

$$\frac{I_s}{K_1} = \frac{\frac{D_w}{W_s}}{\frac{0.5K_1 L}{K_1(D_1 + H_w - 0.5D_w) + K_2 D_2} + \frac{1}{\pi}\ln\frac{a(H_w + D_1)}{\text{w.p.}}} \qquad (26)$$

Equation (26) was used to calculate I_s/K_1 in relation to K_2/K_1 for different values of D_1/W_b for a trapezoidal channel with $H_w = 0.75 W_b$ and $\alpha = 45°$. If $K_2 = K_1$, the system of Fig. 19 reduces to the soil profile for condition B with $D_i = D_1 + D_2$, and the results from Eq. (26) should agree with those presented in Fig. 7. Using the data from Fig. 7 as matching points for this case, the resulting curves are shown in Fig. 21. The curve for $D_1/W_b = 2$ was calculated with $D_2 = 2(D_1 + H_w)$. The other two curves applied to $D_2 = 4(D_1 + H_w)$. Figure 21 is taken from Bouwer [24a].

FIG. 19. Geometry and symbols for channels in two-layered soil underlain by an impermeable layer.

FIG. 20. Values of a as a function of K_2/K_1 for different values of $D_2/(H_w + D_1)$.

FIG. 21. Graph of I_s/K_1 as a function of K_2/K_1 for channel in two-layered profile.

If $K_2/K_1 = 0$, the two-layered soil reduces to the profile for condition B with $D_i = D_1$. Similarly, if $K_2/K_1 = \infty$, condition A is obtained with $D_p = D_1$. The values of I_s/K_1 for these extreme values of K_2/K_1, as evaluated from Fig. 7 and plotted as open circles in Fig. 21, appear to be accurately approached by the curves calculated with Eq. (26).

The curves in Fig. 21 give some indication of how high or how low K_2 should be in relation to K_1 before K_2 can be considered infinity (for condition A) or zero (for condition B). As can be expected, the curves show that the effect of K_2/K_1 on I_s/K_1 increases with decreasing D_1. However, if $D_1 > 2W_b$, condition A is essentially reached when $K_2 > 10K_1$ and condition B when $K_2 < 0.1K_1$.

F. Effect of Channel Geometry on Seepage

1. Effect of Channel Shape

The most important channel factors with respect to seepage are the width W_s of the water surface and the water depth H_w in the center of the channel. For given values of W_s and H_w, seepage increases from a triangular cross section to trapezoidal and rectangular cross sections. The magnitude of the increases depends on the soil and water table conditions. This is illustrated in Table II, which is taken from Bouwer [1] and applies to a channel with $H_w = 0.3W_s$. The effect of channel shape on seepage can be expected to be particularly pronounced for condition A' with relatively low D_p, where a change from a triangular to a rectangular cross section significantly reduces the extent of the soil medium over which the total head $H_w + D_p$ is dissipated. In contrast to this, the effect of channel shape on seepage for condition B with relatively small D_i will be quite minor. For condition C, seepage for the different cross sections was calculated with Eq. (25) for $P_{cr} = 0$ and expressed in terms of the dimensionless parameter $I_s W_s R_a/H_w^2$. It appears that, for this condition, seepage increases approximately in proportion to the cross-sectional area of the channel.

The effect of channel shape on I_s/K for condition A' with infinite D_p can also be evaluated from Fig. 10, where curves for different values of $\cot \alpha$ are shown. Morel-Seytoux [25] applied hodograph techniques, Schwarz-Christoffel transformations, and the Green-Neumann function to obtain solutions of seepage for condition A' with $D_p = \infty$ for channels of different geometry, including shapes deviating from the standard rectangular, trapezoidal, and triangular cross sections. The different standard shapes to which the various shapes could be reduced were characterized by $(A_c)^{1/2}$ and α, and a dimensionless graph relating $I_s W_s/K(A_c)^{1/2}$ to W_s/H_w for different values

TABLE II

EFFECT OF CHANNEL SHAPE ON SEEPAGE AND ON RELATIVE SEEPAGE LOSSES
($H_w/W_s = 0.3$)

Cross section	Seepage condition	D_w/W_s	D_p/W_s	D_i/W_s	I_s/K	Index of relative seepage losses
Triangular	A	0.9	3	—	0.93	156
Trapezoidal	A	0.9	3	—	1.00	100
Rectangular	A	0.9	3	—	1.13	73
Triangular	A'	∞	∞	—	1.56	145
Trapezoidal	A'	∞	∞	—	1.81	100
Rectangular	A'	∞	∞	—	2.31	83
Triangular	A'	0.63	0.33	—	1.68	140
Trapezoidal	A'	0.63	0.33	—	2.03	100
Rectangular	A'	0.63	0.33	—	2.83	91
Triangular	B	0.37	—	0.2	0.064	171
Trapezoidal	B	0.37	—	0.2	0.063	100
Rectangular	B	0.37	—	0.2	0.066	68
Triangular	C	$I_s W_s R_a/H_w^2 = 1.94$				119
Trapezoidal	C	$= 2.74$				100
Rectangular	C	$= 4.33$				103

of α was presented (Fig. 22). For channels of near optimum hydraulic radius, the results could be generalized into the equation

$$\frac{I_s}{K} = \frac{4.2}{K}(A_c)^{1/2} \tag{27}$$

which is in agreement with the data for condition A' with $D_p = \infty$ in Table II.

The minimum value of the coefficient in Eq. (27) as given by Morel-Seytoux is 3.76, which applies to a channel of which the shape was defined by Preissmann [26] as

$$y = \frac{1}{\pi} \int_{-I_s W_s/4K}^{x} \ln \frac{1 - \sin(2\pi K x/I_s W_s)}{1 + \sin(2\pi K x/I_s W_s)} \tag{28}$$

using an approximate solution technique. The resulting geometry, which is the shape of the channel yielding minimum seepage for a given value of A_c, is reproduced from Preissmann [26] in Fig. 23.

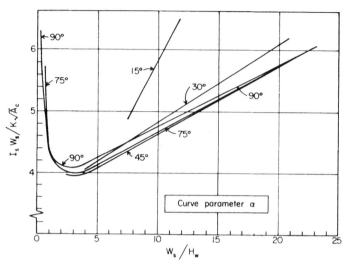

FIG. 22. Graph of $I_s W_s/K(A_c)^{1/2}$ versus W_s/H_w for condition A' with $D_p = \infty$.

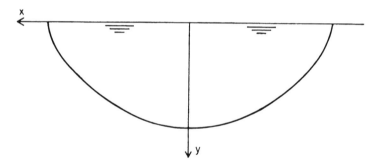

FIG. 23. Channel geometry yielding minimum seepage for a given value of A_c under condition A' with $D_p = \infty$.

To properly assess the effect of channel shape on seepage, the effect of shape on the discharge in the channel should also be taken into account, so that the relative seepage losses can be evaluated. This was done for the data in Table II, where the discharge in the channels was calculated with Manning's formula. The relative seepage losses, i.e., seepage loss divided by channel discharge, for the trapezoidal channel were arbitrarily set at 100 and the other relative losses proportionally adjusted. The data in Table II show that, for a given value of H_w and W_s, rectangular channels are more efficient conveyors of water than trapezoidal and triangular channels.

2. Effect of Water Depth

Seepage increases more with increasing water depth for condition A' with relatively small values of D_p and for condition C than for conditions A and B. This was shown by Bouwer [1], where the effect of H_w/W_b on I_s/K for a trapezoidal channel with $\alpha = 45°$ and a constant water table depth at a horizontal distance of $10W_b$ from the channel was evaluated with the graphs of Fig. 7 for conditions A, A', and B. The effect of H_w on I_s/K for condition C was calculated with Eq. (25). For a proper evaluation of the effect of water depth on seepage, the effect of water depth on the discharge in the channel for uniform flow should also be taken into account in this case. Using again Manning's equation to calculate the discharge in the channel, it appeared that discharge increased more with water depth than did seepage [1]. Thus, a channel with uniform flow tends to be a more efficient conveyor of water when it is deep than when it is shallow.

G. Unsaturated Flow

1. Theoretical Aspects

Use of the water table or the line of zero pressure (gage) as the boundary of the flow system assumes that the hydraulic conductivity of the soil becomes zero when $> _d 0$. In reality, however, the hydraulic conductivity at small negative values of P is essentially the same as the hydraulic conductivity at $P \geq 0$. The reduction in hydraulic conductivity with decreasing P does not take place until P has reached a value whereby drainage of pore space and entry of air into the system begin. Once air has replaced part of the water in the soil, the hydraulic conductivity usually has undergone a large reduction to values that are only a small fraction of the hydraulic conductivity at $P \geq 0$.

As in the previous sections, the hydraulic conductivity at $P \geq 0$ will be called K. This value of K is often referred to as the saturated hydraulic conductivity. However, saturated conditions will not always, and perhaps rarely, prevail in natural seepage flow systems because of entrapped air. Thus, it is preferable to refer to the hydraulic conductivity below the water table, or at the "wet" side of the free boundary, as K at $P \geq 0$. Available data seem to indicate that K at $P \geq 0$ may range from the value at complete saturation to approximately one-half that value ([27] and references therein), depending on the amount of entrapped air in the soil.

The hydraulic conductivity at $P < 0$ will be designated K_p. The relation between K_p and P is a characteristic of the soil, and it must be experimentally determined for the material in question ([13] and references therein). Plotting

K_p versus P on Cartesian coordinates generally yields sigmoid curves (see Fig. 17). Curves of K_p versus P collected from the literature for different soils are shown by Bouwer [13].

The flow in unsaturated soil, or rather soil with negative values of P, is generally treated on the basis of the Darcy equation. Sewell and Van Schilfgaarde [17] furnished proof that the streamlines and equipotential lines in unsaturated soil are orthogonal, provided that the soil is isotropic at saturated conditions.

To include flow at negative values of P in the solution of underground flow systems, the K_p-P relationship of the soil or soils in question must be known. An iterative procedure is then used to obtain such a distribution of K_p that the values of P yielded by the flow system satisfy the K_p-P relationship of the soil at each point in the system. The iterative procedure is tedious and requires execution by resistance network analog [7, 19] or digital computer [16, 17].

An example of a complete solution for a steady-state seepage system under condition A' is presented in Fig. 24. This figure is obtained from two illustrations of Reisenauer's analysis with a digital computer [16]. In addition to streamlines and equipotentials, contours of K_p, expressed as K_p/K, are shown. The values of P and the volumetric water contents corresponding to the

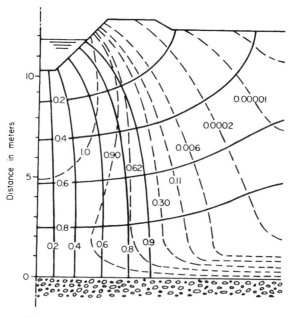

FIG. 24. Seepage flow system for condition A' with flow at negative P values included. The dashed lines are contours of K_p/K.

K_p/K ratios for the soil in question (gravelly sand) are given in Table III. As shown in Fig. 24, there is an oval zone with positive values of P immediately below the channel. The rest of the flow system is at negative values of P. The 90% streamline coincides approximately with the K_p/K contour of 0.5, which shows that most of the seepage occurs in a region with K_p values relatively close to K. Thus, the "wet" end or upper portion of the K_p-P relationship is the most significant part of the curve. The concentration of essentially horizontal contours of K_p/K above the permeable layer at some distance from the centerline indicates the presence of a capillary fringe above the drainage layer.

Solution of seepage flow systems on the basis of the K_p-P relationship yields a higher seepage rate than solving the flow system on the assumption that the hydraulic conductivity is zero when $P < 0$. This is because of the additional medium at saturated or near-saturated hydraulic conductivity resulting from the inclusion of the K_p-P relationship. For systems of mainly downward flow (conditions A and B with relatively large D_w and condition A'), the increase in seepage is due to an increase in the effective width of the zone of mainly downward flow. For systems with predominantly horizontal flow (condition B with relatively small D_i), the seepage increase is caused by an increase in the height of the flow system due to the introduction of a *capillary fringe*.

Inclusion of negative-pressure flow in the solution of seepage flow systems by computer or resistance network analog is not a simple matter. In addition, it will be necessary to use experimentally determined curves of K_p versus P for each soil in the flow system. The determination of these curves is also a

TABLE III

RELATION BETWEEN K_p/K, P, AND VOLUMETRIC WATER CONTENT OF SOIL FOR FLOW SYSTEM OF FIG. 24

K_p/K	P (cm water)	Volumetric water content
1	0	0.33
0.90	−12	0.32
0.62	−31	0.30
0.30	−58	0.26
0.11	−74	0.22
0.006	−110	0.16
0.0002[a]	−170	0.12
0.00001[a]	−280	0.10
	−420	0.09

[a] Extrapolated values.

time-consuming process, and representative K_p-P relationships for field soils are difficult to obtain. Moreover, the K_p-P relationship for a given soil is not unique, but depends on the antecedent moisture conditions. Since the flow at negative pressures often constitutes only part of the total flow system, and since the total system is usually dominated by the flow at positive and/or small negative pressures, accurate solution of the flow at the complete range of negative pressures is normally not necessary. For practical purposes, therefore, a more simple, though less accurate, solution may be more appropriate. Such a solution will be discussed in the next section.

2. Simplified Analysis

A simplification can readily be applied to include the flow above the water table in systems with mainly horizontal flow. This is done by replacing the actual capillary fringe in which K_p continuously decreases with increasing vertical distance from the water table, by an equivalent fringe in which K is constant and equal to K below the water table. At the top of the equivalent fringe, an abrupt reduction of K to zero is then assumed to take place.

The height of the equivalent fringe is calculated so that the flow in this fringe is the same as that in the actual fringe. In accordance with this, the pressure head at the top of the equivalent fringe, called the critical pressure head P_{cr}, was derived by Bouwer [13] as

$$P_{cr} = \frac{\int_0^{P_g} K_p \, dP}{K} \tag{29}$$

where P_g is the P value at field surface and K is the hydraulic conductivity at $P \geq 0$. Ignoring vertical flow in the fringe, as was done in the derivation of this equation, the height of the equivalent fringe will then be $-P_{cr}$.

As shown in Fig. 17, P_{cr} constitutes the width of a step function of height K with the same area as that under the actual K_p-P curve. Although the integration in Eq. (29) should be carried out between 0 and P_g, integration between 0 and a P value whereby K_p is insignificantly small compared to K will be sufficient for practical purposes, provided that the top of the fringe is below field surface. As discussed in Sect. II.D.1, values of P_{cr} may range from -20 cm or more to -200 cm or less [13, 27]. A field method for estimating P_{cr} via the air entry value of the soil is discussed in Sect. III.B.1.

The increase in seepage obtained by the inclusion of the flow in the capillary fringe in the analysis of seepage systems with predominantly horizontal flow will be proportional to the height of the capillary fringe in relation to the height of the flow system below the water table. In formula,

$$\left(\frac{I_s}{K}\right)_{P_{cr}} = \left(\frac{I_s}{K}\right)_0 \left(1 + \frac{-P_{cr}}{H_w + D_i - \frac{1}{2}D_w}\right) \tag{30}$$

where $(I_s/K)_{P_{cr}}$ refers to the seepage with the flow in the capillary fringe included, and $(I_s/K)_0$ to the seepage obtained with the water table as the upper boundary of the flow system. Equation (30) shows that the flow in the fringe is significant only if $-P_{cr}$ is not small compared to $H_w + D_i - \frac{1}{2}D_w$. The validity of Eq. (30) is limited to relatively small values of D_i, where I_s/K is essentially linear with D_i. For larger values of D_i, the effect on seepage of the addition of a capillary fringe can be estimated more accurately by adding the fringe thickness to the depth of the impermeable layer. The resulting value of $D_i - P_{cr}$ is then used as the effective value of D_i in evaluating I_s/K according to one of the procedures in Sects. II.B.2 and II.B.3. The validity of this approach was demonstrated for the case of flow toward parallel subsurface drains [28].

The inclusion of flow at negative P values in the analysis of seepage systems by the use of P_{cr} essentially means that the flow system is solved on the basis of a constant K value throughout the system, but with P_{cr} instead of zero as the pressure condition at the free boundary. The validity of using P_{cr} for systems with mainly downward water movement is not as readily demonstrated for this case as for the systems with mainly horizontal movement, except when the actual K_p-P relationship of the soil already approaches a step function. By comparison, however, it was shown that the P_{cr} concept also yielded accurate solutions for systems of mainly downward flow if K_p exhibited a more gradual reduction with decreasing P. This comparison was made by Bouwer [13] for the seepage system of Fig. 24, which was solved with a resistance network analog for a P_{cr} value of -45.4 cm as calculated with Eq. (29) from the data in Table III. The resulting value of I_s/K was 2.27, which agreed very well with Reisenauer's value of 2.20 obtained with the actual K_p-P relationship. Good agreement also existed regarding the shape of the zone of positive P below the channel, and the location of the effective boundary of the flow system [13]. It seems, therefore, that the P_{cr} concept can also be used to include negative-pressure flow in systems of mainly downward water movement, such as seepage under condition A'.

An illustration of how seepage, expressed as $I_s W_s/KW_b$, is affected by P_{cr} is given in Fig. 25, which was constructed from data obtained with a resistance network analog [1]. Because P_{cr} has the dimension of a length, it can be divided by some other length parameter (W_b in this case) for dimensionless expression. The channel geometry for Fig. 25 is trapezoidal with $\alpha = 45°$. The curve with $H_w/W_b = 0$ was obtained with Risenkampf's analysis [29]. The effect of $-P_{cr}/W_b$ on $I_s W_s/KW_b$, as evidenced by the general slope of the curves, increases with increasing D_p/W_b. This is because the opportunity for full streamline divergence increases with increasing D_p [1]. The curves also show that the relative increase in seepage with increasing $-P_{cr}/W_b$ decreases with increasing H_w/W_b. This effect can be explained by considering that, when H_w is small, the streamline divergence with $P_{cr} = 0$ is also small, so that the relative increase in streamline divergence, and hence in seepage, by letting P_{cr} become increasingly negative is greater than when H_w is relatively large.

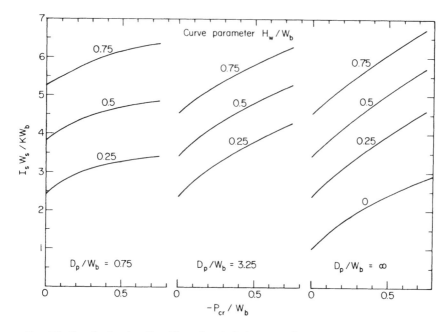

FIG. 25. Graph showing the effect of $-P_{cr}/W_b$ on $I_s W_s/KW_b$ for seepage from a trapezoidal channel with $\alpha = 45°$ under condition A' for different values of H_w/W_b and D_p/W_b.

If $-P_{cr}$ is small compared to W_b, the effect on seepage of flow at negative P values in the soil can be ignored. For a discussion of the effect of P_{cr} on the shape of the oval zone of positive P below the channel and on the location of the free boundary, reference is made to Bouwer [1].

Mathematical analyses of the effect of assigning a negative pressure to the free boundary on seepage for condition A' with $D_p = \infty$ were presented by Vedernikov ([30], see also Polubarinova-Kochina [10]) for trapezoidal canals and by Risenkampf ([29], see also Harr [9], Polubarinova-Kochina [10]) for the flow below an inundated strip of zero water depth at soil surface. It is of interest to note that in this earlier work the pressure at the free boundary was taken as the negative static height of capillary rise, h_{cap}, in the soil. Because the hydraulic conductivity K_p in the zone of capillary rise decreases with height, however, the numerical value of h_{cap} can considerably exceed that of P_{cr} as defined by Eq. (29). The effect of decreasing hydraulic conductivity in the capillary zone was considered by Averjanov, who suggested that h_{cap} be multiplied by 0.3 to obtain a more realistic value of the negative pressure at the free boundary [1, 10, 31].

H. TRANSIENT FLOW SYSTEMS

Transient seepage flow systems can be divided into two basic types. The first type is the seepage occurring after water has entered and then continues to flow in an initially dry channel. The second type is seepage under condition *A* or *B* without external control of the position of the water table at relatively great distance from the channel. The absence of drainage of the seepage water in that case causes the water table to rise until (1) drainage begins to occur somewhere and a control position for the water table is reached, (2) the water table is sufficiently close to ground surface for the seepage flow to be disposed by evaporation from soil or plants, or (3) the water table becomes horizontal at the same elevation of the water surface in the channel and seepage ceases.

Of the transient systems under conditions *A* and *B*, those of condition *B* are of most concern. The permeable layer of condition *A* not only causes the water table to become almost horizontal at a relatively small distance from the channel, but its short-circuiting effect also yields an essentially uniform flux across the water table over its entire near-horizontal position. If the lateral extent of the seepage system is large, the water table will rise only very slowly. Thus, flow systems under condition *A* tend to approach a steady state. Transient systems under condition *A* can be treated as a succession of steady states, assuming that the entire flow is stored above the near-horizontal portion of the water table. In view of this, the discussion of transient flow systems will be limited to the case of seepage from a dry channel following the onset of inundation, and to seepage under condition *B*.

1. Seepage from Initially Dry Channels

Seepage following entry of water into a dry channel should in principle be treated as a problem of two-dimensional infiltration. Computer solutions for the one-dimensional case [32] based on the theory of infiltration of water into soil [33, 34] must then be extended to the two-dimensional case. For a number of situations, however, the water depth will be relatively large for most of the channel perimeter. This condition is conducive to piston-like flow (sharp wet front and constant *K* in the wetted zone) in the soil surrounding the channel, which permits a simplified treatment of the problem using Green and Ampt's approach [35]. The infiltration rate at each point of the channel bottom can be described with this approach as

$$i = K \frac{H_w + L_w - P_w}{L_w} \tag{31}$$

The factor K in this equation refers to K in the wetted zone, which is the K value at $P \geq 0$ for sorption. This value may be taken as one-half the value of K for saturation [27]. The term L_w refers to the depth of the wet front below the channel bottom, and P_w is the water entry value of the soil. This value, which is negative, is the soil-water pressure head just above the wet front, and it may be estimated as one-half the air-entry value, P_a, of the soil [27]. Values of K for sorption and of P_a for field conditions can be obtained with the air-entry permeameter (see Sect. III.B.1).

The rate of advance of the wet front can be described as

$$dL_w/dt = i/f \tag{32}$$

where f is the difference in volumetric moisture content between the soil before and after wetting. Combining Eqs. (31) and (32) and solving for t yields

$$t = \frac{f}{K}\left[L_w - (H_w - P_w) \ln \frac{H_w + L_w - P_w}{H_w - P_w}\right] \tag{33}$$

This equation applies to vertically downward flow, as can be expected below the channel bottom. For the flow into the channel banks, which is assumed to occur in a direction normal to the bank, a similar expression can be derived. The resulting equation for the flow at a given point of the bank is

$$t = \frac{f}{K}\left[\frac{L_w}{\cos \alpha} - \frac{h_w - P_w}{\cos^2 \alpha} \ln \frac{L_w \cos \alpha + h_w - P_w}{h_w - P_w}\right] \tag{34}$$

where h_w is the water depth at the point in question and α refers to the angle of the bank with the horizontal (see Fig. 1, for example).

Using Eqs. (33) and (34), graphs can be constructed relating t and L_w for different values of H_w, P_w, and α. Assuming that the infiltration flow occurs in a direction normal to the channel perimeter, such graphs can then be used to determine the position of the wet front for different values of t. After plotting the successive positions of the wet front, the area A_w of the wetted zone can be evaluated for different values of t. Multiplying A_w by f then yields the volume of water that has entered the soil from the channel.

This procedure was applied to a trapezoidal channel with $\alpha = 45°$ and $H_w = 0.75W_b$. The assumption was made that P_w was negligibly small compared to H_w and to most of the h_w values for the points on the banks. Thus, P_w was taken as zero. The resulting successive positions of the wetting front as characterized by the dimensionless time parameter Kt/fW_s are shown in Fig. 26. Since the infiltration flow is considered to be normal to the channel

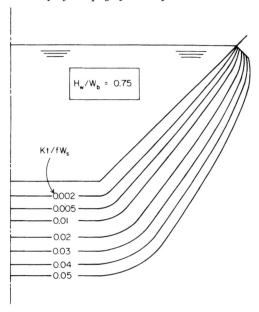

FIG. 26. Positions of wet front in relation to Kt/fW_s due to seepage from an initially dry channel.

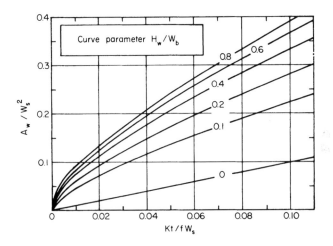

FIG. 27. Graph of A_w/W_s^2 versus Kt/fW_s for different values of H_w/W_b for seepage from an initially dry trapezoidal channel with $\alpha = 45°$.

perimeter, the analysis does not yield the position of the wet front in the zone extending from the toe of the bank between the lines normal to the bank and normal to the channel bottom. The wet fronts in this zone were arbitrarily constructed as the connecting curves between the positions of the wet fronts below the bank and below the bottom. The resulting error is initially small, but increases with increasing Kt/fW_s.

Repeating the procedure used to construct Fig. 26 for other values of H_w, a dimensionless graph could be prepared relating A_w/W_s^2 to Kt/fW_s for different values of H_w/W_b. The resulting graph (Fig. 27) applies to a trapezoidal canal with $\alpha = 45°$ and $P_w = 0$. To use this graph in determining the accumulated seepage losses at time t, the parameter Kt/fW_s is calculated for the desired value of t. The corresponding value of A_w/W_s^2 is read from the graph for the appropriate curve of H_w/W_b. Knowing W_s, A_w is calculated, and multiplying A_w by f yields the accumulated volume of seepage per unit length of canal since water entered the channel.

The calculation of seepage on the basis of Eqs. (33) and (34) is applicable only to relatively small values of t and L_w. As L_w increases, the streamlines no longer are normal to the channel banks but curve downward because of gravity. An estimate of the seepage in relation to time for the more advanced stages of soil wetting below the channel can be obtained by solving the problem as a succession of steady-state systems for condition A' with increasing D_p. The assumptions in this approach are that the bottom of the wetted zone is horizontal and that the wetted zone itself is of uniform K. From results of analyses with a resistance network analog [1], plots of flow system geometries for condition A' could be prepared for increasing values of D_p. This was done for a trapezoidal channel with $\alpha = 45°$ and different values of H_w/W_b. An example of the results is shown in Fig. 28 for $H_w/W_b = 0.75$. The values of Kt/fW_s at the successive geometries in this figure were determined with the procedure discussed in the next paragraph. Because the analog solutions applied to the case of zero pressures at the free boundary and at the top of the permeable layer, P_w in this analysis is also zero.

From Fig. 28 and the graphs for the other values of H_w/W_b, plots of the wetted soil area, A_w, as a function of D_p were constructed. Combining these plots with the relation between I_s/K and D_p from Fig. 7 enabled the construction of curves relating A_w to t. Expressing these curves in dimensionless form then yielded Fig. 29, where A_w/W_s^2 is plotted against Kt/fW_s for different values of H_w/W_b. This graph serves as a continuation of Fig. 27. Because an increase in D_p has a marked effect on seepage only when D_p is relatively small, the curves in Fig. 29 soon approach straight lines.

2. Seepage from Channels in Soils with an Impermeable Floor

An example of transient seepage under condition B is the seepage occurring after the water level in a channel is raised a distance h_0 from a position that was in equilibrium with the water table (Fig. 30). The analysis of the resulting

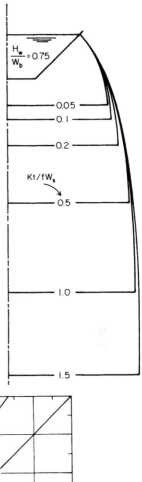

FIG. 28. Use of flow systems for condition A' with increasing D_p as successive positions of the wetted zone for seepage from an initially dry channel.

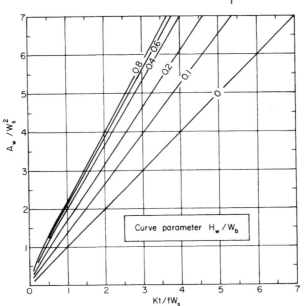

FIG. 29. Graph of A_w/W_s^2 versus Kt/fW_s for different values of H_w/W_b for seepage from an initially dry trapezoidal channel, with larger values of Kt/fW_s.

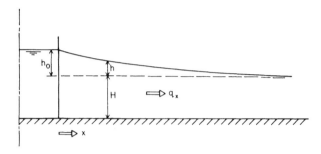

FIG. 30. Geometry and symbols for transient seepage flow under condition B.

transient flow system is based on the Dupuit-Forchheimer assumption and on the continuity principle. Following the analysis by Collis-George and Smiles [36], the vertical distance between the water table and the impermeable layer is divided into a constant H and a variable h. This yields the following equation for the Dupuit-Forchheimer assumption:

$$q_x = k(h + H)\frac{\partial h}{\partial x} \tag{35}$$

where q_x is the flow per unit length of channel at a distance x from the channel. Combining this equation with the continuity equation

$$\frac{\partial h}{\partial t} = \frac{1}{f}\frac{\partial q_x}{\partial x} \tag{36}$$

yields

$$\frac{\partial h}{\partial t} = \frac{K}{f}\frac{\partial}{\partial x}\left[(h + H)\frac{\partial h}{\partial x}\right] \tag{37}$$

If the values of h are relatively small compared to H, Eq. (37) can be reduced to

$$\frac{\partial h}{\partial t} = \frac{KH}{f}\frac{\partial^2 h}{\partial x^2} \tag{38}$$

If the fluxes across the water table are solely because of the increase in the water depth in the channel, the boundary conditions are: at $x = 0$ and $t > t_0$, h is constant; when $t = t_0$, h is constant for $x > 0$; and $0 \leq x \leq \infty$. The solution of Eq. (38) for these conditions is the error function

$$h_{x,t} = h_0 \frac{2}{\pi^{1/2}}\int_{(x/2)(f/KHt)^{1/2}}^{\infty} e^{-\zeta}\, d\zeta \tag{39}$$

for which solutions are tabulated. Collis-George and Smiles [36] differentiated Eq. (39) and combined the result with Eq. (35) to yield

$$q_x = h_0 K \left(\frac{fH}{\pi Kt}\right)^{1/2} \exp\left(-\frac{fx^2}{4KHt}\right) \qquad (40)$$

which, for $x = 0$, gives the following expression:

$$\frac{I_s}{K} = \frac{2h_0}{W_s}\left(\frac{fH}{\pi Kt}\right)^{1/2} \qquad (41)$$

for I_s/K in relation to t after raising the water level in the canal a distance h_0.

The foregoing equations apply in principle only to canals with $\alpha = 90°$ and $D_i = 0$ (Fig. 30), where the D-F assumption gives exact solutions [2]. Based on the discussion of the validity of the D-F assumption in Sect. II.B.1, Eqs. (39) and (41) should also yield reasonably accurate solutions for canals of different shapes and for $D_i > 0$, as long as D_i is relatively small. To apply Eqs. (39) and (41) to seepage systems with relatively large values of D_i for which the D-F assumption is no longer valid (including $D_i = \infty$), the value of H in these equations should be based on the "equivalent" depth of the impermeable layer, as was done by Van Schilfgaarde for transient flow in subsurface drainage systems [37]. To determine the equivalent depth of the impermeable layer for seepage from open channels, I_s/K is evaluated with one of the procedures discussed in Sects. II.B.2 and II.B.3. The value of I_s/K obtained in this manner is then substituted in Eq. (2), and, knowing all parameters in this equation except the effective or equivalent value of D_i, the latter can be calculated. The value of H in Eqs. (39) and (41) is then based on the equivalent position of the impermeable layer.

III. Measurement of Soil Hydraulic Conductivity

Applying the solutions presented in the previous sections toward predicting seepage rates for a given channel requires that the soil and boundary conditions governing the flow system be known. Of critical importance is the adequate evaluation of the hydraulic conductivity profile of the soil because of the direct proportionality between I_s and K and because of the effect of subsoil conditions on I_s/K. It can be stated without reservation that, with the advances in solution techniques, including analogs and computers, the most difficult aspect of seepage prediction for a given channel presently lies in the evaluation of the boundary conditions and of the hydraulic properties of the soil, particularly the latter.

The field techniques for K measurement can be divided into those that measure K of soil below the water table and those that measure K of soil above, or in the absence of, a water table. The latter are of special interest in seepage prediction, because the soil in which the channel is, or will be, imbedded is not always below a water table. The various techniques for measuring K will be briefly reviewed in the following sections.

A. Hydraulic Conductivity Measurements of Soil Below the Water Table

Values of the average hydraulic conductivity conditions over a relatively large soil region can be obtained with pumping-test techniques of wells for evaluating aquifer transmissibility. More local values of K can be obtained with shallower or smaller holes, such as with the auger hole, tube, piezometer, and multiple-well techniques. The general principle of the auger-hole, tube, and piezometer techniques consists of installing a vertical hole in the soil to below the water table, allowing the water level in the hole to become in equilibrium with the water table, removing a volume of water from the hole, and measuring the subsequent rate of rise of the water level in the hole for calculation of K. With the multiple-well technique, several wells are dug a relatively short distance apart, and K is calculated from the flow system created by pumping from one well into another well. For a review and bibliography of below the water table methods, reference is made to Kirkham [38].

B. Hydraulic Conductivity Measurements of Soil Above the Water Table

The general principle of these methods, which enable measurement of K of soil not saturated prior to taking the measurements, consists of wetting a portion of the soil, preferably to positive soil-water pressures, and creating in this wetted zone a flow system of known behavior for evaluation of K. The simplest flow system that can be created in this manner is a system of one-dimensional, vertically downward flow. This system is the basis for the gradient-intake methods such as the infiltrometer techniques and the air-entry permeameter. Axisymmetric flow systems are the basis for the shallow well pump-in method, the double-tube device, and the seepage meter techniques.

Because K is evaluated from a wetted zone that was relatively dry prior to the test, the resulting value of K is less than K at saturation because of entrapped air. If K is measured shortly after the soil has been wetted, the K value obtained may be approximately one-half of K at saturation ([27] and references therein). The water used in the tests should be of about the same quality as the expected seepage water because of the effect of chemical constituents in the water on K [39].

Theory of Seepage from Open Channels

1. Gradient-Intake Techniques

With the infiltrometer techniques, vertically downward flow is established by maintaining equal water levels in two concentric cylinders that are installed in the surface of the soil or the bottom of an auger hole or shallow well. Winger (as discussed by Boersma [40]) used several tensiometers installed just outside the inside cylinder to a depth approximately equal to the depth of penetration of this cylinder to determine the hydraulic gradient. Bouwer and Rice [41] used small piezometers with rapid response, which were pushed into the soil at 1- to 2-cm increments to obtain a plot of pressure head (adjusted to a common infiltration rate in case this rate was not constant during the period of the measurements) versus depth for evaluating the hydraulic gradient. The K value is then calculated with Darcy's equation from the gradient and the infiltration rate for the inner cylinder. With this technique, the occurrence of vertical flow can be verified from the plot of pressure head versus depth, and errors in K due to clogged or compacted surfaces of the soil can be eliminated [41].

The air-entry permeameter [27] consists essentially of a cylinder with clamped-on lid, a vacuum gage, and a standpipe with reservoir (Fig. 31). The diameter of the cylinder is approximately 25 cm, and the depth of penetration into the soil is about 10 cm. Water is applied at a relatively high head (about 100 cm) to create a wetted zone with predominantly positive soil-water pressure. When the wet front is expected to have reached a depth approximately equal to that of the cylinder (as determined by a few trial runs), the

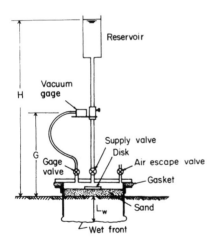

FIG. 31. Sketch of air-entry permeameter.

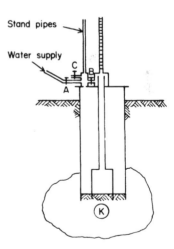

FIG. 32. Sketch of double-tube apparatus.

supply valve is closed. The pressure of the water inside the cylinder will then decrease, reach a minimum when air enters at the bottom of the wetted zone, and increase again as air moves upward through the wetted zone and emerges at the soil surface. Referring the minimum pressure in this water, P_{min}, which is negative and is read with a memory pointer on the vacuum gage, to the bottom of the wetted zone yields the air-entry value P_a of the wetted zone, or $P_a = P_{min} + L_w + G$ (Fig. 31). The depth of the wet front L_w is measured by direct observation or with penetrometer techniques immediately after P_{min} is reached. The factor G is the height of the vacuum gage above soil surface. The term P_a is negative, and it is the soil-water pressure at which air of atmospheric pressure enters a soil with continuous water phase.

During the advance of the wet front, prior to closing the supply valve, the water pressure just above the wet front will be the water-entry (air-exit) value, P_w, which according to available data can be approximated as $\frac{1}{2}P_a$ ([27] and references therein). Knowing H, L_w, and P_w, the hydraulic conductivity of the wetted zone can be calculated if the infiltration rate just before closing the supply valve is measured. The value of K is then computed by applying Darcy's equation to the wetted zone of height L_w with a positive pressure of H at the soil surface and a negative pressure P_w at the wetting front. The resulting value of K applies to K for sorption, which may be taken as one-half the value of K at saturation ([27] and references therein). The air-entry perme-ameter thus yields, in step-function form, the relation between hydraulic conductivity and (negative) soil-water pressure head of the soil for sorption and for desorption [27]. These step functions can be used in the solution of seepage systems using the P_{cr} concept to include the flow at negative water pressures in the soil. Depending on whether the flow system is better described by sorption or by desorption conditions, P_{cr} may be taken as $\frac{1}{2}P_a$ or as P_a, respectively.

The time required per test with the air-entry permeameter may range from 0.3 to 1 hr, depending on type and water content of the soil. Approximately 10 liters of water are required per test. The soil should be initially sufficiently dry to permit development of a well-defined, easily detectable wet front. In its present form of construction, the air-entry permeameter is a surface device. Subsurface measurements can be obtained by placing the device in the bottom of pits or trenches.

2. Techniques Based on Axisymmetric Flow Systems

With the *shallow well pump-in method*, a constant water level is maintained in an auger hole (using perforated casing if necessary) until the rate of flow from the auger hole into the soil has become essentially constant. Nomographs and equations have been developed for different depths of the impermeable layer below the hole bottom to calculate K in the wetted zone around and below the auger hole from the final flow rate and the hole geometry [40, 42]. In layered soils, the resulting value of K reflects primarily the hydraulic conductivity of the more permeable layer penetrated by the auger hole. The time to reach essentially constant outflow from the hole may be several days, and large volumes of water are generally required.

With the *double-tube method*, two concentric tubes are inserted into an auger hole and covered by a lid with a standpipe for each tube (Fig. 32). Water levels are maintained at the top of the standpipes to create a zone of positive water pressure in the soil below the auger hole. The hydraulic conductivity of this zone is evaluated from the reduction in the rate of flow from the inner tube into the soil when the water pressure in the inner tube is allowed to become less than that in the outer tube. This is done by stopping the water supply to the inner tube (closing valve B) and measuring the rate of fall of the water level in the standpipe on the inner tube while keeping the standpipe on the outer tube full to the top. This rate of fall is less than that obtained in a subsequent measurement in which the water level in the standpipe on the outer tube is allowed to fall at the same rate (by manipulating valve C) as that in the standpipe on the inner tube. Using dimensionless factors determined by electric analog [43], K is calculated from the difference between the two rates of fall. This can be done using a graphical procedure [44] or a simplified, tabular calculation [45].

Field studies have shown that, because of soil disturbance in the annular space between the inner tube and the outer tube, the diameter of the outer tube should be larger than the theoretical minimum value [41]. Thus, for an inner tube of 12.7-cm diameter, an outer tube with a diameter of 25.4 cm should be used, instead of the theoretical minimum, which would be approximately 20 cm in this case. If smaller tube sizes are desired, diameters of 10 and 20 cm may be used for the inner and outer tubes, respectively [41].

In soils with different hydraulic conductivity in horizontal and vertical directions, the value of K obtained with the double-tube apparatus is the resultant of horizontal and vertical hydraulic conductivity, K_x and K_z, and lies closer to the latter. If K_z below the auger hole bottom is also measured, for example, by applying the gradient-intake technique to the inner tube, K_x can be computed [23, 41].

The double-tube method operates best if the soil at the bottom of the auger hole is relatively free from stones and gravel. A special hole-cleaning device has been developed to remove disturbed soil from the auger hole bottom, and a layer of sand is used as a protective cover [44]. Depending upon the type of soil and the depth of the hole, tests are usually completed in 1 or 2 hr after the tubes are filled with water. Several hundred liters of water are generally required per test.

C. HYDRAULIC CONDUCTIVITY MEASUREMENTS BELOW SURFACE INUNDATIONS

1. Hydraulic Conductivity of Bottom Material

Hydraulic conductivity of bottom soil in channels or reservoirs filled with water can be measured by applying the double-tube principle to seepage meters. The latter are lid-covered cylindrical devices of approximately 25 cm diameter and 15 cm high, which are placed in the channel or reservoir bottom to measure local seepage rates. The cylinder is connected to a falling-level reservoir mounted at water surface elevation on a rod, and to a manometer that is placed on top of the channel bank. The manometer registers the pressure difference between the water inside the seepage meter and that in the channel. Using a falling-head technique, the local seepage rate, i, is measured as the outflow from the cylindrical device into the channel bottom when the water pressure inside the seepage meter is equal to that outside the meter [46].

When the hydraulic head inside the seepage meter is allowed to fall below that due to the water surface in the channel, a component flow system is created whereby water enters the seepage meter from the channel through the bottom material. This flow results in a reduction of the flow leaving the seepage meter due to seepage, and, if the head inside the seepage meter continues to fall, a point will be reached whereby the net outflow from the seepage meter is zero. The hydraulic-head difference at this point, which is artificially established by closing the water supply to the seepage meter, is called the balanced-flow differential head, H_b [46]. Knowing i and H_b, the value of K of the soil in which the seepage meter is placed can be calculated using the same dimensionless flow factors developed for the double-tube method [46].

2. Hydraulic Impedance of Clogged Soil

The term H_b is proportional to the hydraulic gradient in the bottom material below the seepage meter [46]. If H_b approaches or exceeds the magnitude of the water depth H_w, the hydraulic head due to water depth is mostly dissipated over a relatively small distance of bottom material. This is indicative of seepage under condition C (channels with clogged soil at their perimeter). In that case, the *hydraulic impedance* of the impeding layer is calculated as [46]

$$R_a = H_b/i \qquad (42)$$

If $H_b > H_w$, the pressure below the restricting layer is negative. This condition occurs if R_a is sufficiently large to cause unsaturated flow in the underlying material (see Sect. II.D.1).

IV. Summary and Conclusions

In the past few decades, the pioneering solutions for seepage from open channels by Dachler, Kozeny, and Vedernikov have been followed by contributions from other workers. Today a sufficient number of solutions is available to permit the calculation of seepage rates for a broad spectrum of channel, soil, and water-table conditions. Most of the analyses pertain to steady-state systems. However, solutions are also presented for two types of transient systems. With the advent of digital computers and electric analogs, an almost endless array of soil and boundary conditions can now be handled.

The advances in solution techniques have been paralleled by a better understanding of the physics of the flow systems involved. Progress in the discipline of soil physics has made it possible to include unsaturated or negative-pressure flow in the analysis of the flow system, including systems of seepage from channels with a restricting layer at their perimeter and seepage from initially dry channels. The inclusion of negative-pressure flow by the use of the true unsaturated hydraulic conductivity characteristics of the soil, which involves complicated soil measurements and flow-system analyses, has been followed by a simpler approach based on reducing the unsaturated hydraulic conductivity curve to a step function.

Application of the solution techniques toward the calculation of seepage from actual channels requires that the field conditions be adequately characterized. Recent developments for measuring hydraulic conductivity of soils below the water table and of initially dry soils may be used for such a purpose. However, the field evaluation of pertinent boundary conditions and hydraulic properties of the soil will generally be a much more difficult and time-consuming process than the actual calculation of the seepage once a realistic representation of the field situation has been developed.

Symbols

A_c Wetted cross-sectional area of channel

A_w Area of wetted soil below channel

α Angle between channel bank and horizontal

D_i Vertical distance between channel bottom and impermeable layer

D_p Vertical distance between channel bottom and permeable layer

D_w Vertical distance between water surface in channel and water table in soil at a distance L from channel center

f Fillable pore space as volume fraction of the soil

I_s Seepage rate per unit length of channel and per unit width of water surface (length/time)

- i Rate at which water enters the soil at a given point (length/time)
- H_w Water depth in center of channel or above horizontal bottom
- h_w Water depth at points other than channel center or horizontal bottom ($h_w < H_w$)
- K Hydraulic conductivity of soil at $P \geq 0$ (length/time)
- K_a K of restricting layer at channel perimeter
- K_p Hydraulic conductivity of soil at $P < 0$ (length/time)
- K_x K in horizontal direction of anisotropic soil
- K_z K in vertical direction of anisotropic soil
- L Horizontal distance between channel center and point where water-table position is characterized through D_w
- L_a Thickness of restricting layer at channel perimeter
- L_w Distance of wet front penetration
- P Pressure head of soil water with respect to atmospheric pressure (length)
- P_a Air-entry value of soil (value of P where air of atmospheric pressure enters saturated soil)
- P_{cr} Critical pressure head of soil (length)
- P_w Water-entry value of soil (value of P where water replaces pore air in soil and becomes continuous, also called air-exit value)
- R_a Hydraulic impedance of restricting layer (time)
- W_b Width of channel bottom
- W_s Width of water surface in channel
- w.p.   Wetted perimeter of channel

References

1. Bouwer, H., Theoretical aspects of seepage from open channels. *J. Hydraulics Div. Am. Soc. Civil Engrs.* **91**, No. HY 3, 37–59 (1965).
2. Bouwer, H., Limitation of the Dupuit-Forchheimer assumption in recharge and seepage. *Trans. ASAE* **8**, 512–515 (1965).
3. Dachler, R., "Grundwasserströmung." Springer, Vienna, 1936.
4. Ernst, L. F., Grondwaterstromingen in de verzadigde zone en hun berekening bij de aanwezigheid van horizontale evenwijdige open leidingen. *Verslag. Landbouwk. Onderzoek.* **67**.15, (1962).
5. Muskat, M., "The Flow of Homogeneous Fluids Through Porous Media." Edwards, Ann Arbor, Michigan, 1946.
6. Van Beers, W. F. J., Some nomographs for the calculation of drain spacings. *Bull. Intern. Inst. Land Reclamation Improvement*, **8** (1965).
7. Bouwer, H., Analyzing subsurface flow systems with electric analogs. *Water Resources Res.* **3**, 897–907 (1967).
8. Kozeny, J., Grundwasserbewegung bei freiem Spiegel, Fluss und Kanalversickerung. *Wasserkraft Wasserwirtsch.* **3**, (1931).
8a. Pavlovsky, N. N., "Collected Works." Akad. Nauk SSSR, Leningrad, 1956.
9. Harr, M. E., "Groundwater and Seepage." McGraw-Hill, New York, 1962.
10. Polubarinova-Kochina, P. Ya., "Theory of Ground Water Movement," translated by J. M. R. de Wiest. Princeton Univ. Press, Princeton, New Jersey, 1962.
11. Vedernikov, V. V., Versickerungen aus Kanälen. *Wasserkraft Wasserwirtsch*, **11–13** (1934).
11a. Hammad, H. Y., Seepage losses from irrigation canals. *J. Eng. Mech. Div. Am. Soc. Civil Engrs.* **85**, No. EM 2, 31–36 (1959).
12. Hammad, H. Y., Discussion of Theoretical aspects of seepage from open channels by H. Bouwer. *J. Hydraulics Div. Am. Soc. Civil Engrs.* **91**, No. HY 6, 218–220 (1965).

13. Bouwer, H., Unsaturated flow in ground-water hydraulics. *J. Hydraulics Div. Am. Soc. Civil Engrs.* **90**, No. HY 5, 121–144 (1964).
14. U. P. Irrigation Res. Inst. (Roorkee, India). Steady state saturation line of seepage from unlined canal—its variation with depth of water table and the impervious stratum, and assessment of the waterlogging conditions thereof. *Intern. Comm. Irrigation and Drainage Ann. Bull.* pp. 42–51 (1964).
15. Southwell, R. V., "Relaxation Methods in Theoretical Physics." Oxford Univ. Press, London and New York, 1946.
16. Reisenauer, A. E., Methods for solving problems of multidimensional, partially saturated steady flow in soils. *J. Geophys. Res.* **68**, 5725–5733 (1963).
17. Sewell, J. I., and Van Schilfgaarde, J., Digital computer solutions of partially unsaturated steady-state drainage and subirrigation problems. *Trans. ASAE* **6**, 292–296 (1963).
18. Taylor, G. S., and Luthin, J. N., The use of electronic computers to solve subsurface drainage problems. *Hilgardia* **34**, 543–558 (1963).
19. Bouwer, H., and Little, W. C., A unifying numerical solution for steady two-dimensional flow problems in porous media with an electrical resistance network. *Soil Sci. Soc. Am. Proc.* **23**, 91–96 (1959).
20. Todd, D. K., and Bear, J., Seepage through layered anisotropic porous media. *J. Hydraulics Div. Am. Soc. Civil Engrs.* **87**, No. HY 3, 31–57 (1961).
21. Santing, G., Ground water models (Dutch with English summary). *Comm. Hydrol. Onderzoek T.N.O. Verslag. Mededel.* **17**, 23–48 (1963).
22. Sternberg, Y. M., and Scott, V. H., The Hele-Shaw model—a research device in ground-water studies. *Ground Water* **2**, 33–38 (1964).
23. Bouwer, H., Measuring horizontal and vertical hydraulic conductivity of soil with the double-tube method. *Soil Sci. Soc. Am. Proc.* **28**, 19–23 (1964).
24. Maasland, M., Soil anisotropy and land drainage. In "Drainage of Agricultural Lands" (J. N. Luthin, ed.), Agron. Monograph No. 7, pp. 216–287. *Am. Soc. Agron.*, Madison, Wisconsin 1957.
24a. Bouwer, H., Closure of discussion of Theoretical aspects of seepage from open channels. *J. Hydraulics Div. Am. Soc. Civil Engrs.* **92**, No. HY 3, 90–95 (1966).
25. Morel-Seytoux, H. J., Domain variations in channel seepage flow. *J. Hydraulics Div. Am. Soc. Civil Engrs.* **90**, No. HY 2, 55–79 (1964).
26. Preissmann, A., A propos de la filtration au-dessous des canaux. *La Houille Blanche* **12**, 181–188 (1957).
27. Bouwer, H., Rapid field measurement of air entry value and hydraulic conductivity of soil as significant parameters in flow system analysis. *Water Resources Res.* **2**, 729–738 (1966).
28. Bouwer, H., Theoretical aspects of flow above the water table in tile drainage of shallow homogeneous soils. *Soil Sci. Soc. Am. Proc.* **23**, 260–263 (1959).
29. Risenkampf, B. K., Hydraulics of groundwater. Part III. *Proc. State Univ. Saratovsky*, *1940*, **15**, No. 25.
30. Vedernikov, V. V., Account of soil capillarity on seepage from a canal. *Dokl. Akad. Nauk SSSR* **28**, No. 5 (1940).
31. Averjanov, S. F., Approximate appraisal of the role of seepage in the zone of the capillary fringe. *Dokl. Akad. Nauk SSSR* **69**, 309–311 (1949).
32. Whisler, F. D., and Klute, A., The numerical analysis of infiltration, considering hysteresis, into a vertical soil column at equilibrium under gravity. *Soil Sci. Soc. Am. Proc.* **29**, 489–494 (1965).
33. Philip, J. R., The theory of infiltration: 1. *Soil Sci.* **83**, 345–357 (1957).

34. Watson, K. K., Infiltration: The physical process. *Civil Eng. Trans. Inst. of Eng. Australia* **8**, 13–20 (1966).
35. Green, W. H., and Ampt, G. A., Studies on soil physics. I. The flow of air and water through soils. *J. Agr. Sci.* **4**, 1–24 (1911).
36. Collis-George, N., and Smiles, D. E., A study of some aspects of the hydrology of some irrigated soils of western New South Wales. *Australian J. Soil Res.* **1**, 17–27 (1963).
37. Van Schilfgaarde, J., Design of tile drainage for falling water tables. *J. Irrigation Drainage Div. Am. Soc. Civil Engrs.* **89**, No. IR 2, 1–12 (1963).
38. Kirkham, D., Saturated conductivity as a character of soil for drainage design. *Proc. Intersoc. Conf. Drainage for Efficient Crop Prod., Chicago, Illinois*, pp. 24–32. Am. Soc. Agr. Eng., 1965.
39. Doering, E. J., Change of permeability with time. *Proc. Intersoc. Conf. Drainage for Efficient Crop Prod., Chicago, Illinois*, pp. 32–36. Am. Soc. Agr. Eng., 1965.
40. Boersma, L., Field measurement of hydraulic conductivity above a water table. *In* "Methods of Soil Analysis, Part I" (C. A. Black, ed.), Agron. Monograph No. 9, pp. 234–252. Am. Soc. Agron., Madison, Wisconsin, 1965.
41. Bouwer, H., and Rice, R. C., Modified tube diameters for the double-tube apparatus. *Soil Sci. Soc. Am. Proc.* **31**, 437–439 (1967).
42. Zangar, C. N., "Theory and Problems of Water Percolation," Eng. Monograph No. 8. Bur. of Reclamation, Denver, Colorado, 1953.
43. Bouwer, H., A double-tube method for measuring hydraulic conductivity of soil *in situ* above a water table. *Soil Sci. Soc. Am. Proc.* **25**, 334–339 (1961).
44. Bouwer, H., Field determination of hydraulic conductivity of soil *in situ* above a water table. *Soil Sci. Soc. Am. Proc.* **26**, 330–335 (1962).
45. Bouwer, H., and Rice, R. C., Simplified procedure for calculation of hydraulic conductivity with the double-tube method. *Soil Sci. Soc. Am. Proc.* **28**, 133–134 (1964).
46. Bouwer, H., and Rice, R. C., Seepage meters in seepage and recharge studies. *J. Irrigation Drainage Div. Am. Soc. Civil Engrs.* **89**, No. IR 1, 17–42 (1963).

RECENT STUDIES ON SNOW PROPERTIES

YIN-CHAO YEN

U.S. Army Terrestrial Sciences Center, Hanover, New Hampshire

I. Introduction	173
II. Metamorphism and Intergranular Bonds	174
A. Snow Metamorphism	174
B. Model Studies on Constructive Metamorphism	176
C. Development of Intergranular Bonds	178
III. Thermal Properties of Snow	183
A. Thermal Conductivity	183
B. Methods for Determining Thermal Conductivity of Snow . . .	184
C. Experimental Results on Thermal Conductivity of Snow	187
D. Theoretical Expressions for Thermal Conductivity of Snow . . .	188
E. Thermal Diffusivity of Snow	190
F. Vapor Diffusion in Snow	192
G. Theoretical Expressions for Effective Thermal Diffusivity	195
H. Effect of Vapor Diffusion on Temperature Propagation	196
I. Heat and Vapor Transfer with Forced Convection	197
IV. Radiation Properties of Snow	201
A. Fundamentals of Radiation Transport	201
B. Concepts of Radiation Interaction	202
C. Absorption and Scattering of Radiation in a Snow Mass	202
D. Reflection from Snow	206
E. Analytical Study of Radiant Energy Distribution and Albedo of Snow	209
Symbols	212
References	213

I. Introduction

The purpose of this review is to summarize, up to date, the findings and developments in regard to the process of snow metamorphism and sintering, thermal and radiation properties of snow, and to provide information for the use of workers in the field of snow and ice research.

In the second section, various phenomena of snow metamorphism, sintering, and development of intergranular bonds are discussed. The understanding of these phenomena is necessary in order to explain the physical processes occurring in a natural snow mass and their effect on the physical properties of snow.

In the third section, a few methods for determining thermal conductivity, water vapor diffusivity, and some experimental results are discussed. Analytical expressions for thermal conductivity are derived. The effects of vapor diffusion under natural and forced convection in heat conduction are discussed.

Finally, in the last section of this article, the essential concepts of radiation interaction with a snow medium are explained. Values for the extinction coefficient and reflectance for radiation in the visible range are reported, and theoretical expressions for albedo and the energy density distribution in a snow layer are given.

It should be pointed out that there is a large amount of research and progress being made on the properties of snow, but it would be impossible to cover them adequately because of the limited length of this article. The reader can refer to the proceedings of the *Intern. Conf. on Low Temp. Sci.* [1] held in 1966 at the Hokkaido University in Japan. At this conference, papers on the viscosity, compressability, and deformability of snow were presented. It also contains studies on the mechanisms of avalanches and the mechanics of snowstorms and blowing snow. An additional reference is the book edited by Kingery [1a]. This book covers studies on the properties of snow densification, snow stabilization, snow compaction techniques, and engineering properties of polar snow. Another important work is the book by Yosida [1b], in which detailed discussions on the various thermal, mechanical, and electrical properties of snow are included.

II. Metamorphism and Intergranular Bonds

A. Snow Metamorphism

Snow is a porous, permeable aggregate of ice grains with its pores filled with air and water vapor. Its physical and mechanical properties are strongly time and temperature dependent. Many of the unique properties of snow are due to the fact that snow, unlike most of the other solid materials, is normally encountered at temperatures very close to its melting point. Water molecules are, therefore, comparatively free to migrate within the ice lattice by volume diffusion, and to evaporate readily into the vapor phase. As a result of this molecular mobility, water molecules will always be redistributed in a given volume of snow in such a way as to decrease the total surface area of the crystals and hence to reduce the surface free energy of the system. This process is called *snow metamorphism* and will take place as soon as snow is deposited and accumulated. At the initial stage, the process of metamorphism changes the variously shaped snow crystals into the spherical particles, as observed in natural powdered snow.

The mechanics and thermodynamics of snow metamorphism and their dependence on environmental factors are rather complicated. Bader [1c] has reported these natural processes in great detail, and so this paper will only briefly describe the four fairly distinct kinds of snow metamorphism and the essential factors contributing to the transformation of snow to ice.

1. Destructive Metamorphism

A few days after the deposition of the original dry snow, the original crystal shape is almost completely lost. The end product of destructive metamorphism usually consists of a fine-grained snow of density (including the pores) between 0.15 and 0.25 gm/cm^3. Grains are rounded and weakly bounded. Very few grains are smaller than 0.2 mm, the predominating size being from 0.5 to 1 mm.

2. Constructive Metamorphism

In this process, the grain size of dry snow increases significantly more than 1 mm. This process is slow if the density is high. If the density is below about 0.3 gm/cm^3, the process can become very rapid when vapor transfer is accelerated by convective air flow in the large pore volume. The end product of this process is a very distinctive snow type known as *depth hoar* with grain size between 2 and 8 mm. Bonding of grains in depth hoar is very poor.

3. Melt Metamorphism

This mechanism characterizes the changes produced in snow by the presence of liquid water. As the climate becomes warmer, snow becomes moist, and temperature gradients and their effect vanish, since both phases are in equilibrium. As melt water seeps down through the snow to colder regions, refreezing occurs, and crystals grow to a maximum size of about 3 mm and composite grains to about 15 mm. Within a few days, this metamorphism produces the well-known *rotten* snow of the thaw season.

4. Pressure Metamorphism

This is a new term to characterize the densification of dry neve (granulated snow accumulated and subsequently compacted to glacier ice). This process of densification due to pressure is very slow, and it may take many decades to change snow of density of 0.45 gm/cm^3 to ice of density of 0.53 gm/cm^3. Thermodynamic processes, such as grain and grain-bond growth by vapor or surface migration, appear to be of secondary importance during this process.

In summary, we can say that melting and refreezing are the major processes in producing drastic changes in snow condition. With moderate melting, much of the melt water may be retained in pores by the action of surface tension, so that subsequent refreezing can provide strong ice bonds between grains. If intensive melting persists, free water will percolate down through the snow and refreeze when it reaches colder layers of snow to form ice sheets.

B. Model Studies on Constructive Metamorphism

Recently, de Quervain [2] devised a few structural models aimed to demonstrate the phenomenon of constructive metamorphism. The following is an outline of his analysis. Let us consider a model made of ice, perforated by parallel cylindrical pores as shown in Fig. 1. A negative temperature gradient is applied in the direction of z. For small pores, we may assume the air in the pores is saturated with water vapor at the ice temperature. By neglecting thermodiffusion and temperature dependence of the diffusion coefficient, we can write

$$m = -nD_0 \, dC/dz \tag{1}$$

where m is the mass flux of water vapor in grams per square centimeter-second, $n = 1 - (\rho_m/\rho_i)$ is the porosity in which ρ_m and ρ_i are the densities of the medium and ice, respectively, D_0 denotes the diffusion coefficient of water vapor in air, and C is the vapor concentration in grams per cubic centimeter. By writing $C = P_s/R_v T$ and expressing $P_s = P_0 \exp[f(T)]$, Eq. (1) becomes

$$m = -\frac{nD_0 P_0}{R_v T}\left[f'(T) - \frac{1}{T}\right]\exp[f(T)]\frac{dT}{dz} \tag{2}$$

where T is the absolute temperature, P_s the saturation pressure of water vapor, and R_v the gas constant for water vapor. From conservation of mass, we have $-dm/dz = d\rho_m/dt$. Using Eq. (2), we obtain

$$\frac{d\rho_m}{dt} = \frac{nD_0 P_0}{R_v T}\left\{\left[f''(T) - \frac{2f'(T)}{T} + \frac{2}{T^2} + (f'(T))^2\right]\left(\frac{dT}{dz}\right)^2 \right.$$
$$\left. + \left[f'(T) - \frac{1}{T}\right]\frac{d^2 T}{dz^2}\right\}\exp[f(T)] \tag{3}$$

Assuming a linear temperature distribution $T = T_0 + az$ with $a = -0.2°C/cm$ and $T_0 = 273°K$, $n = 0.6$, $D_0 = 0.202 \text{ cm}^2/\text{sec}$, and

$$P_s = P_0 \exp[B \ln T - A/T - CT + DT^2]$$

(see Dorsey [3]), de Quervain computed the variations of m and $d\rho_m/dt$ and C. It can be seen from Fig. 2 that the density is increasing everywhere throughout the height of the model. To explain the conservation of mass, it is necessary to assume a fictitious warm surface of infinitesimal thickness at the bottom where this mass is supplied. This phenomenon is contrary

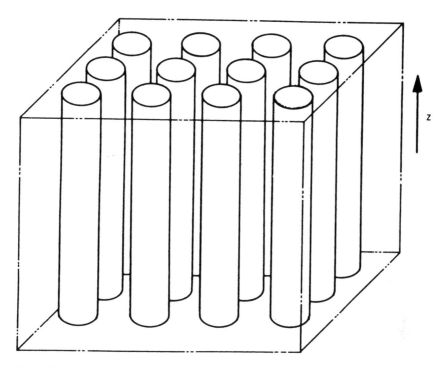

FIG. 1. Structural model of snow with pores parallel to the temperature gradient [2].

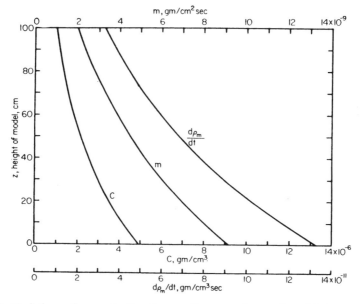

FIG. 2. Variations of concentration C, mass flux m, and rate of model density change with model height z [2].

to the naturally observed fact that, in a natural snow cover, the bottom layer that has more than an infinitesimal thickness usually loses its mass. Furthermore, from the preceding calculation, the increase in density appears rather small (about 0.001 gm /cm^3 in 80 days). According to de Quervain, the reason for this is that the following effects are neglected: (a) thermodiffusion, (b) temperature dependence of the diffusion coefficient, and (c) mechanism of mass transfer between the walls and the pores.

Since the foregoing model was not satisfactory, he constructed another model having the same porosity as the first but consisting of parallel ice lamellae perpendicular to the temperature gradient as shown in Fig. 3. This model was aimed to distribute the loss of mass to a certain zone of the warm end instead of being restricted to the warm surface. Since thermal conductivity of ice is about 100 times larger than that of air, it is reasonable to assume that the temperature gradient is concentrated mostly on the air lamellae, and its value is $(dT/dz)_{m2} = (dT/dz)_{m1}(1/n)$, where subscripts $m1$ and $m2$ refer to the first and second model, respectively. In model $m1$, $(dT/dz)_{m1} = -0.2°C/$cm, and, therefore, $(dT/dz)_{m2} = -(0.2/0.6) = -0.33°C/cm$. It can be noted clearly from Eq. (2) that the mass flux increased by a factor of 1.67. Another factor that causes the increase in the mass flux is the increased cross-sectional area for vapor diffusion which also amounts to 1.67. Thus, the mass flux in the second model is 2.79 times higher than that in the first. However, this model also deviates from reality in various respects. Because of the evaporation and condensation, the temperature gradient is reduced in the air gaps and increased in the ice, and, therefore, the mass flux of water vapor is reduced. Also, as in the first model, vapor transfer due to thermodiffusion and convection has not been taken into account. The values of m obtained using this second model are still small compared to the observed values under natural conditions.

Finally, de Quervain proposed a model shown in Fig. 4 which is a combination of the previous two models. This model is supposed to give more satisfactory values for m because it resembles the internal snow structure. In this model, there are vapor currents that pass through the model as well as short-range currents that originate and end in the structure. This is similar to the "hand-to-hand delivery" vapor transfer mechanism proposed by Yosida [4]. However, the analysis of this model is very difficult, and therefore it is not yet done.

C. DEVELOPMENT OF INTERGRANULAR BONDS

When a mass of snow grains is deposited in cold, dry conditions, there is little cohesion between grains at the beginning, but they develop *intergranular bonds* right away. This phenomenon can take place in an isothermal snow

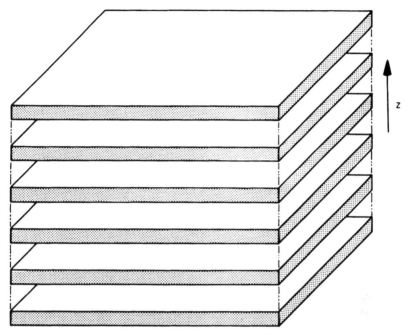

Fig. 3. Structural model of snow with ice lamellae perpendicular to the temperature gradient [2].

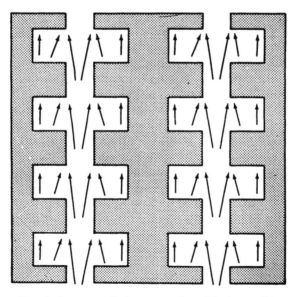

Fig. 4. A more realistic structural model of snow [2].

mass. However, this process will be accelerated by the presence of a temperature gradient and convection in snow pores, which usually exist in natural snow covers. The development of intergranular bonds can be contributed to two main causes, i.e., sublimation due to temperature gradient and sintering in isothermal condition.

Vapor pressure difference along the grain surfaces is the cause of *sublimation transfer* of water molecules. Surfaces with high positive curvature (small grains and sharp corners) have higher vapor pressures than flatter grain surfaces. Therefore, there is evaporation from areas of high convex curvature (positive curvature) and condensation on relatively flat or concave surfaces (negative curvature). Since vapor pressure increases with temperature, there will be mass transfer through the evaporation-condensation mechanism toward the cold end of the snow mass subjected to a temperature gradient.

Sintering is now commonly used in snow technology to describe the process by which snow particles tend to stick to each other and ice bonds (necks) form among them even below the melting temperature. This phenomenon has been known for a long time in powder metallurgy. Many investigators have attempted to study the cause of ice-bonding mechanism between snow grains or small ice spheres. Faraday [5] was the first one to observe the fact that, when two pieces of solid ice are brought together at the melting point, they become firmly cemented; this cementation occurs in air, *in vacuo*, and even under water. Tyndall [6] used the term *regelation* to describe this process. In order to explain the phenomenon of adhesion of ice, Faraday [5] and Tyndall [6] suggested the hypothesis that a thin layer of water exists on the surface of ice and it would freeze if two pieces of ice were brought together. Thomson [7], on the other hand, interpreted the phenomenon as due to melting of ice when pressure is applied (melting point depression of ice with pressure) and resolidification upon removal of pressure. These early investigators observed that ice did not readily adhere at temperatures well below the melting point. They concluded that the effect was present only for ice at the melting point and for *thawing* ice with a superficial coating of liquid. Nakaya and Matsumoto [8], however, observed the adhesion of ice in the temperature range of $0°$ to $-16°C$, and measured the force required to separate two spheres brought into contact. They observed that the adhesive force of ice increases with increasing temperature and found that two ice spheres suspended by this thin filament are caused to rotate before separation as the inclination of the filaments from vertical is increased. From this fact, they concluded that the adhesion of ice was due to a liquid-water film existing on the ice surface even at temperatures below $0°C$. Hosler *et al.* [9] supported the liquid-film hypothesis. They also indicated that the adhesive force of ice is strongly dependent on ambient vapor pressure. They found that the force required to separate ice spheres is strongly dependent on temperature and the environmental conditions (see Fig. 5). Jellinek [10] also measured the adhesive force between ice and various kinds of materials such as metals, quartz, and high polymeric substances, and interpreted his results in terms of *liquid-like film* at the surface.

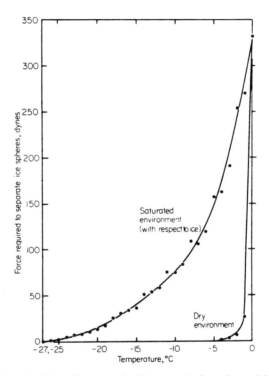

FIG. 5. Mean force required to separate ice spheres [9].

It should be pointed out that the pressure melting theory is incapable of explaining the adhesion of ice-bonding, which is actually observed when no pressure is applied, and that the liquid-film hypothesis at the temperature below the melting point has no rigorous thermodynamic justification.

Recently, Kingery [11] treated the ice-bonding problem from the viewpoint of *sintering*. He observed the growth rate of ice-bond as a function of temperature and concluded, in contrast to the liquid-film hypothesis, that *ice-bonding* or *ice-adhesion* takes place as a result of mass transfer to the contact area arising from surface diffusion. On the other hand, Kuroiwa [12] measured the growth rate of necks between ice spheres (radius 100 μ) in both saturated air and kerosene environments. He reported that ice-bonding takes place even in the medium of kerosene where no vapor phase is present, although the growth velocity of the neck is lower. Kuroiwa concluded that volume diffusion is dominant above $-10°C$ and surface diffusion becomes dominant below $-15°C$. Ramseier and Sander [13] studied the time dependence of the unconfined compressive strength of snow, making the assumption that the strength of snow is a function of both the size and number of intergranular bonds. By comparing activation energies and mass transport coefficients, they supported the evaporation-condensation hypothesis. To substantiate this finding, Ramseier and Keeler [14] further measured the unconfined compressive strengths of two groups of identical snow samples. One group was allowed to sinter under atmospheric conditions, whereas the other group was kept immersed in silicone oil. They found the latter group of snow had a much lower rate of strengthening and suggested that evaporation-condensation must be the major mechanism of mass transport in snow under atmospheric conditions.

The conclusion drawn by Kingery and Kuroiwa is based on theoretical models of sintering used in powder metallurgy (see Kuczynski [15], Kingery and Berg [16]). In the case of two spheric particles of radius r, the model suggested four possible mechanisms for material transport to the point of contact or neck-growth area and can be expressed by

$$\left(\frac{x}{r}\right)^n = \frac{A(T)}{r^m} t \qquad (4)$$

where x is the radius of the neck at time t. $A(T)$ is a temperature-dependent constant, and n and m are integers, the values of which depend on actual mechanism of mass transport. According to Kuczynski [15], $n = 2, m = 1$ for plastic or viscous flow (a concept used in powder metallurgy); $n = 3, m = 2$ for material transfer through vapor phase or evaporation-condensation; $n = 5, m = 3$ for volume diffusion through the particle; and $n = 7, m = 4$ for surface diffusion on the particles.

In the derivation of Eq. (4) for the case of mass transfer by evaporation and condensation, the assumption was made that the mean free path of the vapor molecules in the environment is large in comparison with the distance over which transfer takes place. Hobbs and Mason [17] have pointed out that, in the case of the sintering of ice, this assumption is incorrect, and therefore Eq. (4) is not valid for this particular mode of mass transfer. They show instead that when water molecules are transferred to the neck by diffusion through the environmental gas, the radius of neck x at time t is given by

$$\left(\frac{x}{r}\right)^5 = \frac{20\sigma\delta^3}{hTr^3}\left(\frac{hT\rho_i}{P_0 m D_0} + \frac{L_s^2 m \rho_i}{k_a hT^2}\right)^{-1} t \qquad (5)$$

where σ is the surface tension of ice, δ the intermolecular spacing, h Boltzmann's constant, T the temperature of sintering, L_s the latent heat of sublimation of ice, m the mass of water molecules, P_0 the equilibrium vapor pressure over a plain surface of ice, and k_a the thermal conductivity of air. They also concluded that mass transfer through the vapor phase is the main mechanism for neck growth and that surface and volume diffusion play only an insignificant role in the process of sintering. In a recent study, Hobbs [18] further showed that experimental results for the variation with time of compressive strength, work of disaggregation, hardness, and Young's modulus of deposited snow can be explained quantitatively in terms of theory of sintering of ice as indicated in Eq. (5). Most recently, Itagaki [19] conducted a rather ingeneous experiment on the ice-bonding problem. He observed the actual process of neck-formation by placing randomly oriented single crystal ice spheres close to (about 10 μ) but without making contact with newly microtomed flat ice surface. From this study, he concluded that the initial phase of the neck-formation must be due to vapor transfer through the atmosphere because there was no contact between the ice sphere and the plane. However, he pointed out that the driving force for the evaporation-condensation mechanism was not only the negative curvature as mentioned by the previous investigators but also the presumable vapor pressure difference between different crystallographic surfaces due to the difference of defect concentration.

III. Thermal Properties of Snow

A. Thermal Conductivity

Thermal conductivity is a physical property of the material. Usually it varies with temperature. For unidirectional steady heat conduction in homogeneous isotropic solids, thermal conductivity k is the coefficient of the temperature gradient in the well-known Fourier's law:

$$q = -k \, d\theta/dz \qquad (6)$$

where q is the heat flux along the axis z. Heat flow in dry snow is a complicated process. We shall designate k_e° (effective thermal conductivity) instead of k in Eq. (6) to indicate the effect of the following mechanisms: (a) conduction through the grains in contact, (b) conduction through the void space or the interstitial air, (c) radiant energy exchange via the void space, and (d) molecular vapor diffusion through the void space. In most cases, because of the low temperature, radiant heat exchange can be neglected. Thermal conductivity of ice is about 100 times larger than that of the air; therefore, the contribution due to conduction through interstitial air can also be considered insignificant. However, the contribution to the overall heat flow due to vapor diffusion can be significant, especially when the snow density is low (Sect. III.H).

B. METHODS FOR DETERMINING THERMAL CONDUCTIVITY OF SNOW

Since Ångström's classical experiment a century ago, numerous methods have been devised for measuring the thermal conductivity or thermal diffusivity of substances. These techniques can usually be grouped into two classes, i.e., steady-state and transient methods. In most of these methods, direct measurement of the internal temperature of the medium under investigation is needed. For snow, it should be pointed out that placing thermocouples or thermometers into the snow sample may change its properties and disturb the heat flow pattern. Snow on ground, especially, is extremely susceptible to damage. In the following, two transient techniques that do not require the knowledge of the internal temperature of snow will be discussed.

1. Method of Yosida and Iwai

Yosida and Iwai [20] use the transient technique to measure the thermal conductivity of snow. Snow is placed in an air-tight container, the temperature of the container is changed, and changes in the air pressure that is saturated with water vapor in the container are measured. This pressure is proportional to the mean internal temperature $\bar{\theta}$ of the snow. The changes of the mean temperature lag behind the temperature changes of the container. This lag is determined by measuring the air pressure as a function of time, and from these values the thermal diffusivity is determined. Since this measurement includes the effect of vapor pressure, the relationship between mean temperature $\bar{\theta}$ and mean pressure $\bar{P}$ will not be linear when the vapor pressure of the substance under investigation is high. However, for snow, changes in vapor pressure with temperature are only one-tenth of those of dry air pressure, and the mean temperature-pressure relationship is approximately linear.

a. MATHEMATICAL ESSENCE OF YOSIDA AND IWAI'S METHOD. Consider a layer of snow of thickness l. Let θ_1 be the initial temperature of the sample and θ_2 be the temperature imposed on the top and bottom boundaries at $t = 0$. The temperature θ at any point within the layer is

$$\frac{\theta - \theta_2}{\theta_1 - \theta_2} = \frac{4}{\pi} \sum_{n=1}^{\infty} \frac{1}{2n - 1} \sin(2n - 1)\frac{\pi z}{l} \exp\left[\frac{-\alpha(2n - 1)^2 \pi^2 t}{l^2}\right] \quad (7)$$

This series converges rapidly. For large t, the expression can be approximated to

$$\frac{\theta - \theta_2}{\theta_1 - \theta_2} = \frac{4}{\pi} \sin\frac{\pi z}{l} \exp\left(\frac{-\alpha \pi^2 t}{l^2}\right) \quad (8)$$

b. ESTABLISHMENT OF A LINEAR RELATION BETWEEN MEAN PRESSURE $\bar{P}$ AND TEMPERATURE $\bar{\theta}$. The interstitial air and the snow will have the same temperature as given by Eq. (8). If ρ_a is the density of air in grams per cubic centimeter and T is the absolute temperature in degrees Kelvin ($T = \theta + 273$), we have

$$P = R\rho_a T$$

where P is the dry air pressure and R is the gas law constant. Between $0°$ to $-10°C$, the variation of vapor pressure of ice can be approximated to

$$P_i = -a + bT$$

where $a = 67.8$, $b = 0.262$. Then the pressure $\bar{P}$ that is registered by the manometer is

$$\bar{P} = R\rho_a T - a + bT \tag{9}$$

Integrating this equation for the entire volume of the sample container, we have

$$(\bar{P} + a) \int dv/T = R \int \rho_a \, dv + b \int dv \tag{10}$$

where $\int \rho_a \, dv = M$ is the total mass of the air, and $\int dv = V$ is the volume of the sample container. In order to integrate the left side of Eq. (10), $(273 + \theta)^{-1}$ is expanded in the Maclaurin's series for small θ. Considering only the first two terms of the series, the following result is obtained:

$$(\bar{P} + a) \frac{V}{273} \left(1 - \frac{\bar{\theta}}{273}\right) = RM + bV \tag{11}$$

or

$$(\bar{P} + a) = C\left(1 + \frac{\bar{\theta}}{273}\right) \tag{12}$$

where $\bar{\theta}$ is the mean temperature defined as $\bar{\theta} = \int \theta \, dv/V$, and C is a constant. Equation (12) can be reformulated as follows:

$$\bar{\theta} - \bar{\theta}_2 = (\bar{\theta}_1 - \bar{\theta}_2) \frac{\bar{P} - \bar{P}_2}{\bar{P}_1 - \bar{P}_2} \tag{13}$$

where the initial and boundary temperatures and pressures are taken as $\bar{\theta}_1, \bar{\theta}_2, \bar{P}_1$, and $\bar{P}_2$, respectively. From Eq. (8), we have

$$\bar{\theta} - \bar{\theta}_2 = \frac{8(\bar{\theta}_1 - \bar{\theta}_2)}{\pi^2} \exp\left(\frac{-\alpha \pi^2 t}{l^2}\right) \tag{14}$$

Hence, according to Eq. (13), we can write

$$(\bar{P} - \bar{P}_2) \propto \exp(-\alpha \pi^2 t/l^2) \tag{15}$$

from which the thermal diffusivity α can be determined from the slope of a plot of $\ln(\bar{P} - \bar{P}_2)$ versus t.

2. Method of van Wijk

van Wijk [21] has recently reported a transient technique for determining thermal conductivity k and heat capacity per unit volume C at a soil surface. The essential feature of this method is that a block of reference material with appropriate size $10 \times 10 \times 4$ cm, having a uniform initial temperature, is placed on the soil sample surface. Let the temperature within the upper soil layer be $\theta_{20} + Ez$ at $t = 0$, where z is the depth measured downward from the surface, E is the temperature gradient at the interface, and θ_{20} is the initial surface temperature of the soil. The temperature near the center of the interface of two semi-infinite bodies that are suddenly brought into contact along the plane $z = 0$ is

$$\theta(0, t) = \frac{\theta_{10}(k_1 C_1)^{1/2} + \theta_{20}(k_2 C_2)^{1/2}}{(k_1 C_1)^{1/2} + (k_2 C_2)^{1/2}} + \frac{2}{\pi^{1/2}} \frac{Ek_2}{(k_1 C_1)^{1/2} + (k_2 C_2)^{1/2}} t^{1/2} \quad (16)$$

where θ_{10} is the initial temperature of the reference material. By recording $\theta(0, t)$ and knowing the values of θ_{10}, θ_{20}, and $k_1 C_1$, the value of $k_2 C_2$ can be obtained from the slope of a plot of $\theta(0, t)$ versus t.

Equation (16) is valid for perfect thermal contact, that is, $\theta(0, t) = \theta_1(0, t) = \theta_2(0, t)$, where θ_1 and θ_2 are the temperature variations at the interface of reference material and the soil, respectively. In reality, this is not true, and a more realistic assumption is $\theta_1(0, t) = \theta_2(0, t) + \sigma H(t)$, in which $H(t)$ is the heat flux and σ a positive constant indicating the quality of contact. If the quality of the contact is considered, Eq. (16) becomes

$$\theta(0, t) = \frac{\theta_{10}(k_1 C_1)^{1/2} + \theta_{20}(k_2 C_2)^{1/2}}{M} - \frac{NEk_2}{M^2}$$
$$+ \left[\frac{(\theta_{10} - \theta_{20})(k_2 C_2)^{1/2}}{M} + \frac{NEk_2}{M^2} \right] \exp\left(\frac{M^2}{N^2} t\right) \mathrm{erfc}\left(\frac{M}{N} t^{1/2}\right)$$
$$+ \frac{2}{\pi^{1/2}} \frac{Ek_2}{M} t^{1/2} \quad (17)$$

where $M = (k_1 C_1)^{1/2} + (k_2 C_2)^{1/2}$, and $N = \sigma(k_1 C_1 k_2 C_2)^{1/2}$. By knowing the values of θ_{10}, θ_{20}, and $k_1 C_1$, values of $k_2 C_2$, Ek_2, and σ can be obtained.

This method is developed primarily for soils, but it can be used for determining thermal conductivity of snow with proper selection of a reference material. The other good feature of this method is that a perfect contact between the reference material and the sample is not necessary.

Recent Studies on Snow Properties

C. Experimental Results on Thermal Conductivity of Snow

Thermal conductivity of dry snow is hard to determine because of its continuous change of state due to metamorphism. Table I summarizes the

TABLE I

Thermal Diffusivity α, Thermal Conductivity k_e°
(Temperature Range -1 to $-6°C$)

Run no.	Density (gm/cm³)	Snow temperature at time of selection (°C)	$\alpha \times 10^3$ (cm²/sec)	$k_e^\circ \times 10^4$ (cal/°C-sec-cm)
1[a]	0.40	—	6.9 2.5	13.2 4.9
2	0.39	-6.2	3.9 3.9	6.4 6.4
3	0.28	-1.2	1.2 2.49 3.22	1.61 3.34 4.31
4	0.14	-3.0	1.89 0.60 1.84 6.62	1.54 0.49 1.50 5.39
5	0.085	-4.5	3.26 4.53 4.22	1.34 1.86 1.73
6	0.072	-1.0	12.0 6.3	4.2 2.2

[a] In measuring snow sample no. 1, the temperature of top side of the sample container was kept at $-1°C$, whereas the temperature of bottom only was changed to $-6°C$.

results obtained by Yosida and Iwai [20]. During measurements with new snow, sublimation causes remarkable changes in its physical structure. Microscopic examinations reveal that sharp edges and corners of the snow grains are rounded off during the sublimation process. Thus, several values of α and k_e° can be found for the same sample of new snow. For old coarse-grained snow in an advanced stage of sublimation, there is only slight alteration in its structure, and therefore α or k_e° does not change during measurement (see run no. 2, Table I). Sublimation and evaporation with the formation of water vapor occur in the relatively warmer parts of the snow sample; the water vapor diffuses to the cooler parts and, by sublimation and condensation, causes a transfer of heat. It should also be noted in Table I that for each snow sample the values of $k_e°$ determined for the case of warming up from $-6°$ to $-1°C$ are indicated by italic numbers and are quite different from corresponding values of α or $k_e°$ determined by cooling the sample from $-1°$ to $-6°C$. There seems to be no satisfactory explanation for this phenomenon.

Quite a few other investigators have determined the thermal conductivity of snow. Without exception, they all empirically correlated their results with snow density as the sole parameter. Some of these empirical equations are listed in Table II [22]. When these equations are plotted as shown in Fig. 6,

TABLE II

THERMAL CONDUCTIVITY OF SNOW[a]

Investigator	Empirical expression	Density range
Abels	$0.0068\rho_s^2$	$0.14 < \rho_s < 0.34$
Jansson	$0.00005 + 0.0019\rho_s + 0.006\rho_s^4$	$0.08 < \rho_s < 0.05$
Van Dusen	$0.00005 + 0.0010\rho_s + 0.0052\rho_s^3$	
DeVaux	$0.00007 + 0.007\rho_s^2$	$0.1 < \rho_s < 0.6$
Kondrat'eva	$0.0085\rho_s^2$	$0.35 < \rho_s$
Bracht	$0.0049\rho_s^2$	$0.19 < \rho_s < 0.35$
Sulakvelidze	$0.0012\rho_s$	$\rho_s < 0.35$
Proskuriakov	$0.000005 + 0.00242\rho_s$	

[a] From Mellor [22].

it is obvious that there is considerable difference among them. Apparently, experiments were conducted under conditions where the effects of vapor diffusion were different. Thermal conductivity is also dependent on temperature, but for snow this dependence is known to be very weak.

D. THEORETICAL EXPRESSIONS FOR THERMAL CONDUCTIVITY OF SNOW

Schwerdtfeger [23] has recently derived a theoretical expression for the thermal conductivity of snow based on Maxwell's work [23a] on the electrical conductivity of inhomogeneous media. For dense snow, the thermal conductivity expression can be written as

$$k_{ia} = \frac{2k_i + k_a - 2n(k_i - k_a)}{2k_i + k_a + n(k_i - k_a)} k_i \quad (18)$$

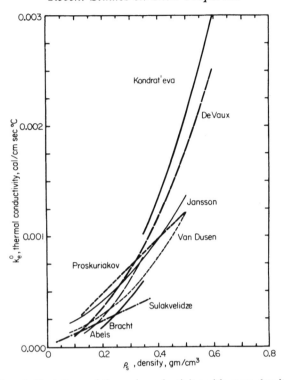

FIG. 6. Variations of thermal conductivity with snow density.

where k_{ia} denotes thermal conductivity of dense snow or ice containing air bubbles, k_i and k_a are, respectively, thermal conductivities of ice and air, and n is *porosity* defined as

$$n = 1 - (\rho_s/\rho_i) \tag{19}$$

where ρ_s and ρ_i are the densities of snow and ice, respectively. Observing $k_a \ll k_i$ and substituting Eq. (19) in Eq. (18), Eq. (18) can be approximated to

$$k_{ia} = \frac{2\rho_s}{3\rho_i - \rho_s} k_i \tag{20}$$

For low-density snow, the model (see Fig. 7) showing the spatial arrangement of the air space parallelepipeds within the ice medium is used to derive the thermal conductivity. For simplicity, the constants, s_1, s_2, and s_3 are all chosen to be equal to s. It can be seen that the thermal conductivities will be different in each direction and, as can be seen from the figure, will be a minimum in the x direction and a maximum along z. For the case of

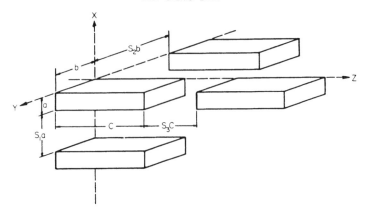

FIG. 7. The orientation and relative dimensions of the spaces in the model for light snow [23].

$a \ll b \ll c$, the following relation for the thermal conductivity of this medium in the x direction is derived:

$$k_{sa} = \frac{[1 + s(1 + s)^2]k_a + s(2 + s)k_i}{(1 + s)^2(sk_a + k_i)} k_i \qquad (21)$$

where k_{sa} denotes the thermal conductivity of snow containing air spaces. The porosity of this medium can be shown to be

$$n = [(1 + s)^3]^{-1} \qquad (22)$$

For snow densities down to about 0.15 gm/cm³, the effect of still air on the conductivity of the compound medium may often be neglected, in which case Eq. (21) reduces to

$$k_{sa} = \frac{s(2 + s)}{(1 + s)^2} k_i \qquad (23)$$

For very low densities of snow, another snow model consisting of a suspension of small spherical ice particles in air not in contact with each other is considered. This leads to a formula of the type of Eq. (18) in which k_i and k_a are interchanged. The conductivity of this fictitious medium is denoted by k_{as}. Figure 8 shows the thermal conductivities of snow calculated from these theoretical expressions. It should be noted that, in all of these derivations, the effect of vapor diffusion is not included.

E. THERMAL DIFFUSIVITY OF SNOW

Thermal diffusivity is defined as the ratio of thermal conductivity to the product of specific heat and density ($\alpha = k/\rho c$). The thermal diffusivity is far less density dependent than the conductivity. Figure 9 shows the variation of diffusivities calculated from the conductivities in Fig. 8. It can be seen that, for snow densities ranging from that of ice (0.917 gm/cm^3) to about 0.6 gm/cm^3, the diffusivity lies within ±4% of the mean value. Since the densities in this range are those commonly encountered in large snow-ice caps such as in Greenland and Antarctica, this more or less constant value of diffusivity, in spite of varying conductivity, should simplify temperature field analysis considerably.

It should be noted that thermal properties are usually temperature dependent. The conductivity k and specific heat c of ice undergo a slight decrease and increase, respectively, with rising temperature. Over the range of 0° to −20°C, their variations are within 0.2%/°C. This leads to a diffusivity change of less than 0.4%/°C. Therefore, in Schwerdtfeger's analysis the temperature dependence of the thermal properties has not been included.

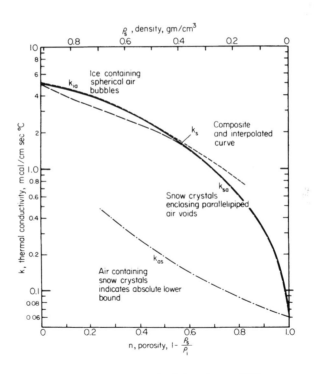

FIG. 8. The thermal conductivity of snow [23].

F. Vapor Diffusion in Snow

Diffusion coefficient D for a binary system is a function of temperature, pressure, and composition, For gas mixtures of low pressure or dilute solutions, D can reasonably be taken as a constant. In one-dimensional and steady diffusion processes, according to the Fick's law, the mass flux m is

$$m = -D \, dC/dz \tag{24}$$

where D is the diffusion coefficient and dC/dz is the concentration gradient in the direction of diffusion

Yosida [24] was the first one to investigate the macroscopic diffusion coefficient D of water vapor through snow. His experimental setup is shown in Fig 10. It consists of four small circular cans with bottoms replaced by fine wire screens. These cans, after having been filled with snow, were set on top of each other within a metal tube. The top and bottom of the tube were maintained at desired temperatures. To collect the sublimate from the bottom of snow can C_4, a circular disk of the same diameter as the snow was placed as shown in Fig. 10.

Under a temperature gradient, sublimation occurs on the snow grains at higher temperatures, and water vapor is produced. This water vapor diffuses and condenses on the nearby snow grains, which have lower temperatures. As a result of this mass-transfer process, there is redistribution of the snow density. At the warmer end the snow cans will lose weight, whereas at the colder end the snow cans will increase in weight. In order to obtain the diffusion coefficient D, we write Eq. (24) in the following form:

$$m = -\beta_P D \, dp/dz \tag{25}$$

It should be pointed out that m is a function of z as well as of time. The diffusion coefficient D can be experimentally determined as follows. Let m_{i-1} and m_i be the mass flux at the top and bottom of snow can C_i, respectively. Representing the mean value of dp/dz for snow can C_i by $\frac{1}{2}[(dp/dz)_{i-1} + (dp/dz)_i]$, the following relation can be derived from Eq. (25):

$$\sum_{i=1}^{4} \frac{m_{i-1} + m_i}{2} = \beta_P D \sum_{i=1}^{4} \frac{p_{i-1} - p_i}{l/4} \tag{26}$$

where l is the total height of the four snow cans. It is clear that Eq. (26) holds at any instant during the experiment or for experiments lasting for a certain period of time as long as the time average values of m_i's and p_i's are used. The left side of Eq. (26), for the case of four cans, can be written as

$$Q = \tfrac{1}{8}(7q_1 + 5q_2 + 3q_3 + q_4) \tag{27}$$

where $q_i = m_i - m_{i-1}$ for the time average value of decrease in weight of

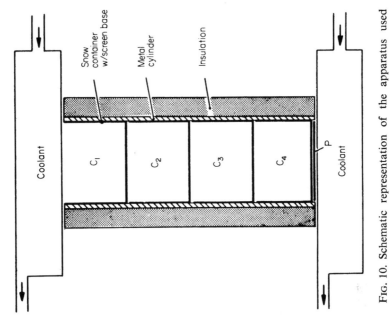

FIG. 10. Schematic representation of the apparatus used in determining the macroscopic diffusion coefficient D of water vapor through snow [24].

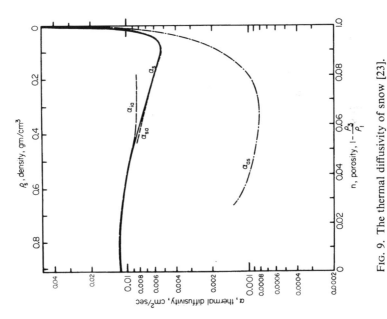

FIG. 9. The thermal diffusivity of snow [23].

snow can C_i which can be determined by weighing the snow cans at the beginning and at the completion of the experiment. It should be noted that $m_0 = 0$, and m_4 is the mass flux at the bottom of can C_4 which sublimes on the disk. D can be found from Eqs. (26) and (27) by

$$D = \frac{lQ}{4\beta_P(p_0 - p_4)} \tag{28}$$

where p_0 and p_4 are the time average values of the saturation water vapor pressure corresponding to temperatures maintained in the upper and lower boundaries, respectively. Using Eq. (28), Yosida [24] calculated D values for different conditions as listed in Table III. It appears that the diffusion coefficient of water vapor through snow is independent of snow density, and its value is four or five times larger than the diffusion coefficient D_0

TABLE III

DIFFUSION COEFFICIENT OF WATER VAPOR THROUGH SNOW

Experiments	1	2	3	4	5
Snow density (gm/cm³)	0.08	0.14	0.25	0.34	0.51
Temperature at top of snow sample (°C)	−1.2	−0.4	−0.3	−9.1	−1.9
Temperature at bottom of snow sample (°C)	−6.7	−7.0	−4.9	−1.4	−8.8
$P_0 - P_4$ (mm-Hg)	1.58	1.90	1.43	1.96	1.75
Duration of experiment (10^3) sec	16.5	16.9	20.1	18.00	16.2
D (cm²/sec)	0.7	1.0	0.9	0.8	0.9

of water vapor through air (which is about 0.22 cm²/sec at 0°C and atmospheric pressure). It seems that natural convection has no apparent effect on water vapor diffusion through snow (see experiment no. 4 in Table III, in which the warmer temperature is maintained at the lower boundary).

To explain the fact that $D > D_0$, Yosida pointed out that ice grains do not act as mere obstacles on diffusion of water vapor as sand grains do, but they produce water vapor themselves, thereby facilitating the mass-transfer process and condensation.

Now let us estimate the contribution of molecular vapor diffusion to heat conduction. Referring to Eq. (24), this can be written as

$$q_v = -\beta_T \, DL_s \, dT/dz \tag{29}$$

where q_v is the heat flux due to vapor diffusion and β_T is the ratio of water vapor density (concentration) to temperature. The product of $\beta_T \, DL_s$ can be considered as k_v°, equivalent thermal conductivity due to molecular vapor diffusion. Taking $L_s = 676$ cal/gm, $\beta_T = 0.39 \times 10^{-6}$ gm/cm^3 °C, and $D = 0.85$ cm^2/sec, Eq. (29) becomes

$$q_v = -k_v^\circ \, dT/dz = -2.2 \times 10^{-4} \, dT/dz \quad \text{cal/cm}^2\text{-sec} \tag{30}$$

Therefore, the molecular vapor diffusion plays a major role in the process of heat conduction in snow of low density.

G. Theoretical Expressions for Effective Thermal Diffusivity

Sulakvelidze [25] formulated a heat-transfer equation for porous media containing saturated vapor, water, or ice at temperatures close to the temperature of phase transitions. This was done by including a term due to evaporation-condensation in the Fourier's heat-conduction equation. For one-dimensional heat transfer, the heat conduction equation becomes

$$\frac{\partial T}{\partial t} = \alpha \frac{\partial^2 T}{\partial z^2} - \frac{L_s}{c_i \rho_s} q \tag{31}$$

where q is the intensity of evaporation (or condensation) and can be expressed in terms of vapor concentration C by

$$q = \frac{\partial C}{\partial t} - D \frac{\partial^2 C}{\partial z^2} \tag{32}$$

Substituting for C a prescribed function $f(T)$, we have

$$\frac{\partial f}{\partial t} = D \frac{\partial^2 f}{\partial z^2} + q \tag{33}$$

Combining Eqs. (31) and (33), we obtain

$$\frac{\partial T}{\partial t} = \alpha \left[\frac{1 + DL_s/(\alpha c_i \rho_s) f'}{1 + L_s/(c_i \rho_s) f'} \right] \frac{\partial^2 T}{\partial z^2} + \frac{L_s D}{c_i \rho_s} f'' \left(\frac{\partial T}{\partial z} \right)^2 \tag{34}$$

where f' and f'' are first and second derivatives of f to T, respectively. Equation (34) describes the general processes of heat transfer taking place in a moist porous medium. When there is no vapor diffusion, $L_s = 0$, then Eq. (34) reduces to the well-known *Fourier heat-conduction equation*

$$\frac{\partial T}{\partial t} = \alpha \frac{\partial^2 T}{\partial z^2} \tag{35}$$

The *effective thermal diffusivity* is defined as the coefficient of $(\partial^2 T/\partial z^2)$, and it is

$$\alpha_e = \alpha \left[\frac{1 + (DL_s/\alpha c_i \rho_s) f'}{1 + (L_s/c_i \rho_s) f'} \right] \tag{36}$$

which can be rewritten as

$$\alpha_e = \alpha \left\{ \frac{1 + [DL_s^2 C_0/(\alpha c_i \rho_s J R_v T^2)] \exp[L_s(T - T_0)/(JR_v TT_0)]}{1 + [L_s^2 C_0/(c_i \rho_s J R_v T^2)] \exp[L_s(T - T_0)/(JR_v TT_0)]} \right\} \quad (37)$$

where use is made of the relation

$$C = C_0 \exp[L_s(T - T_0)/(JR_v TT_0)] \quad (38)$$

where T_0 is the freezing temperature of water on the absolute scale, J is the mechanical equivalent of heat, and C_0 is the vapor concentration at 0°C. From Eq. (37), it can be seen that, for $T \ll T_0$, the exponential term becomes very small so that, at temperatures between -20 and -30°C, $\alpha_e = \alpha$.

H. Effect of Vapor Diffusion on Temperature Propagation

In a recent study, Eq. (34) was solved numerically [26] for studying the combined effect of snow density and water vapor transfer on the rate of temperature propagation. For this purpose, a step increase in temperature T_s was imposed on one end of a snow column having uniform initial temperature T_i. Expressing saturation vapor pressure of snow $p_s = A \exp[B(T - T_i)]$, in which A and B are empirical constants, and noting $C = p_s/R_v T$, we obtain $C = [A/(R_v T)] \exp[B(T - T_i)]$. Equation (34) becomes

$$\frac{\partial T}{\partial t} = \alpha \left\{ \frac{1 + [DL_s A(TB - 1)/(\alpha c_i \rho_s R_v T^2)] \exp[B(T - T_i)]}{1 + [L_s A(TB - 1)/c_i \rho_s R_v T^2)] \exp[B(T - T_i)]} \right\} \frac{\partial^2 T}{\partial z^2}$$

$$+ \frac{DL_s}{c_i \rho_s R_v} \left\{ \frac{[A(T^2 B^2 - 2TB + 2)/T^3] \exp[B(T - T_i)]}{1 + [L_s A(TB - 1)/(c_i \rho_s R_v T^2)] \exp[B(T - T_i)]} \right\} \left(\frac{\partial T}{\partial z} \right)^2 \quad (39)$$

For practical purposes, the term $[L_s A(TB - 1)/(\rho_s R_v T^2)] \exp[B(T - T_i)]$ is much smaller than c_i; thus Eq. (39) becomes

$$\frac{\partial T}{\partial t} = \alpha \{1 + [DL_s A(TB - 1)/(\alpha c_i \rho_s R_v T^2)] \exp[B(T - T_i)]\} \frac{\partial T^2}{\partial z^2}$$

$$+ \frac{DL_s}{c_i \rho_s R_v} \{[A(T^2 B^2 - 2TB + 2)/T^3] \exp[B(T - T_i)]\} \left(\frac{\partial T}{\partial z} \right)^2 \quad (40)$$

By introducing dimensionless variables $\theta = (T - T_s)/(T_i - T_s)$, $\tau = \alpha t/l^2$, and $\xi = z/l$, in which l is a characteristic length in the medium, and by replacing $\partial T/\partial t$, $\partial^2 T/\partial z^2$, and $(\partial T/\partial z)^2$ with finite difference forms of

$$\frac{\partial \theta}{\partial \tau} = \frac{\theta_{m,n+1} - \theta_{m,n}}{\Delta \tau}$$

$$\frac{\partial^2 \theta}{\partial \xi^2} = \frac{\theta_{m+1,n} - 2\theta_{m,n} + \theta_{m-1,n}}{(\Delta \xi)^2}$$

$$\left(\frac{\partial \theta}{\partial \xi}\right)^2 = \frac{(\theta_{m+1,n} - \theta_{m-1,n})^2}{4(\Delta \xi)^2}$$

Eq. (40) is numerically computed with the stability criterion $\Delta \tau / \Delta \xi^2 = \frac{1}{4}$ for various values of α calculated from $\alpha = 6.8 \times \rho_s/c_i$ and with the following constants: $L_s = 676$ cal/gm, $A = 0.457$ cm-Hg, $B = 0.0857°K^{-1}$, $R_v = 346$ cm^3-cm-Hg/gm°K, $D = 0.8$ cm^2/sec, and $T_i =$ initial snow temperature, 273°K. Figure 11 illustrates the computed results for $\rho_s = 0.1$ gm/cm^3.

Figure 12 shows the dimensionless temperature distribution in a medium without the effect of water vapor diffusion. This is computed for the same problem from the analytical solution of Eq. (35). The solution is

$$\theta = \xi + (2/\pi) \sum_{n=1}^{\infty} (\sin n\pi\xi) \exp(-n^2\pi\xi) \tag{41}$$

A comparison of Figs. 11 and 12 indicates that, at low snow density, the effect of water vapor diffusion has considerable effect on the rate of temperature propagation. As the density of snow increases, the effect of the water vapor diffusion in temperature change decreases [26], because, in these cases, the conduction through the ice grains dominates the overall heat-transfer process. Figures 11 and 12 will also provide a means of determining the effect of vapor diffusion on the thermal conductivity of snow. For example, to accomplish a change of θ from 1 to 0.25, for $\xi = 0.2$, it takes only $\tau = 0.06$ in Fig. 11, whereas it takes $\tau = 0.2$ in Fig. 12. Therefore, the conductivity of snow of density of 0.1 gm/cm^3 is 3.33 times larger than that of the fictitious snow, where the vapor diffusion is not considered.

I. HEAT AND VAPOR TRANSFER WITH FORCED CONVECTION

The effects of air flow through snow on effective thermal conductivity and vapor diffusivity have been studied experimentally and theoretically and reported in a series of paper by Yen [27–29]. For one-dimensional and steady-state cases, the governing equations for heat and vapor transfer can be summarized as follows:

For heat transfer,

$$Gc_a \frac{dT}{dz} + \frac{GM_w L_s}{M\pi} \frac{dP_s}{dz} + k_e \frac{d^2 T}{dz^2} = 0 \tag{42}$$

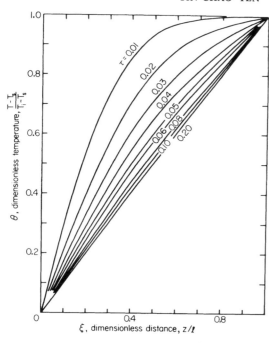

FIG. 11. Relation between dimensionless temperature and dimensionless distance with water vapor diffusion [26].

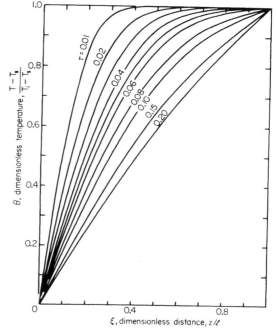

FIG. 12. Relation between dimensionless temperature and dimensionless distance without water vapor diffusion [26].

For vapor transfer,

$$\frac{dP_s}{dz^2} + \left(\beta - \frac{R_v T_m}{D_e \, dP_s/dz} \frac{\partial \rho_s}{\partial t}\right) \frac{dP_s}{dz} = 0 \tag{43}$$

where G is the mass flow rate of dry air, M_w and M are the molecular weight of water and dry air, π the total pressure of the system, k_e and D_e the effective thermal conductivity and vapor diffusivity, T_m the mean temperature of the system, and $\beta = G R_v T_m M_w / M \pi D_e$. As shown in the preceding references, for calculation of k_e, only the steady-state temperature distribution is needed, whereas, for obtaining D_e, the density distribution before and after the completion of the experiment as well as the temperature distribution are required.

For unconsolidated snow of density in the range of 0.376 to 0.472 gm/cm^3, the values of k_e which were calculated by Yen [27] from the temperature data are plotted in Fig. 13. These values can be approximated to $k_e = k_e^\circ + 0.58G$, where $k_e^\circ = 0.0014$ and G varies from 10×10^{-4} to 40×10^{-4} gm/cm^2-sec. The value of 0.0014 for k_e° is in good agreement with the data reported by Kondrat'eva (see Mellor [22]). In the same manner, the values of effective diffusivity of water vapor which were calculated by Yen [28] are plotted in Fig. 14 and can be approximated to $D_e = 95.38(G + 0.465 \times 10^{-4})^{1/2}$. When $G = 0$, D_e becomes D (diffusion coefficient of water vapor through snow), the value of which is found from the preceding expression to be equal to 0.65 cm^2/sec. Using a completely different approach, as reported earlier in this section, Yosida [24] reported an average value of 0.85 cm^2/sec for D with snow density between 0.08 and 0.51 gm/cm^3. The difference between these two values is believed to be partially due to the difference in temperature levels used in my experiments and those of Yosida.

Now, let us estimate the contribution of the vapor movement to the overall heat-transfer process. Using Eq. (25) and noting that $C/P_s = 1/R_v T$, and replacing D by D_e since we are concerned with forced convection, we have

$$q_v = -\frac{L_s D_e}{R_v T} \frac{dP_s}{dz} \tag{44}$$

Substituting $P_s = 4.58 \exp[0.0857(T - 273)]$, we have

$$q_v = -\frac{0.393 L_s D_e}{R_v T} \exp[0.0857(T - T_0)] \frac{dT}{dz} \tag{45}$$

The coefficient of dT/dz in Eq. (45) is equivalent to k_v, equivalent thermal conductivity due to vapor transfer in forced convection.

For naturally compacted snow (see Yen [29]), the following empirical correlations for k_e and D_e are found to be valid for snow density varying

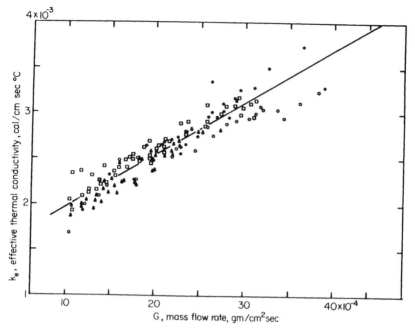

Fig. 13. Relation between the effective thermal conductivity and mass flow rate [27]. Snow particle (nominal diam.): $\bigcirc = 0.07$ cm; $\bullet = 0.13$ cm; $\square = 0.19$ cm; $\blacktriangle = 0.22$ cm.

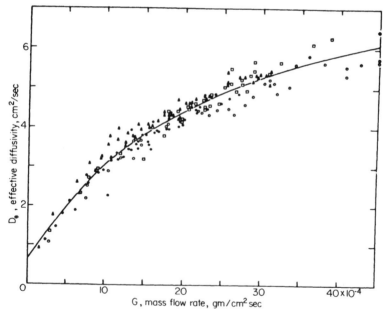

Fig. 14. Relation between the effective water vapor diffusivity and the mass flow rate [28]. Snow particle (nominal diam.): $\bigcirc = 0.13$ cm; $\bullet = 0.07$ cm; $\square = 0.19$ cm; $\blacktriangle = 0.22$ cm.

from 0.50 to 0.59 gm/cm^3, and G ranging from 5×10^{-4} to 32×10^{-4} gm/cm^2-sec:

$$k_e = 0.0077(\rho_s)^2 + 0.60G \qquad (46)$$

and

$$D_e = 0.65 + 2113(\rho_s)^{3.20} G^{0.615} \qquad (47)$$

Under no-flow conditions, k_e reduces to $k_e°$, which is $0.0077(\rho_s)^2$ cal/cm-sec-°C, and D_e becomes D, which is 0.65 cm^2/sec. Both values are in agreement with previously reported experimental results [22, 28].

From these investigations, it can be concluded that air flow has considerable effect on the values of k_e and D_e of both unconsolidated and naturally compacted snow. The increase in thermal conductivity and vapor diffusivity due to air ventilation is held responsible for the small variations in snow density and temperature gradient near the surface layers of a snow cover.

IV. Radiation Properties of Snow

A. Fundamentals of Radiation Transport

All substances, solids as well as liquids and gases, emit and absorb energy in the form of radiation at normal and especially at high temperatures. The mechanism of heat transfer by radiation differs fundamentally from that of conduction and convection. In both conduction and convection, the transfer of heat relies on the existence of a material medium. For heat conduction to occur, there must be temperature inequalities at neighboring points in the conducting medium, whereas for convection there must be a fluid that is free to move and transport energy with it. For the case of radiation, the energy is transferred by electromagnetic waves at different frequencies ($v = C/\lambda$, where C is speed of light in vacuum and λ is the wavelength) shown in Table IV. In this section, we are mostly concerned with the visible radiation.

The radiation flux emitted from a surface can be expressed by

$$q = e\sigma T^4 \qquad (48)$$

known as *Stefan-Boltzmann law*, where e is the emissivity (a property of the radiating material, in the case of a blackbody, $e = 1$), σ is Stefan-Boltzmann constant, and T is the absolute temperature of the radiating body. It becomes obvious, then, that the radiation received at a given point is entirely independent of the temperature of the material present at that point. This is clearly shown by the transfer of the solar energy to the earth's surface.

TABLE IV

Modes of Radiation

Type of wave	Wavelength, λ^a
Cosmic rays	0.5 $\mu\mu$
Gamma rays	0.05 $\mu\mu$–10 $\mu\mu$
X rays	1 $\mu\mu$–10 mμ
Ultraviolet radiation	20 mμ–0.4 μ
Visible radiation (light)	0.4 μ–0.8 μ
Thermal (infrared) radiation	0.8 μ–0.8 mm
Electrical waves (radio, etc.)	0.2 mm–x km

^{a} 1 $\mu = 10^{-6}$ m $= 10^{-3}$ mm, 1 m$\mu = 10^{-6}$ mm $=$ 10 Å, 1 $\mu\mu = 10^{-9}$ mm.

B. Concepts of Radiation Interaction

When a radiation ray falls on an interface between two homogeneous media, its energy is usually disposed in three different modes. One portion is reflected from the interface, whereas the nonreflected portion of the radiation, which penetrates into the second medium, can be either absorbed or transmitted, or partially absorbed and transmitted. Thus, we can write

$$I = R + A + T \qquad (49)$$

where I is the energy of the incident radiation, R is the energy reflected, A is the energy absorbed, and T is the energy transmitted. The quantity A will be the essential factor affecting such problems as the heat economy of, and the temperature distribution in, a snow cover. It is almost impossible to measure the absorbed radiation directly, and, therefore, I, R, and T are measured, and A is computed from Eq. (49). Since most of the radiation is absorbed within the top few centimeters of the snow layer, we can assume that $T = 0$, and only I and R need to be determined.

C. Absorption and Scattering of Radiation in a Snow Mass

In any consideration of the earth's total heat budget as well as for short- and long-range weather forecasting, the effect of snow cannot be ignored. Solar radiation is an important form of energy flux to the world's snow surfaces. The amount of radiation absorbed and reflected depends very largely on the physical properties of the snow below the surface. For simplicity, it is usually assumed that snow behaves as a homogeneous diffusing medium and follows the *Bouger-Lambert law*. The attenuation of the intensity for monochromatic radiation at a point z measured downward from the surface may be written as

$$I = I_0 \exp(-vz) \tag{50}$$

where I is the radiation intensity, I_0 is the net radiation intensity at surface $z = 0$, and v is usually called the *extinction coefficient*, which is related to the absorption coefficient a and scattering coefficient s by $v = a(1 + 2s/a)^{1/2}$. The extinction coefficient normally varies with wavelength.

It is easy to visualize that, even if the Bouger-Lambert law held for a homogeneous snow pack (artificial or natural), one would not expect the law to hold rigorously for a natural layered snow cover where density, crystal structure, and grain size change with depth. Furthermore, it should be pointed out that radiation properties of snow, like thermal properties, are strongly dependent on the surface characteristics and the internal structure of the snow mass, and, therefore, it is very difficult to obtain consistent values of *spectral transmittance* τ (defined as the ratio of transmitted energy to incident energy), spectral reflectance r, and albedo α (both will be defined in Sect. IV.D). These inconsistencies are substantiated by the data compiled, covering a period of several decades, at the University of Minnesota [30].

Some of the old work on extinction coefficients was compiled by Mellor [22], as shown in Figs. 15 and 16. According to Fig. 15, both absorption (for dry powdered snow in the range of 0.4 to 1 μ) and extinction coefficients (for fresh and old snow in the range of 0.4 to 0.66 μ) increase with wavelength. Figure 16 indicates that the extinction coefficient decreases linearly with increasing density. It is obvious that such a relation cannot be extended to lower densities, as v must approach zero as a limiting value.

Some of the most recent results on extinction coefficients were reported by Mellor [31] in a study on optical properties of snow, by using homogeneous snow samples prepared under controlled conditions. He found that the extinction coefficient is a function of wavelength (Fig. 17) and density (Fig. 18). The variations of extinction coefficients for fine-grained and coarse-grained snow samples are within 0.8 and 1.7 cm^{-1} and 0.16 and 0.37 cm^{-1}, respectively.

It can be seen from Fig. 17 that there is a general decline in for fine-grained snow as λ increases from 0.4 to 0.7 μ. Since this trend is contrary to that which might be expected in an absorption-controlled attenuation, he concludes that scattering is the dominant mechanism for attenuation. In coarse-grained snow, however, he believes absorption is more significant. Previous reports (Fig. 15) show an opposite trend [22].

The results shown in Fig. 18 for snow having a mean grain diameter of 0.2 mm indicate an interesting trend. Since both air and ice are more

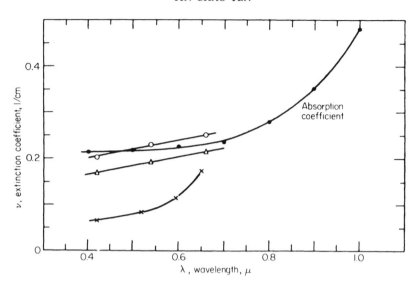

Fig. 15. Extinction and absorption coefficients as a function of wavelength [22]. ●, Ambach and Habicht (powder, $\rho_s = 0.33$ gm cm^{-3}); ×, Liljequist (dry polar snow, $\rho_s = 0.42$ gm cm^{-3}); ○, Thomas (old snow, $\rho_s = 0.33$ gm cm^{-3}); △, Thomas (old snow, $\rho_s = 0.42$ gm cm^{-3}).

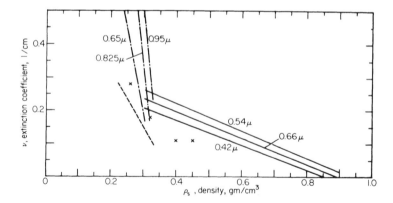

Fig. 16. Extinction coefficient as a function of snow density for various wavelengths [22]. —, Thomas, old snow (data plotted by Mellor; correlation coefficients 0.94–0.97, std. deviation 0.17–0.23). ---, Ambach and Habicht, mixed light, powder snow. —·—, Ambach and Habicht, powder snow. ×, Gerdel, mixed light.

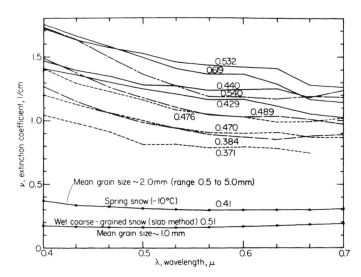

FIG. 17. Extinction coefficient as a function of wavelength [31]. Mean grain size: —, 0.2 mm (very uniform); —·—, 0.3 mm (range 0.2–0.7 mm); ---, 0.6 mm (range 0.2–1.1 mm).

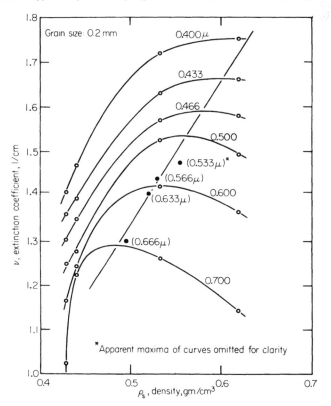

FIG. 18. Extinction coefficient as a function of snow density for fine-grained snow [31].

205

transparent than snow, the curve relating v and density must exhibit a maximum. From the limited data available, it appears that the maximum v occurs between 0.45 and 0.6 gm/cm^3 and also depends on λ.

The relation between extinction coefficient and grain size is shown in Fig. 19. It clearly demonstrates that, for some snow density (0.45 gm/cm^3), v is inversely related to grain size and wavelength λ. It is interesting to note that the curves appear to converge as the grain size both increases and decreases. The data indicate that v increases sharply as grain size falls below

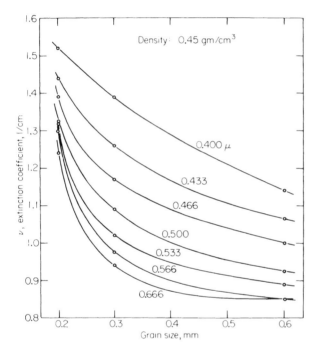

FIG. 19. Extinction coefficient as a function of grain size [31].

0.2 mm. Furthermore, it also suggests that spectral selection may disappear for very small particle sizes. As grain size increases, absorption must assume the dominant role in extinction, and the absorption coefficients for ice can probably be taken as the limiting value for extinction as grain size approaches infinity.

D. REFLECTION FROM SNOW

The passage of radiation through snow differs fundamentally from that of radiation through a continuous medium because internal reflection takes place in snow. Therefore, energy reflection from a snow surface consists of two parts: one is surface reflection, and the other is energy return from beneath the surface due to multiple reflections occurring within the mass. The reflecting property of the surface alone is expressed as surface reflectivity, whereas the overall reflection characteristic of the snow mass is described as *reflectance*.

Surface structure is the major factor affecting the magnitude of reflectivity, and it varies with the characteristics of the incident light. In the diffuse light of a cloudy day, the snow surface can be regarded as a diffuse reflector; however, in direct sunlight only smooth fresh snow can be considered as a diffuse reflector. (This means that snow will hold *Lambert's cosine law*: intensity of reflected radiation is proportional to the cosine of the angle between the direction of reflection and the normal to the surface.) There are various factors affecting reflectivity, such as grain form and solid impurities.

Energy returns from the subsurface depend on scattering and absorption within the snow, which in turn depend on density, grain form, and the presence of subsurface ice-lenses and inhomogeneities.

The overall reflectance covering a wide spectrum of the electromagnetic waves is usually interpreted as the *albedo* of snow, defined mathematically by

$$\alpha = (\int_{\lambda_1}^{\lambda_2} rI \, d\lambda)/(\int_{\lambda_1}^{\lambda_2} I \, d\lambda) \tag{51}$$

where α is the albedo, r is the *spectral reflectance* defined as R/I, and the integration limits are the extremes of the spectrum concerned (about 0.4 to 0.7 μ for the visible range). If r is a constant for all wavelengths, then, from Eq. (51), $\alpha = r$. The white color of snow indicates that r is not a strong function of wavelength in the visible spectrum, but there is evidence that some dependence of r on wavelength λ and snow type might be anticipated as snow in transformed to ice.

Mellor [22] compiled the available published data on reflectance as a function of wavelength in Fig. 20. Curves A and B are for snow illuminated by light diffused through a cloud cover, whereas curves C–F are for snow in direct sunlight. It can be seen that reflectance decreases as wavelength increases from 0.4 to 0.8 μ (curve A). Curve B indicates very little change between 0.42 and 0.595 μ, but curve F shows an appreciable increase of reflectance as wavelength increases from 0.45 to 0.66 μ.

Curves C and D are similar through the visible spectrum, but they deviate from each other for wavelengths above 0.7 μ. From the foregoing curves, it can be seen that no general trend can be defined in regard to snow reflectance and wavelength.

Snow density is a widely used parameter to describe various physical properties of snow, but the effect of density on reflectance does not appear to have been studied extensively. Using the data available [32], no significant correlation can be found between reflectance and density for wavelengths of 0.42, 0.54, and 0.66 μ and for unfiltered light.

Figure 21 shows the recent results obtained by Mellor [31]. Without

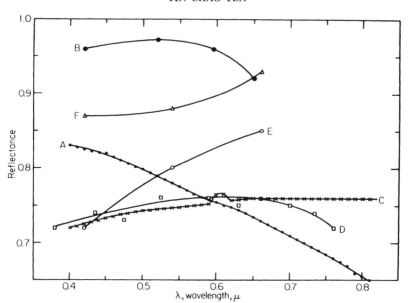

FIG. 20. Reflectance as a function of wavelength [22]. ○, Thomas (MM analysis); ●, Liljequist (cloud); □, Sauberer; △, Thomas (fresh snow); ●, Krinov (fresh snow; cloud); ×, Krinov (iced snow).

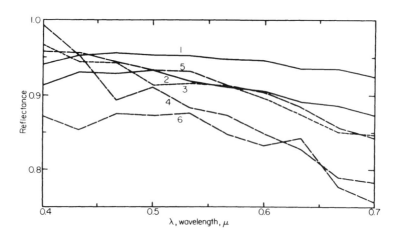

FIG. 21. Summary of results from field measurements of reflectance [31]: (1) fresh snow (dry), 0.28 gm/cm^3, 0°C; (2) 1–2 cm fresh snow (0.1 gm/cm^3) on older snow (0.4 gm/cm^3), °C; (3) metamorphosed snow, 0.43 gm/cm^3, 0°C; (4) slightly metamorphosed new snow, 0.2 gm/cm^3, 0°C; (5) wet snow, 2 days old, 0.4 gm/cm^3, melting during test; (6) same as (5) after 5 hr more melting.

exception, the lowest reflectance was found at 0.700 μ (the limit of the range studied), but the trend of spectral selection differed according to the lighting conditions. It can be seen that the reflectance of fine-grained snow does not vary much with λ, since surface reflectivity is high, and backscatter from the first layer of grain is apparently not very selective with respect to wavelength. As grain size increases, the reflectance and the surface reflectivity are expected to decrease; a relatively large portion of the reflected light is backscattered from beneath the surface, so that reflectance becomes inversely dependent on λ.

E. Analytical Study of Radiant Energy Distribution and Albedo of Snow

The most recent theoretical study concerning radiant energy distribution and albedo of snow was conducted by Giddings and LaChapelle [33]. In their analysis, the penetration of radiation into snow is assumed to follow a *random walk model*. In this model, radiation may be considered to travel a distance l_z in the z direction and then by a random choice either continue or reverse its direction. Assuming that the radiation that ends its random walk step within a thin lamina at depth z originates at a distance l_z above and below, we have, for the steady-state equation for radiative transfer,

$$kI_z = \gamma/l_z(I_{z-l_z} + I_{z+l_z} - 2I_z) \tag{52}$$

Expanding I_{z-l_z} and I_{z+l_z} in a series about z, it follows that

$$I = \frac{\gamma}{kl_z}\left(\frac{d^2I}{dz^2}l_z^2 + \frac{1}{12}\frac{d^4I}{dz^4}l_z^4 + \cdots\right) \tag{53}$$

where k is the absorption constant. γ is defined as the ratio of radiation flux to the energy intensity at that plane.

Dropping the fourth-power term and higher, Eq. (53) becomes

$$d^2I/dz^2 = b^2 I \tag{54}$$

where $b^2 = k/\gamma l_z$. It should be noted that Eq. (54) can also be obtained from the differential equation,

$$\frac{\partial I}{\partial t} = D\frac{\partial^2 I}{\partial z^2} - kI \tag{55}$$

if the steady-state conditions are assumed. Equation (55) is obtained by assuming that the mechanism of penetration of radiation into snow follows *Fick's second law of diffusion* and the snow medium is homogeneous and isotropic. D in Eq. (55) is defined as the *diffusion coefficient*. The solution of Eq. (54) is

$$I = C_1 \exp(-bz) + C_2 \exp(bz) \tag{56}$$

For a semi-infinite medium, Eq. (56) reduces to

$$I = I_0 \exp(-bz) \tag{57}$$

Equation (56) can be generalized for n successive snow layers having different optical properties, as outlined by Giddings and LaChapelle. The solution for a snow layer of depth h with boundary conditions $I = I_0$ at $z = 0$ and $-D(dI/dz)_h = A\gamma I_{z=h}$ at $z = h$, where A is the absorptivity of the layer, is

$$\frac{I}{I_0} = \exp(-bz) + \frac{\exp(-bh)(\omega - A)\sinh bz}{A \sinh bh + \omega \cosh bh} \tag{58}$$

where $\omega = bD/\gamma = (kD)^{1/2}/\gamma$. For a small grain size snow cover, D is small, and ω is small by comparison with unity. If the surface is not a strong reflector, A is near unity, and if h is at a depth comparable to $1/b$ or larger, Eq. (58) can be reduced to

$$\frac{I}{I_0} = \exp(-bz) - \frac{\exp(-bh)\sinh bz}{\sinh bh} \tag{59}$$

An analytical expression for spectral albedo can be derived as

$$\alpha = \frac{\gamma I_0}{-D(dI/dz)_0 + \gamma I_0} \tag{60}$$

where the term $-D(dI/dz)_0$ is the net downward flux at the surface, and γI_0 is the reflected radiation flux at $z = 0$. The preceding expression is valid irrespective of the internal snow structure or the presence of absorbing surfaces. By substituting $(dI/dz)_0$ from Eq. (57) into Eq. (60), we have

$$\alpha = (1 + \omega)^{-1} \tag{61}$$

which is valid for a semi-infinite, homogeneous, and isotropic snow cover. In practice, the medium can be considered as semi-infinite as long as h is $\gg 1/b$.

Similarly, the albedo of snow with an internal absorbing surface can be obtained by combining Eqs. (58) and (60):

$$\alpha = \left\{ 1 + \omega \left[1 - \frac{\exp(-bh)(\omega - A)}{A \sinh bh + \omega \cosh bh} \right] \right\}^{-1} \tag{62}$$

We now wish to find the error incurred when the second term in (Eq. 53) was neglected. An approximation to the relative error r_1 is defined as

$$r_1 = (d^4 I/dz^4) l_z^2 / 12 (d^2 I/dz^2) \tag{63}$$

For $I = I_0 \exp(-bz)$, then $r_1 = b^2 l_z^2/12 = \omega^2/12$. If r_1 is expressed in terms of albedo, it becomes $r_1 = (1 - \alpha)^2/12\alpha^2$. It can be seen that for $\alpha \leq 0.5$ the relative error becomes important (about 10% and larger).

In the foregoing analysis, the radiation mechanism was assumed to be by diffusion. In the following, we shall consider a case where radiation is not entirely by diffusion; however, since the treatment will still be based on diffusion theory, the deviations from diffuseness cannot be too large. In this approach, the radiation that is passing through a plane is assumed, on the average, to be halfway through its step of length l_z, and thus can be considered as originating at a distance $l_z/2$ above and below. Therefore,

$$\text{net flux} = \gamma(I_{z-lz/2} - I_{z+lz/2}) \tag{64}$$

By expanding the terms in the parentheses, we get $-\gamma l_z(dI/dz)$ for the net flux. Therefore, the relative error caused by writing the flux through plane z as γI in the diffuse radiation instead of $\gamma[I - (dI/dz)(l_z/2)]$ is approximately

$$r_2 = -(d \ln I/dz)(l_z/2) \tag{65}$$

Assuming again $I = I_0 \exp(-bz)$, and expressing r_2 in terms of albedo, $r_2 = (1 - \alpha)/2\alpha$. It can be seen that the errors become important for $\alpha \leqslant 0.80$. Therefore, Eqs. (58), (60), (61), and (62) are valid for $\alpha \geqslant 0.80$. For lower values of α, γI in Eqs. (58), (60), (61), and (62) must be replaced by $\gamma I \pm \gamma(dI/dz)(l_z/2)$. Since $\gamma l_z = D$, the equation for albedo becomes

$$\alpha = \frac{I_0 + (D/2)(dI/dz)_0}{I_0 - (D/2)(dI/dz)_0} \tag{66}$$

Utilizing Eq. (66), Eqs. (61) and (62) can be rewritten as

$$\alpha = \frac{1 - \omega/2}{1 + \omega/2} \tag{67}$$

and

$$\alpha = \frac{1 - \omega(1 - y)/2}{1 + \omega(1 - y)/2} \tag{68}$$

where

$$y = \frac{\exp(-bh)[A + \omega(A/2 - 1)]}{\omega(A/2 - 1) \cosh bh - A \sinh bh}$$

and Eq. (58) is changed to

$$\frac{I}{I_0} = \exp(-bz) + \frac{\exp(-bh)[A + \omega(A/2 - 1)] \sinh bz}{\omega(A/2 - 1) \cosh bh - A \sinh bh} \tag{69}$$

The preceding equations are expected to be more accurate and probably will not incur serious errors until the albedo drops to 0.50.

Acknowledgment

The author is grateful to Dr. Fuat Odar for his helpful suggestions and critical review during the preparation of this manuscript.

Symbols

- a Absorption coefficient
- A Absorptivity constant
- $A(T)$ Temperature-dependent constant
- b A constant
- c Specific heat at constant pressure
- C Vapor concentration
- D Diffusion coefficient of water vapor through snow; also for radiation
- D_e Effective diffusion coefficient with forced convection
- D_0 Diffusion coefficient of water vapor in air
- e Emissivity
- E Temperature gradient
- G Mass flow rate
- h Boltzmann's constant; also depth of a snow layer
- I Radiation intensity
- J Mechanical equivalent of heat
- k Thermal conductivity; also absorption constant
- k_e Effective thermal conductivity with forced convection
- k_e° Effective thermal conductivity of snow with stagnant air
- l Characteristic length
- L_s Latent heat of sublimation
- m Mass flux, molecular weight; also exponent
- M Parameter; also molecular weight
- n Porosity; also exponent
- N Parameter
- P Pressure
- P_i Saturation vapor pressure of ice
- P_0 Equilibrium vapor pressure over a plain surface of ice
- P_s Saturation vapor pressure of snow
- q Heat flux; also evaporation or condensation intensity
- Q Parameter
- q_v Heat flux due to vapor transfer
- r Radius of a sphere; also spectral reflectance
- R Gas constant for dry air
- R_v Gas constant for water vapor
- s Scattering coefficient
- t Time
- T Absolute temperature
- x Radius of a neck
- z Rectangular coordinate

Greek Letters

- ρ Density
- σ Surface tension, Stefan-Boltzmann constant; also a constant
- δ Intermolecular spacing
- λ Wavelength (μ)
- θ Temperature (°C); also dimensionless temperature
- α Thermal diffusivity; also albedo
- π 3.1416; also pressure of a system
- β_P Ratio of vapor concentration to pressure
- β_T Ratio of vapor concentration to temperature
- τ Dimensionless time
- ξ Dimensionless distance
- γ A constant
- ω A constant
- ν Extinction coefficient

Subscripts

- 1 Material 1
- 2 Material 2
- a Air
- as Air containing snow
- e Effective
- i Ice
- ia Ice containing air bubbles
- m Model, mean
- 0 Surface
- s Snow; also constant
- sa Snow containing air
- v Vapor

Recent Studies on Snow Properties

REFERENCES

1. *Proc. Intern. Conf. Low Temp. Sci., Inst. Low Temp. Sci., Hokkaido Univ., Japan,* 1966.
1a. Kingery, W. D., ed., "Ice and Snow." M.I.T. Press, Cambridge, Massachusetts, 1963.
1b. Yosida, Z., ed., "Physical Studies on Deposited Snow" (Z. Yosida, ed.). Inst. of Low Temp. Sci., Hokkaido Univ., Japan, 1955.
1c. Bader, H., The physics and mechanics of snow as a material. "Cold Regions Science and Engineering, Part II: Physical Science," Sect. B. Monograph. U.S. Army Cold Regions Res. and Eng. Lab., Hanover, New Hampshire, 1962.
2. de Quervain, M. R., On the metamorphism of snow. *In* "Ice and Snow" (W. D. Kingery, ed,), pp. 377–390. M.I.T. Press, Cambridge, Massachusetts, 1963.
3. Dorsey, N. E., "Properties of Ordinary Water Substances." Reinhold, New York, 1940.
4. Yosida, Z., Thermal properties, *In* "Physical Studies on Deposited Snow" (Z. Yosida, ed.), pp. 19–74. Inst. of Low Temp. Sci., Hokkaido Univ., Japan, 1955.
5. Faraday, M., Note on regelation. *Proc. Roy. Soc.* **10**, 440–450 (1860).
6. Tyndall, J., On some properties of ice. *Phil. Trans. Roy. Soc. London* **148**, 211–229 (1858).
7. Thomson, J., On the theories and experiments regarding ice at or near its melting point. *Proc. Roy. Soc.* **A10**, 152–160 (1859).
8. Nakaya, U., and Matsumoto, A., Simple experiment showing the existence of liquid-water film in the ice surface. *J. Colloid Sci.* **9**, 41–49 (1954).
9. Hosler, C. L., Jensen, D. C., and Goldshlak, L., On the aggregation of ice crystals to form snow. *J. Geophys. Res.* **70**, 3903–3907 (1965).
10. Jellinek, H. H. G., Adhesive properties of ice. U.S. Army Snow Ice and Permafrost Res. Estab., Res. Rep. 62 (1960).
11. Kingery, W. D., Regelation surface diffusion, and ice sintering. *J. Appl. Phys.* **31**, 833–838 (1960).
12. Kuroiwa, D., A study of ice sintering. *Tellus* **13**, 252–259 (1961).
13. Ramseier, R. O., and Sander, G. W., Sintering of snow as a function of temperature. *Intern. Assoc. Sci. Hydrology, Symp. Davos,* 1965, *Publ.* No. 69, pp. 119–127.
14. Ramseier, R. O., and Keeler, C. M., Sintering of snow. *J. Glaciol.* **6**, 421–424 (1966).
15. Kuczynski, G. C., Self-diffusion in sintering of metallic particles. *J. Metals* **1**, 169–178 (1949).
16. Kingery, W. D., and Berg, M., Study of the initial stages of sintering solids by viscous flow, evaporation-condensation, and self-diffusion. *J. Appl. Phys.* **26**, 1205–1212 (1955).
17. Hobbs, P. V., and Mason, B. J., The sintering and adhesion of ice. *Phil. Mag.* **9**, 181–197 (1964).
18. Hobbs, P. V., The effect of time on the physical properties of deposited snow. *J. Geophys. Res.* **70**, 3903–3907 (1965).
19. Itagaki, K., personal communication, 1967.
20. Yosida, Z., and Iwai, H., Measurement of the thermal conductivity of snow cover (Sekisetsu no netsu dendō-ritsu no sokutei). U.S. Army Snow Ice and Permafrost Res. Estab., Transl. 30 (1954); also *Seppyo.* **8**, 48–53 (1946).
21. van Wijk, W. R., New method for measuring heat flux density at the surface of soils or of other bodies. *Nature* **213**, 214 (1967).
22. Mellor, M., Properties of snow. "Cold Regions Science and Engineering, Part III," Sect. AI. Monograph. U.S. Army Cold Regions Res. and Eng. Lab., Hanover, New Hampshire, 1964.

23. Schwerdtfeger, P., Theoretical derivation of the thermal conductivity and diffusivity of snow. *Intern. Assoc. Sci. Hydrology, Gen. Assembly, Berkeley*, 1963, Publ. No. 61, pp. 75–81.
23a. Maxwell, J. C., "Electricity and Magnetism," 3rd Ed., Vol. 1, Dover, New York, 1891.
24. Yosida, Z., Heat transfer by water vapor in a snow cover. *Teion Kagaku A* **5**, 93–100 (1950).
25. Sulakvelidze, G. K., Thermoconductivity equation for porous media containing saturated vapor, water or ice. *Bull. Acad. Sci. USSR, Geophys. Ser. (English Transl.)* pp. 186–188 (1959); also *Izv. Akad. Nauk SSSR Ser. Geofiz.* pp. 284–287 (1959).
26. Yen, Y. C., The rate of temperature propagation in moist porous mediums with particular reference to snow. *J. Geophys. Res.* **72**, 1283–1288 (1967).
27. Yen, Y. C., Effective thermal conductivity of ventilated snow. *J. Geophys. Res.* **67**, 1091–1098 (1962).
28. Yen, Y. C., Heat transfer by vapor transfer in ventilated snow. *J. Geophys. Res.* **68**, 1093–1101 (1963).
29. Yen, Y. C., Effective thermal conductivity and water vapor diffusivity of naturally compacted snow. *J. Geophys. Res.* **70**, 1821–1825 (1965).
30. University of Minnesota, Review of the properties of snow and ice. U. S. Army Snow Ice and Permafrost Res. Estab., Special Rep. 4 (1951).
31. Mellor, M., Some optical properties of snow. *Intern. Assoc. Sci. Hydrology, Symp. Davos*, 1965, Publ. No. 69, 128–140.
32. Thomas, C. W., On the transfer of visible radiation through sea ice and snow. *J. Glaciol.* **4**, 481–484 (1963).
33. Giddings, J. C., and LaChapelle, E., Diffusion theory applied to radiant energy distribution and albedo of snow. *J. Geophys. Res.* **66**, 181–189 (1961).

THEORY OF INFILTRATION

J. R. PHILIP

Division of Plant Industry
Commonwealth Scientific and Industrial Research Organization
Canberra, Australia

I. Introduction 216
II. Water Transfer in Unsaturated Soils 217
   A. Hydraulic Conductivity of Unsaturated Soils 217
   B. Moisture Potential and Total Potential of Water in Unsaturated Soils 220
   C. General Partial Differential Equation of Soil-Water Transfer . 220
   D. Limits to Applicability of Approach 224
III. Mathematical Preliminaries 229
   A. Particular Forms of Eq. (12) 229
   B. Initial and Boundary Conditions 232
   C. Relevance to Hydrologic Processes Other than Absorption and Infiltration 233
   D. Techniques for Solving Nonlinear Diffusion and Fokker–Planck Equations 233
IV. Exact Solutions of Absorption Equations 235
   A. One-Dimensional Quasi-Analytical Solution 235
   B. One-Dimensional Analytical Solution 236
   C. The Sorptivity of Porous Media and Soils 237
   D. Illustrative Example 239
   E. Two- and Three-Dimensional Series Solutions 240
   F. Three-Dimensional Quasi-Analytical Large-Time Solution . 243
   G. Illustrative Example: Two- and Three-Dimensional Absorption 243
   H. Two- and Three-Dimensional Arbitrary Geometry: Quasi-Analytical Large-Time and Steady Solutions 244
   I. Similarity Solutions in Absorption Problems 246
V. Exact Solutions of Infiltration Equations 248
   A. One-Dimensional Series Solution 248
   B. One-Dimensional Asymptotic (Large-Time) Solution . . 251
   C. Illustrative Example: One-Dimensional Infiltration . . 252
   D. Two- and Three-Dimensional Small-Time Solutions . . 253
   E. Two- and Three-Dimensional Arbitrary Geometry: Quasi-Analytical Large-Time and Steady Solutions 255
VI. Linearized Solutions 258
   A. Principles of Method 260
   B. Absorption Solutions 260
   C. Infiltration Solutions 262
   D. Large-Time Solutions in Two and Three Dimensions . . 265
VII. Delta-Function Solutions 266
   A. Principles of Method 266

B. Absorption Solutions 267
  C. Infiltration Solutions 268
VIII. COMPARISON TECHNIQUE FOR ESTIMATING INTEGRAL PROPERTIES OF
  SOLUTIONS 270
  A. Principles of Method 270
  B. Absorption Solutions 271
  C. Infiltration Solutions 274
  D. Aggregated Media 275
IX. EXACT SOLUTIONS OF MORE COMPLICATED PROBLEMS . . . 275
  A. Processes Subject to Flux Boundary Conditions . . . 276
  B. Absorption and Infiltration in Heterogeneous Media . . 278
X. PHYSICS OF ONE-DIMENSIONAL INFILTRATION 279
  A. Moisture Profile 280
  B. Comparison with Experiment 280
  C. Effect of Initial Moisture Content 282
  D. Effect of Water Depth 283
  E. Algebraic Infiltration Equations 283
XI. EFFECTS OF GEOMETRY AND GRAVITY ON INFILTRATION . . . 284
  A. Moisture Distribution 284
  B. Existence of Steady State and Character of Ultimate Wetting . 285
  C. Infiltration Rate 286
SYMBOLS 289
REFERENCES 291

I. Introduction

A very large fraction of the water falling as rain on the land surfaces of the earth moves through *unsaturated* soil during the subsequent processes of infiltration, drainage, evaporation, and the absorption of soil-water by plant roots. Hydrologists, and their textbooks and handbooks, have tended, nevertheless, to pay relatively little attention to the phenomenon of water movement in unsaturated soils. Most research on this topic has been done by soil physicists, concerned ultimately with agronomic or ecological aspects of hydrology; but their colleagues in engineering hydrology have exhibited an increasing interest in this field in recent years.

The present article deals with the theory of infiltration which is one important outcome of the mathematical-physical approach to the study of water movement in unsaturated soils which has been developed over the past 15 years or so, principally in the U.S., England, and Australia. We shall consider briefly the physical basis for the general formalism of this approach and the limits of its applicability; but we shall be concerned principally with developing the general flow equation (a nonlinear Fokker–Planck equation), with describing methods for its solution, and with presenting the solutions and discussing their physical significance.

Theory of Infiltration

We define *infiltration* as the process of the entry into the soil of water made available (under appropriately defined conditions) at its surface. This "surface" may be the natural, more or less horizontal upper surface of the soil; or it may be the bed of a natural or artificial furrow or stream, or the walls of a natural or artificial tunnel or cavity. As we shall see, a fundamental point of departure for the study of infiltration is the study of *absorption*, the particular case of infiltration when the effect of gravity may be neglected, as in horizontal systems, in the early stages of infiltration, and in fine-textured soils in which the influence of moisture gradients dominates that of gravity.

II. Water Transfer in Unsaturated Soils

A. Hydraulic Conductivity of Unsaturated Soils

1. Darcy's Law for Saturated Media

We may write Darcy's law [1] for the flow of water in the liquid phase through saturated porous media, including soils, in the form

$$\mathbf{U} = -K \nabla \Phi \quad (1)$$

where $\mathbf{U}$ is the vector flow velocity, Φ is the total potential, and K is the hydraulic conductivity of the medium.

$\mathbf{U}$ has the character of a macroscopic mean. It is a flow rate per unit cross section of the medium formed by averaging over an area large compared with the cross section of individual pores and grains of the medium. $\mathbf{U}$ has the dimensions [length][time]$^{-1}$, and we shall use the specific unit cm-sec^{-1}.

Φ is the potential defined by the equation

$$\Phi = (P/\rho g) + \Omega \quad (2)$$

where P is the pressure (more precisely, the resultant stress due to the hydrostatic and adsorptive force fields), ρ the density of water, g the acceleration due to gravity, and Ω the potential of the external forces per unit weight of water. Like $\mathbf{U}$, P is a macroscopic quantity and is the outcome of smoothing the actual microscopic distribution of pressure in the water over a volume rather larger than that of the individual pores. Usually the only external force on the system is that of gravity, and then $\Omega = z$, the vertical ordinate. As we have defined it, Φ has the dimensions [length]. We shall use the unit centimeters for Φ.

We employ this definition of Φ, which corresponds to the *hydraulic head* used by engineers in their flow calculations, because it is convenient and simple for the purposes of our exposition. The formalism could be developed equally well in terms of a potential based on unit mass. It is an elementary matter to transpose our various results into this form—at the expense of the intrusion of the factor g into various expressions, and the use of a more cumbersome set of units.

It then follows that K has the dimensions $[\text{length}][\text{time}]^{-1}$, and that we shall specifically use the unit cm-sec^{-1}.

If interaction between medium and water is negligible, K is related to the *intrinsic permeability* of the medium, κ, by the expression

$$K = \kappa \rho g / \mu \tag{3}$$

where μ is the dynamic viscosity, and κ, which depends solely on the internal geometry of the medium, has the dimensions $[\text{length}]^2$.

Equation (1) holds for an isotropic medium, the extension to anisotropic media being

$$\mathbf{U} = -\mathbf{K}\,\nabla\Phi \tag{4}$$

where $\mathbf{K}$ is the hydraulic conductivity tensor.

A great deal has been written in recent years on the theoretical basis of Darcy's law, although much of it seems to constitute self-education rather than serious research. It is sufficient here to remark that there are both hydrodynamic and statistical elements to the basis of Darcy's law. The hydrodynamic one is as follows: the Navier-Stokes equation governing the flow of a viscous, incompressible, Newtonian fluid (e.g., Lamb [2]) is linear in the limit of small Reynolds number. We have already hinted at the statistical one in defining $\mathbf{U}$ and P: it is that, although it is not feasible to know details of the internal geometry of the medium and of the *microscopic* distribution of velocity and potential, local mean quantities such as $\mathbf{U}$ and Φ can be defined and exist.

2. Darcy's Law for Unsaturated Media

It has long been assumed [3], and was confirmed experimentally by Childs and Collis-George [4] and others, that Darcy's law may hold for the flow of liquid water in unsaturated media in a modified form in which K is a function of the volumetric moisture content, θ. The theoretical validity of this concept depends on the (usually very reasonable) assumption that the drag at the air-water interfaces in the soil is negligibly small [5]. The general behavior of the $K(\theta)$ function is now fairly well established, thanks to the work of Richards [3], Moore [6], Childs and Collis-George [4], and other workers in soil physics and petroleum engineering research. K is found to decrease very rapidly as θ decreases from its saturation value. This is not surprising, for the following reasons [7]:

(a) The total cross section available for flow decreases with θ.

Theory of Infiltration

(b) The largest pores are emptied first as θ decreases. Since the contribution to K per unit area varies roughly as the square of pore "radius," K can be expected to decrease much more rapidly than θ.

(c) As θ decreases, the probability increases that water will occur in pores and wedges isolated from the general three-dimensional network of water films and channels. Once continuity fails, there can be no flow in the liquid phase, other than flow through liquid "islands" in series-parallel with the vapor system [8]. (Flow of this type is usually negligible in the absence of temperature gradients, and need not detain us here.)

K may, in fact, vary through six or more decades over the θ-range of interest [9]. Figure 1 shows a typical $K(\theta)$ relationship.

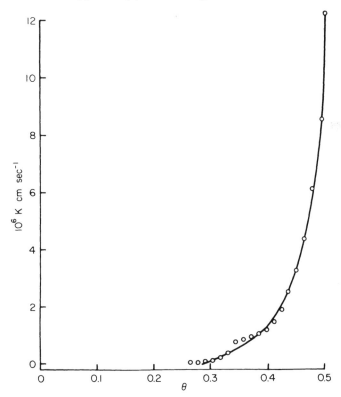

FIG. 1. Relationship between hydraulic conductivity, K, and moisture content, θ, for Yolo light clay [6].

Although the possibility of hysteresis in $K(\theta)$, analogous to that in $\Psi(\theta)$ discussed in Sect. D.6 below, cannot be excluded *a priori*, present indications are that it is relatively unimportant [10–13]. It would not, in any case, enter into the nonhysteretic class of soil-moisture phenomena which is our principal theme here.

We thus express Darcy's law for the flow of water in unsaturated porous media, including soils, in the modified form of Eq. (1):

$$\mathbf{U} = -K(\theta) \nabla \Phi \tag{5}$$

B. Moisture Potential and Total Potential of Water in Unsaturated Soils

The water in unsaturated soils is not "free" in the thermodynamic sense, because of capillarity and adsorption, the latter tending to be significant in dry soils [14]. The energy state of the water is commonly expressed by the *moisture potential*, Ψ, which has, as well, been known variously as *capillary potential, capillary pressure, moisture tension, moisture suction, negative pressure, pressure head*, etc. [15–19]. We here define Ψ as a potential based on unit weight of water, so that, like Φ, it has the dimensions [length] and is conveniently expressed in the unit centimeters. The convenient datum of Ψ is the potential of free water at atmospheric pressure. The appropriate differential specific Gibbs function is then $g\Psi$. It follows that, if the effect of solutes is negligible, the liquid and vapor water systems are connected by the relation

$$h = \exp[g\Psi/\mathbf{RT}] \tag{6}$$

where h is the relative humidity, $\mathbf{R}$ (erg-gm^{-1}-°K^{-1}) is the gas constant for water vapor ($= 4.65 \times 10^6$), and $\mathbf{T}$(°K) is the absolute temperature. It is important to note that, in unsaturated media that are wet by water, Ψ is negative. Figure 2 depicts the $\Psi(\theta)$ relation for the soil for which $K(\theta)$ is given in Fig. 1. Ψ decreases rapidly as θ decreases. When θ is so small that water is retained only in the smallest pores, or in thin adsorbed films, the forces holding it are very great indeed.

For unsaturated media, the total potential Φ may be regarded as comprising Ψ and the gravitational component z, the height above some datum level. Thus, under these conditions, Ψ takes the place of $P/\rho g$ in Eq. (2), Ω is set equal to z, and we have

$$\Phi = \Psi(\theta) + z \tag{7}$$

C. General Partial Differential Equation of Soil-Water Transfer

Combining Eq. (5) with the requirement of continuity that

$$\partial \theta/\partial t = -\nabla \cdot \mathbf{U} \tag{8}$$

we obtain the equation

$$\partial \theta/\partial t = \nabla \cdot (K \nabla \Phi) \tag{9}$$

Theory of Infiltration

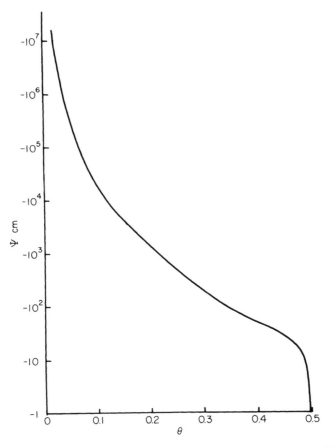

FIG. 2. Relationship between moisture potential, Ψ, and moisture content, θ, for the soil of Fig. 1 [6].

Here t denotes time (sec). Substituting Eq. (7) in Eq. (9), we then have

$$\partial\theta/\partial t = \nabla \cdot (K\,\nabla\Psi) + (\partial K/\partial z) \tag{10}$$

Equations (9) and (10) hold quite generally in the sense that they apply equally well to both homogeneous and heterogeneous soils, and do not depend on any requirement that relations between K, Ψ, and θ be single-valued. For a homogeneous soil, however, we may express Eq. (10) in more tractable form in the two following ways. If K and Ψ are single-valued functions of θ, we may introduce the quantity D, also a single-valued function of θ, such that

$$D = K(d\Psi/d\theta) \tag{11}$$

Then Eq. (10) may be written as

$$\frac{\partial \theta}{\partial t} = \nabla \cdot (D \nabla \theta) + \frac{dK}{d\theta} \cdot \frac{\partial \theta}{\partial z} \tag{12}$$

The coefficient $dK/d\theta$ is evidently a function of θ. Alternatively, if K and θ are single-valued functions of Ψ, Ψ may be taken as the dependent variable and Eq. (10) rewritten as

$$\frac{d\theta}{d\Psi} \cdot \frac{\partial \Psi}{\partial t} = \nabla \cdot (K \nabla \Psi) + \frac{dK}{d\Psi} \cdot \frac{\partial \Psi}{\partial z} \tag{13}$$

The coefficients $d\theta/d\Psi$ and $dK/d\Psi$ are functions of Ψ.

Contrary to the assertion of a recent review [19], Eqs. (12) and (13) are not completely equivalent. For example, Eq. (13) may still apply when Ψ exceeds the *air-entry value* (i.e., the value at which air enters an initially saturated porous body [20]) or is positive (as it may be when a depth of free water is ponded over the soil), whereas Eq. (12) cannot [21, 22]. In general, however, Eq. (12) is the more tractable form. We are here concerned almost exclusively with solutions of Eq. (12); but it should be understood that, whenever a solution of Eq. (12) can be found, a corresponding solution of Eq. (13) can be found by the same (or a slightly modified) method (see Sects. IV.A, V.A, and V.B).

Equation (12) is a *nonlinear Fokker–Planck equation*. Linear forms of the Fokker–Planck equation arise in such physical phenomena as heat conduction from sources moving relative to the medium [23] and diffusion under an external force field (e.g., sedimentation with Brownian motion) and in the theory of Markov processes in mathematical probability (e.g., Chandrasekhar [24], Bailey [25]). It has entered some recent studies of turbulent diffusion [26, 27]. Nonlinear forms of the Fokker–Planck equation do not seem to have been studied other than in the present context, although work on the closely related nonlinear diffusion equation

$$\partial \theta / \partial t = \nabla \cdot (D \nabla \theta) \tag{14}$$

goes back at least to Boltzmann [28]. Equation (12) reduces to Eq. (14) for horizontal systems and in other instances where gravity may be neglected.

It will be understood that Eq. (12) is of the diffusion, or heat-conduction, form but with two complications that add very greatly to the difficulty of its solution. Firstly, the equation contains on its right side the first-order term representing the influence of gravity on the process. Secondly, the coefficients D and $dK/d\theta$ are both markedly dependent on θ. D may vary typically through three or more decades (and $dK/d\theta$ through several more) between the wet and dry ends of the moisture range. Neither of these complications can be ignored, so that the quantitative study of the physics of soil-water transfer depends quite centrally on the availability of solutions to the nonlinear diffusion and Fokker–Planck equations.

D is called the *moisture diffusivity*. It has the dimensions $[\text{length}]^2 [\text{time}]^{-1}$, and the unit $\text{cm}^2\text{-sec}^{-1}$ is appropriate. Figure 3 is a graph of $D(\theta)$ for the soil with the $K(\theta)$ and $\Psi(\theta)$ functions of Figs. 1 and 2.

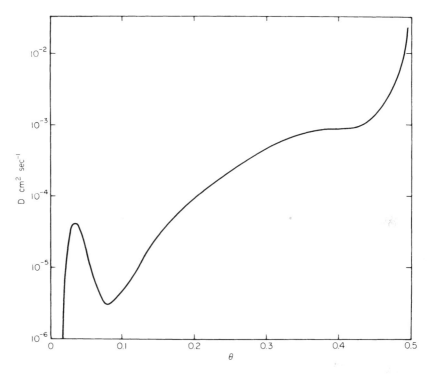

FIG. 3. Relationship between moisture diffusivity, D, and moisture content, θ, for the soil of Fig. 1 [21]. (Note that D for $\theta \leq 0.06$ includes dominant contribution in vapor phase.)

Many of the concepts leading to Eq. (12) were implicit in the 1907 monograph of Buckingham [29]. In 1931, Richards [3] presented equations formally equivalent to Eqs. (10) and (13). In 1948, Childs and George [30] recognized the diffusion character of the particular form of Eq. (14) which applies to a one-dimensional horizontal system. In 1952, Klute [31, 32] explicitly derived Eq. (12). Philip [21, 33, 34] extended the approach to include moisture transfer in the vapor and adsorbed phases within the same mathematical formalism. These extensions depended, essentially, on the use of Eq. (6) and the chain rule for differentiation.

D. Limits to Applicability of Approach

The remainder of this article is, for the most part, concerned with the development of solutions of Eq. (12) which are relevant to the process of infiltration, and with the physical implications of these solutions. In a sense, our study can be looked on as an essay on infiltration as the outcome of physical interactions between capillarity, gravity, and geometry. (Here, and in what follows, the term *capillarity* is convenient to describe the consequence of gradients of Ψ, although we recognize that the principal determinant of Ψ in very dry soils and in soils of high colloid content may not be the curvature of air-water interfaces.)

It is, however, desirable that we discuss the limits to the applicability of the approach, not only so that the reader will not overestimate it, but equally so that he will not underrate it (as he might well do from a superficial reading of some of the literature). We therefore present here a very brief but, it is hoped, fairly inclusive review of the ways in which deviations from the formalism of Sect. II.C may arise.

It is convenient to develop the discussion under seven headings, although it will be understood that the complications of the subject are such that these are not all mutually exclusive.

1. Inadequacy of Specification on Darcy Scale. Local Disequilibrium in Aggregated Media

The possibility must be considered that the specification of the medium "on the Darcy scale" (i.e., in terms of quantities such as K, Ψ, and θ based on volume elements large compared with that of the individual pore) may not be adequate. Where the medium is aggregated, i.e., where it contains aggregates microporous that form the "grains" of a macroporous (or a "cracked") structure, it must be admitted *a priori* that, during transient phenomena, local disequilibria between Ψ values in the microporosity and in the macroporosity may invalidate any description of the process at the Darcy level. A recent detailed analysis of the problem [35, 36], however, indicates that deviations produced in this way will usually be negligible; for the exceptional case of very large aggregates of low sorptivity (see Sect. IV for a definition of this term), the approaches developed by Philip [35, 36] provide the necessary extensions of the methods of analysis discussed in this article.

2. Deviations from Darcy's Law

We discuss two possible causes of deviations from Darcy's law.

Theory of Infiltration

a. INERTIAL EFFECTS. As we have already remarked, the Navier–Stokes equation underlying Darcy's law ceases to be linear when the "inertia terms" are important. For steady flows in porous media, this has the consequence that Darcy's law fails when the Reynolds number of the flow (based on $|U|$ and on a characteristic pore dimension) exceeds a value that tends to lie in the range 1 to 10. Some authors have mistakenly ascribed deviations from proportionality between U and $\nabla\Phi$ to turbulence, although it is definitely a nonlinear laminar regime that emerges as the Reynolds number increases, and not a turbulent one [37, 38]. The Reynolds numbers characteristic of steady soil-water flows are very small indeed, and the possibility of the failure of Darcy's law due to inertial effects cannot be entertained in such cases.

The influences of inertia on transient flows are somewhat different, and the matter requires more careful examination. Analyses by Philip, however, indicate that the deviations from Darcy's law due to inertia are unlikely to be of much significance during transient flows in either saturated [39] or unsaturated [40] soils.

b. NON-NEWTONIAN BEHAVIOR. The applicability of the Navier–Stokes equation depends also on the soil-water behaving as a *Newtonian fluid* (in which the stresses are proportional to the rates of strain). *Bulk water* is a Newtonian fluid, and it would seem that the flow behavior of soil-water can differ significantly from that of bulk water only if a large fraction of the soil-water is subject to surface forces capable of modifying its rheological properties in the required way. Deviations from Darcy's law due to non-Newtonian behavior may therefore be possible in principle in soils of high colloid content and large specific surface; and observations of "non-Darcy" behavior tend to be on soils of this type or, indeed, on pastes made of water and colloidal extracts. The work of Olsen [41] suggests that at least some of these observations are the spurious outcome of poor experimental technique. Swartzendruber [42] has considered various physical explanations for these "non-Darcy" observations and favors the non-Newtonian explanation.

An aspect arises in connection with unsaturated systems which does not seem to have been discussed (although it is perhaps suggested by Olsen's paper). This concerns the influence of the *surface viscosity* of air-water interfaces, and of the abnormal rheology of such surfaces, which are, of course, natural regions of accumulation of contaminants.

These ideas lead us into areas of surface physical chemistry well beyond the scope of this article. We mention them only to indicate that the question of the reality or otherwise of non-Newtonian behavior seems to be a matter for rheologists and surface physical chemists. We shall not allow it to detain us here, as it does not appear to have much bearing on the movement of soil-water in general, or, in particular, on the process of infiltration.

3. Other Complications due to Colloidal Behavior

a. COLLOID SWELLING. The processes of swelling and shrinkage in soils of high colloid content involve the motion of the soil particles. The movement of water in one-dimensional systems of this type may be analyzed through the equation obtained by combining the continuity requirement and Darcy's law applied (as it should be) to flow *relative to the soil particles*. Appropriate methods of developing the physical theory, which are similar in principle to those to be discussed in this article, are presently under active development by Dr. D. E. Smiles and the author. This work, which is too recent for an adequate account of it to be possible in this article, points up the importance of mass flow (i.e., of water convected with the soil particles) in such systems.

b. EFFECTS OF ELECTROLYTE CONCENTRATION. The variation of the structure and hydraulic conductivity of soils of high colloid content with electrolyte concentration [43] is, of course, not specifically taken into account in the present study.

4. Possible Complications due to Soil-Air

The analysis developed in Sect. II.C neglects any effect of pressure differences in the soil-air. We may develop a set of equations similar to the foregoing, which embrace flows of both soil-water and soil-air. This, slightly more general, two-component formalism is, in fact, used in petroleum engineering. In most applications to soil-water movement, this elaboration is unnecessary, as the pressure differences within the soil-atmosphere are trivially small. We discuss below certain instances where this is not so, and certain other complications that may arise from interactions between the soil-air and the soil-water.

a. AIR ESCAPE IMPEDED OR PREVENTED. The dynamics of processes such as the absorption and infiltration of water may be modified significantly if the air initially in the soil is not free to escape as it is displaced by water. The process may be studied, to some extent, by use of the two-component formalism mentioned previously (see Elrick [44] for a linearized attack on a problem of this type). As Youngs and Peck have pointed out [45], however, capillary hysteresis may lead to unexpected complications. Also, as the experiments of Peck [46, 47] indicate, air pressures may increase to the point where the pressure buildup is partly relieved by a bubbling air escape at the water-supply surface. Such a phenomenon is, of course, beyond the scope of a Darcy-type analysis.

These considerations do not appear to be generally of much importance in the field, although, as this author is well aware from his personal experiences in the Riverina of Australia, limits to air escape may well affect infiltration into large inundated areas. In fact, soil-air pressures have been developed which are great enough to lift the pavements of highways passing through the flooded region.

b. ENTRAPPED AIR. As Philip [48] pointed out, air is more likely to be trapped in pores remote from the water-supply surface than in those near the surface. Such spatial differences in air-entrapment will be reflected in spatial differences in the $\Psi(\theta)$ relation. The effect is thus essentially one of spatial heterogeneity (Sect. IX), but it is apparently of minor significance. It is a possible explanation of the sometimes observed *transition zone* in moisture profiles during infiltration (Sect. X).

c. THE SOLUTION OF SOIL-AIR. Relatively little consideration appears to have been given to the diffusion of air dissolved in the soil-water, and to the associated processes of air going into, and coming out of, solution. Such processes clearly lead to changes in the incidence of entrapped air, and may thus affect the stability of the $K(\theta)$ and $\Psi(\theta)$ relations in the medium. It has been remarked elsewhere [48] that the use of air-free water, often a (supposed) precaution in laboratory experiments, leads to the gradual solution of trapped air, with consequent changes in medium properties. It seems preferable to use water in equilibrium with the atmosphere, since the experimental conditions are then closer to nature, and the complications of progressive solution of entrapped air are avoided. Peck [49] has suggested that the use of initially de-aired water may be the cause of the deviant behavior in absorption experiments by Nielsen *et al.* [50].

5. Thermal Effects

The analysis of Sect. II.C refers, as it stands, to isothermal systems. The extensions to nonisothermal systems have been made by Philip and de Vries [8, 21, 51–53]; analysis of such systems involves, in general, solution of simultaneous equations for both heat and moisture transfer. Recent attempts to study these matters through the formalism of the thermodynamics of irreversible processes do not seem to have yielded any real progress. The principal specific result [54] seems to reduce to the identification of the well-known proportionality of water vapor and latent heat transfer in an evaporation-condensation system as an Onsager reciprocity relation.

It has been shown (e.g., Philip and de Vries [8, 51]) that thermally induced moisture movement tends to be most important in rather dry systems where vapor diffusion is a dominant mechanism. The processes of absorption and infiltration treated in this article take place almost exclusively in the liquid phase, and we may safely limit our attention to isothermal systems.

The possibility that thermal effects arising from the heat of wetting might have a significant influence on processes such as absorption and infiltration has recently been studied experimentally [55, 56]. The authors concluded that such effects are quite negligible even in soils initially so dry that Ψ is significantly less than $-15,000$ cm. A similar conclusion was reached in an earlier, theoretical investigation of the energetics of absorption and infiltration [57, 58].

6. Capillary Hysteresis

Perhaps the most important and challenging limitation to the applicability of Eqs. (12) and (13) arises from the fact of hysteresis in the $\Psi(\theta)$ relation, first discussed by Haines [59]. As Philip [21] remarked, hysteresis does not in itself invalidate the use of Eqs. (12) and (13), so long as the $\Psi(\theta)$ relation adopted is appropriate to the phenomenon under study. A wetting curve is thus the $\Psi(\theta)$ function appropriate to the analysis of absorption or infiltration into a homogeneous soil of uniform initial moisture content; and, conversely, a drying curve should be used in the analysis of suitably uniform processes of removal of soil-water. However, as Childs first pointed out [21], it is not, in general, useful to treat in this way phenomena, such as redistribution and drainage after infiltration ceases, wherein wetting and drying occur simultaneously in the one soil mass. In general, the "diffusion" analysis fails in such cases: different points in the soil mass follow different hysteresis scanning curves, so that there is no definite relationship between gradients of Ψ and gradients of θ. In such cases, a knowledge of the complete family of scanning curves within the main $\Psi(\theta)$ hysteresis loop is prerequisite to any quantitative treatment of the transfer process. (Some special cases of simultaneous wetting and drying to which the diffusion analysis is, nevertheless, relevant have been treated by Philip [7] and Youngs [60].)

Collis-George [61] suggested that the independent domain model of hysteresis due to Everett and others [62–64] might be applicable to capillary hysteresis. Miller and Miller [65] gave an original and perceptive treatment of capillary hysteresis, but not a quantitative one. Poulovassilis [66, 67] gave the first published account of capillary hysteresis in terms of the independent domain model and presented experimental evidence for its validity. Poulovassilis used an improved form of the model [68–70]. As Philip [71, 72] showed, the approach is most simply described without any specific reference to independent domain theory. It depends solely on the supposition that any infinitesimal volume element of the pore space may be characterized by two values of Ψ: α, the (arithmetically) smallest value at which, under equilibrium conditions, the element may be occupied by air; and β, the (arithmetically) largest value at which the element may, under equilibrium conditions, be occupied by water. If, then, each volume element of the total pore space is characterized by its α and β values, there exists a bivariate distribution density function $f(\alpha, \beta)$ such that the fraction

$$f(\alpha_0, \beta_0)\, d\alpha\, d\beta$$

of the total pore volume consists of elements for which α lies between the values α_0, $\alpha_0 + d\alpha$, and β lies between the values β_0, $\beta_0 + d\beta$. Then $f(\alpha, \beta)$ completely defines the hysteretic behavior of the medium. Philip [72] showed that a similarity hypothesis about $f(\alpha, \beta)$ (which implied, loosely, that the distribution of geometrical relationships between "wetting" and "drying" air-water interfaces is independent of pore size) holds fairly well for Poulovassilis' data, and that the use of this hypothesis enables $f(\alpha, \beta)$ to be estimated from the main hysteresis loop, without the laborious mapping of scanning curves. Topp and Miller [13] consider that their experiments do not support the independent domain model. Should it prove that the independent domain model is not reasonably adequate, this will seemingly imply that a more complicated formalism is needed. This would not augur well for the prospects of developing a manageable quantitative analysis of hysteretic soil-water processes.

Fortunately, as we have observed previously, capillary hysteresis does not enter the class of phenomena we consider in the following sections. We have discussed the matter at this length because it is a fascinating aspect of the general problem of soil-water behavior, and one that is currently exciting the interest of soil physicists in various parts of the world.

7. Discussion

The theory we develop in the succeeding parts of this article represents the *basic fluid-mechanical theory of infiltration*. Those of the complications we have raised, but have not been able to dismiss out of hand as negligible, will, in some circumstances, impose more or less severe perturbations on the basic behavior that the theory predicts. Our theory, nevertheless, provides a natural point of departure for the study of any such perturbations; and it seems clear that it must remain a central and integral part of any more elaborate formalism that aims to take specific account of one or more of these complications.

III. Mathematical Preliminaries

A. Particular Forms of Eq. (12)

In considering specific infiltration phenomena, we shall make use of various particular forms of Eq. (12). It is useful to set out these forms here. There is, of course, an analogous set of specific forms of Eq. (13).

1. Absorption Equations

We have remarked already that Eq. (12) reduces to the nonlinear diffusion equation

$$\partial \theta / \partial t = \nabla \cdot (D \nabla \theta) \tag{14}$$

when gravity may be neglected. This then is the equation describing absorption, i.e., infiltration into horizontal systems, or into fine-textured soils in which the influence of the moisture gradients is much more important than that of gravity. As we shall see, absorption solutions have the additional theoretical importance that they yield the limiting small-time behavior of transient infiltration processes even when gravity cannot be neglected, and so provide a basic point of departure for the solution of the (more complicated) infiltration equations.

a. ONE-DIMENSIONAL FORM. The one-dimensional form of Eq. (14) is

$$\frac{\partial \theta}{\partial t} = \frac{\partial}{\partial x}\left(D \frac{\partial \theta}{\partial x}\right) \tag{15}$$

with x the space coordinate.

b. TWO-DIMENSIONAL FORMS. The Cartesian two-dimensional form of Eq. (14) is

$$\frac{\partial \theta}{\partial t} = \frac{\partial}{\partial x}\left(D \frac{\partial \theta}{\partial x}\right) + \frac{\partial}{\partial y}\left(D \frac{\partial \theta}{\partial y}\right) \tag{16}$$

where x, y are the two (Cartesian) space coordinates.

In what follows, we shall be specially concerned with two-dimensional systems exhibiting cylindrical radial symmetry. In such cases, the appropriate form of the equation is

$$\frac{\partial \theta}{\partial t} = \frac{1}{r}\frac{\partial}{\partial r}\left(Dr \frac{\partial \theta}{\partial r}\right) \tag{17}$$

where r is the radial coordinate.

c. THREE-DIMENSIONAL FORMS. The Cartesian three-dimensional form of Eq. (14) is

$$\frac{\partial \theta}{\partial t} = \frac{\partial}{\partial x}\left(D \frac{\partial \theta}{\partial x}\right) + \frac{\partial}{\partial y}\left(D \frac{\partial \theta}{\partial y}\right) + \frac{\partial}{\partial z}\left(D \frac{\partial \theta}{\partial z}\right) \tag{18}$$

For three-dimensional systems exhibiting spherical radial symmetry, the equation reduces to

$$\frac{\partial \theta}{\partial t} = \frac{1}{r^2}\frac{\partial}{\partial r}\left(Dr^2 \frac{\partial \theta}{\partial r}\right) \tag{19}$$

*d. m-*DIMENSIONAL RADIALLY SYMMETRICAL FORM. Equations (15), (17), and (19) are all of the form

$$\frac{\partial \theta}{\partial t} = r^{1-m} \frac{\partial}{\partial r}\left(Dr^{m-1} \frac{\partial \theta}{\partial r}\right) \qquad (20)$$

where *m* is the number of dimensions of the system. This form is useful in Sect. IV.

2. Infiltration Equations

It turns out that the calculation of solutions of the infiltration equation may be made more naturally and a little more conveniently if we adopt the convention that *z* is taken *positive downward*. In deference to this convention, we rewrite Eq. (12) as

$$\frac{\partial \theta}{\partial t} = \nabla \cdot (D \nabla \theta) - \frac{dK}{d\theta} \cdot \frac{\partial \theta}{\partial z} \qquad (21)$$

a. ONE-DIMENSIONAL FORM. The one-dimensional form of Eq. (21) is

$$\frac{\partial \theta}{\partial t} = \frac{\partial}{\partial z}\left(D \frac{\partial \theta}{\partial z}\right) - \frac{dK}{d\theta} \cdot \frac{\partial \theta}{\partial z} \qquad (22)$$

b. TWO-DIMENSIONAL FORM. The two-dimensional form is

$$\frac{\partial \theta}{\partial t} = \frac{\partial}{\partial x}\left(D \frac{\partial \theta}{\partial x}\right) + \frac{\partial}{\partial z}\left(D \frac{\partial \theta}{\partial z}\right) - \frac{dK}{d\theta} \cdot \frac{\partial \theta}{\partial z} \qquad (23)$$

with *x* the horizontal coordinate.

c. THREE-DIMENSIONAL FORMS. The general Cartesian three-dimensional form is

$$\frac{\partial \theta}{\partial t} = \frac{\partial}{\partial x}\left(D \frac{\partial \theta}{\partial x}\right) + \frac{\partial}{\partial y}\left(D \frac{\partial \theta}{\partial y}\right) + \frac{\partial}{\partial z}\left(D \frac{\partial \theta}{\partial z}\right) - \frac{dK}{d\theta} \cdot \frac{\partial \theta}{\partial z} \qquad (24)$$

with *x, y* the horizontal coordinates. For three-dimensional systems exhibiting cylindrical radial symmetry about a vertical axis, this becomes

$$\frac{\partial \theta}{\partial t} = \frac{1}{r}\frac{\partial}{\partial r}\left(Dr \frac{\partial \theta}{\partial r}\right) + \frac{\partial}{\partial z}\left(D \frac{\partial \theta}{\partial z}\right) - \frac{dK}{d\theta} \cdot \frac{\partial \theta}{\partial z} \qquad (25)$$

where the horizontal radial coordinate $r = (x^2 + y^2)^{1/2}$.

B. Initial and Boundary Conditions

1. Initial Conditions

We shall be concerned almost exclusively with infiltration into a soil mass at uniform initial moisture content θ_0. In such cases, the initial condition governing the appropriate form of Eq. (12) is

$$t = 0, \quad \theta = \theta_0 \quad \text{in all} \quad \mathbf{V} \tag{26}$$

where $\mathbf{V}$ signifies the whole volume of the soil mass. This type of initial condition is a simple and natural one, but some care is needed in its application to the infiltration equation. If θ_0 is not so small that $K_0[= K(\theta_0)]$ is zero, or at least negligibly small, there will be an initial, gravity-caused flow throughout the system and, in infinite systems, a "flow at infinity" throughout the course of the phenomenon. Once this is recognized, it produces no real difficulty beyond a slight complication in certain calculations [34, 58, 73].

2. Boundary Conditions at the Water-Supply Surface

a. CONCENTRATION CONDITIONS. Most of the following analysis deals with the process of absorption or infiltration which follows the continuous supply of water from time $t = 0$ onward to surface $\mathbf{W}$ at moisture potential Ψ_1. The boundary condition at the water-supply surface is then

$$t \geq 0, \quad \Psi = \Psi_1 \quad \text{at} \quad \mathbf{W} \tag{27}$$

When Eq. (12) is used (as in most of the present article), it is necessary to rewrite this in the form

$$t \geq 0, \quad \theta = \theta_1 \quad \text{at} \quad \mathbf{W} \tag{28}$$

where $\theta_1 = \theta(\Psi_1)$.

Commonly, the water supplied is "free" and under little or no hydrostatic pressure. We may then take $\Psi_1 = 0$, and θ_1 is then the "saturated" moisture content (which may be rather less than the moisture content with all pores water-filled, because of air-entrapment). Where the water depth over the supply surface (or the hydrostatic pressure under which water is supplied) is not negligibly small, and also where the *air-entry value* of the soil is (arithmetically) large, Eq. (13) is the appropriate equation and (27) the appropriate form of the condition at $\mathbf{W}$.

In keeping with terminology in the mathematics of diffusion, we designate this type of boundary condition a *concentration condition*.

b. FLUX CONDITIONS. For a second class of infiltration phenomena, water reaches the supply surface at a definite rate (as in infiltration from rainfall, or from sprinkler irrigation). If the soil is able to take up the arriving water for some period $0 \leq t < t_0$, the appropriate boundary conditions (for one-dimensional systems) are as follows.

For **W** a horizontal surface,

$$0 \leq t < t_0, \quad K\, \partial\Psi/\partial z \quad \text{or} \quad D\, \partial\theta/\partial z = K - U_0 \quad \text{at} \quad \mathbf{W} \quad (29)$$

For **W** a vertical surface,

$$0 \leq t < t_0, \quad K\, \partial\Psi/\partial x \quad \text{or} \quad D\, \partial\theta/\partial x = -U_0 \quad \text{at} \quad \mathbf{W} \quad (30)$$

x is here the ordinate normal to **W**. U_0 is the supply rate per unit area of **W**. The first form of the left side of the equalities is appropriate to equations with Ψ the dependent variable and the second to ones with θ the dependent variable.

We refer to conditions like (29) and (30) as *flux conditions*, in accordance with diffusion terminology.

In some circumstances, $t_0 = \infty$, i.e., a condition like (29) and (30) holds for all $t \geq 0$; but, when t_0 is finite, a second condition at **W** is needed for $t \geq t_0$. Its form depends on whether the excess water at **W** ponds, runs off, etc.

C. Relevance to Hydrologic Processes Other than Absorption and Infiltration

Although the methods and solutions discussed below are presented specifically in connection with absorption and infiltration, the reader will understand that, with appropriate redefinitions of symbols (and, in some cases, changes of sign), certain of them are relevant also to other processes, such as desorption, drainage, capillary rise, and evaporation from soil [34, 51, 74].

D. Techniques for Solving Nonlinear Diffusion and Fokker–Planck Equations

1. Computer Solutions

The various nonlinear flow equations we have discussed, subject to the conditions we have considered, may, of course, be solved by the brute-force use of finite-difference methods on high-speed computers. There is no special obstacle to using such methods, even for much more elaborate flow problems with one or more of the complications of spatial heterogeneity, more involved initial and boundary conditions, and the operation of capillary hysteresis.

A computer attack on problems of this class will always be feasible provided (i) the problem is well-posed in the mathematical sense; (ii) the (well-established) criteria for computational stability are satisfied by the finite-difference scheme adopted. It should be added that the efficiency of the technique demands that the calculator avail himself of analytical methods to establish the behavior of his solutions in the neighborhood of any singularities.

The ready availability of computer solutions is not, however, so fruitful a potential source of understanding of the theory of infiltration as superficial consideration might suggest. The difficulty is that, in general, each of these nonlinear problems requires for each soil [with its particular hydrological characteristics such as $K(\theta)$ and $\Psi(\theta)$] its own *ad hoc* solution, so that the approach is necessarily a piecemeal one. The possibilities of developing general statements about the phenomena on the basis of calculations of this type are not very good, because they tell us relatively little about the *fundamental structure of the solutions*; and it is upon this which the prospects of useful generalization depend. (It is not my intention to denigrate the use of computers for soil-water calculations; but their most effective rôle in this context seems to me to be as the immensely helpful tools of *ad hoc* investigations.) The emphasis in this article is therefore on quasi-analytical and analytical methods of solution.

2. Quasi-Analytical and Analytical Solutions

We shall describe as *quasi-analytical solutions* those in which we may use the methods of mathematical analysis to establish the basic form of the solution, even though some "coefficients" that appear in the solution require to be determined by numerical methods (conceivably with the aid of computers!). Most of the "exact" solutions we shall discuss are of this type.

In some special cases, the solutions may be found completely by mathematical analysis; we shall call such solutions *analytical solutions*.

It is convenient to describe both quasi-analytical and analytical solutions as *exact solutions*. This usage is well established in fluid mechanics [75], although it differs from that of Philip [76].

3. Analytically Based Estimates of Integral Properties of Solutions

We shall have occasion to make use of two special classes of analytical solutions, *linearized solutions* and *delta-function solutions*. In particular, we use integral forms of these solutions and a comparison technique (Sect. VIII) to establish estimates of integral properties of certain solutions.

Theory of Infiltration

IV. Exact Solutions of Absorption Equations

A. One-Dimensional Quasi-Analytical Solution

Absorption into an effectively semi-infinite, homogeneous, one-dimensional system of uniform initial moisture content is described by Eq. (31) subject to conditions (32):

$$\frac{\partial \theta}{\partial t} = \frac{\partial}{\partial x}\left(D \frac{\partial \theta}{\partial x}\right) \quad (31) \text{ [also Eq. (15)]}$$

$$\begin{aligned} t = 0, &\quad x > 0, \quad \theta = \theta_0 \\ t \geq 0, &\quad x = 0, \quad \theta = \theta_1 \end{aligned} \quad (32)$$

The similarity substitution

$$\phi = xt^{-1/2} \quad (33)$$

apparently first used by Boltzmann [28], reduces Eqs. (31) and (32) to Eqs. (34) and (35):

$$-\frac{\phi}{2}\frac{d\theta}{d\phi} = \frac{d}{d\phi}\left(D \frac{d\theta}{d\phi}\right) \quad (34)$$

$$\begin{aligned} \phi = 0, &\quad \theta = \theta_1 \\ \phi \to \infty, &\quad \theta = \theta_0 \end{aligned} \quad (35)$$

The solution of Eq. (31) subject to conditions (32) is therefore

$$x(\theta, t) = \phi(\theta) \cdot t^{1/2} \quad (36)$$

where $\phi(\theta)$ is the solution of Eq. (34) subject to conditions (35).

Klute [31, 32] used a numerical method due to Crank and Henry [77, 78] to find $\phi(\theta)$. Philip [33, 34, 79] subsequently developed a much more rapid and accurate method. Equation (34) may be rewritten as

$$-\frac{\phi}{2} = \frac{d}{d\theta}\left(D \frac{d\theta}{d\phi}\right)$$

and integration yields

$$\int_{\theta_0}^{\theta} \phi \, d\theta = -2D \, d\theta/d\phi \quad (37)$$

subject to the condition

$$\theta = \theta_1, \quad \phi = 0 \quad (38)$$

The lower limit of the integral is θ_0 because the second of conditions (35) implies $d\theta/d\phi = 0$ as $\theta \to \theta_0$.

Philip's method uses an appropriate finite difference form of Eq. (37) in an iterative procedure that is initiated by an estimate of $\int_{\theta_0}^{\theta_1} \phi \, d\theta$. A prescription for making the first estimate is given by Philip [79]. The method employs the Richardson h^2-extrapolation [80] to reduce truncation errors, and avoids difficulties connected with the infinite tail (which occurs as $\theta \to \theta_0$ unless $D(\theta_0) \to 0$ sufficiently rapidly [48]) by using an analytical solution in that region.

The method applies in slightly modified form [81] when water is supplied under positive hydrostatic pressure, and Eq. (31) must be replaced by the relevant form of Eq. (13), at least in the region of Ψ positive.

B. One-Dimensional Analytical Solution

Subsequently, Philip [76, 82, 83] showed that there exists an indefinitely large class of functional forms of $D(\theta)$ for which $\phi(\theta)$ in Eq. (36) [and therefore the solution of Eq. (31) subject to conditions (32)] may be found analytically. Equation (37) may be rewritten as

$$D = -\tfrac{1}{2} d\phi/d\theta \cdot \int_{\theta_0}^{\theta} \phi \, d\theta \qquad (39)$$

It follows that any functional form of ϕ,

$$\phi = F(\theta) \qquad (40)$$

which satisfies the simple conditions we state below, must be the solution of Eq. (34) subject to conditions (35) for the case where

$$D(\theta) = -\tfrac{1}{2} dF/d\theta \cdot \int_{\theta_0}^{\theta} F \, d\theta \qquad (41)$$

D is evidently expressible in terms of known functions if F is integrable in terms of known functions. F must satisfy three conditions:

(i) The condition imposed by (35) is that $F(\theta_1) = 0$.

(ii) The condition imposed by the requirement that $D(\theta)$ exists for $\theta_0 \leq \theta \leq \theta_1$ is that both $\int_{\theta_0}^{\theta} F \, d\theta$ and $dF/d\theta$ exist throughout $\theta_0 < \theta \leq \theta_1$. (But, if a finite number of discontinuities in D are allowable, or if D is permitted to be infinite at a finite number of points in the range $\theta_0 < \theta \leq \theta_1$, $dF/d\theta$ may either not exist or be infinite at the appropriate number of points in the θ range.)

(iii) The condition imposed by the nonnegativity of D is that $dF/d\theta$ be nonpositive in the range $\theta_0 \leq \theta \leq \theta_1$.

Theory of Infiltration

Some restriction to the utility of this method is imposed by the result that $D(\theta_0)$ is finite, nonzero, only if

$$\lim_{\theta \to \theta_0} F(\theta)/[-\log(\theta - \theta_0)]^{1/2}$$

is finite, nonzero. The most readily generated *D-F* analytical pairs satisfying this criterion are rather complicated in form, and are not especially well adapted to fitting empirical data on $D(\theta)$. More careful examination of the whole class of *F* functions, with special reference to those that behave like $[-\log(\theta - \theta_0)]^{1/2}$ as $\theta \to \theta_0$, is needed before the potentialities of this approach will have been explored and exploited.

C. The Sorptivity of Porous Media and Soils

The one-dimensional absorption result discussed in Sects. IV.A and IV.B is the simplest transient solution of the nonlinear Fokker–Planck equation. As will be seen from the succeeding developments here, it plays a central rôle in the study of more difficult problems involving geometrical complications and the effect of gravity. As we shall show, the early stages of the infiltration process are virtually independent of both gravity and geometry, and are essentially the same as one-dimensional absorption: it follows that the preceding solution represents the leading term of perturbation-type analyses of more complicated problems. It is for this reason that the sorptivity, S, which arises naturally in the one-dimensional absorption problem and which we define below, is so important in the development of the theory of infiltration.

1. Sorptivity

Reverting to the solution (36), we introduce the symbol i (dimension [length]; convenient unit centimeters) to designate the cumulative absorption into the medium at time t. It follows at once that

$$i = \int_{\theta_0}^{\theta_1} \phi \, d\theta \cdot t^{1/2}$$

i.e.,

$$i = St^{1/2} \tag{42}$$

where

$$S = \int_{\theta_0}^{\theta_1} \phi \, d\theta \tag{43}$$

Furthermore, we denote by v_0 the absorption (or infiltration) rate, which is also the flow velocity at $x = 0$. Then

$$v_0 = di/dt = \tfrac{1}{2}St^{-1/2} \tag{44}$$

It is seen that the integral properties of the process [by which we mean functions such as $i(t)$, $v_0(t)$] are completely determined by the quantity, S, which we call the *sorptivity* of the porous medium or soil [84]. Evidently, S embodies in a single parameter the influence of capillarity on the transient flow processes that follow a step-function change in θ (or, more precisely, Ψ) at the surface of a porous mass. Strictly, we should write $S(\theta_0, \theta_1)$ or $S(\Psi_0, \Psi_1)$, since S has meaning only in relation to an initial state of the medium and an imposed boundary condition; however, we may usually omit the arguments of S without ambiguity. The dimensions of S are [length] $\times$ [time]$^{-1/2}$, and we use here the unit cm-sec$^{-1/2}$.

Clearly, $S(\theta_0, \theta_1)$ can be found by integration of the solution (36), so that the methods of Sect. IV.A enable $S(\theta_0, \theta_1)$ to be calculated for any porous medium or soil for which $D(\theta)$ is known in the θ range $\theta_0 \leq \theta \leq \theta_1$. Philip [85] discussed the variation of S with θ_0 with the aid of calculations of this type and of the "delta-function" model treated below (Sect. VII). There is no special difficulty in calculating S in cases (such as when water is supplied under positive pressure) where the analysis must be made, at least in part, with Ψ as the dependent variable. Philip [81] has studied the effect of the depth of the supplied (free) water on S, using the required extension to the method of Sect. IV.A; and he has related these results to the delta-function model [22].

2. Intrinsic Sorptivity

Under the assumption that absorption is the consequence of viscous flow induced by capillarity (which is certainly the primary mechanism of the absorption of water, unless it is supplied at very large [negative] Ψ values, or the soil is extremely colloidal in character), we may define the *intrinsic sorptivity*, $\mathscr{S}$, by the equation

$$\mathscr{S} = (\mu/\sigma)^{1/2} S \qquad (45)$$

μ is the dynamic viscosity, and σ is the surface tension of the absorbate.

$\mathscr{S}$ is, then, like the intrinsic permeability κ, a characteristic of the internal geometry of the medium, and is independent of the properties of the absorbate. It is interesting to note that the dimensions of $\mathscr{S}$ are [length]$^{1/2}$, whereas those of κ are [length]^2. Different contact angles produce different geometrical relations of air-water interfaces, however, so that (for given θ_0 and θ_1) $\mathscr{S}$ depends on the contact angle in a very complicated way. The inclusion of the contact angle in the original definition of $\mathscr{S}$ by Philip [84] is not justified.

Theory of Infiltration

D. ILLUSTRATIVE EXAMPLE

It is convenient for the purposes of exposition to introduce as an example the case of horizontal absorption into a particular soil, and to carry the illustration through the later developments below by presenting results for more complicated infiltration processes into the same soil. We choose for illustration the Yolo light clay of Moore [6]; we have already presented in Figs. 1–3 $K(\theta)$, $\Psi(\theta)$, and $D(\theta)$ functions for this soil. Figure 4 gives the $\phi(\theta)$

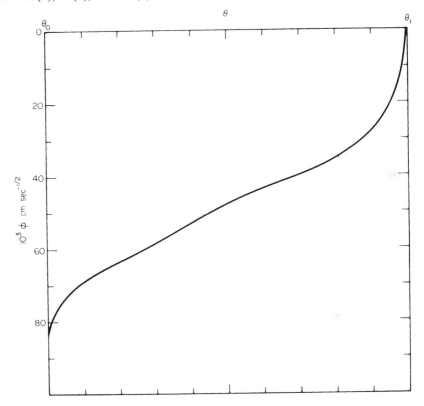

FIG. 4. Moisture profile during one-dimensional absorption in the dimensionless form $\phi(\theta)$, calculated for illustrative example of Yolo light clay of Figs. 1–3 with $\theta_0 = 0.2376$ and $\theta_1 = 0.4950$ [34].

function for the case where $\theta_0 = 0.2376$ and θ_1 is the "saturated" moisture content, 0.4950. For this system, the sorptivity, S, has the value 1.254×10^{-2} cm-sec$^{-1/2}$.

E. Two- and Three-Dimensional Series Solutions

Radially symmetrical m-dimensional absorption into a uniform soil with uniform initial moisture content is described by Eq. (46) subject to conditions (47):

$$\frac{\partial \theta}{\partial t} = r^{1-m} \frac{\partial}{\partial r}\left(Dr^{m-1} \frac{\partial \theta}{\partial r}\right) \qquad (46) \text{ [also Eq. (20)]}$$

$$\begin{aligned} t = 0, & \quad \theta = \theta_0, \quad r > r_0 \\ t \geq 0, & \quad \theta = \theta_1, \quad r = r_0 \end{aligned} \qquad (47)$$

$r = r_0$ is the surface of the cylindrical ($m = 2$) or spherical ($m = 3$) cavity from which water is supplied.

The following method of solution is due to Philip [86]. It is convenient to introduce the new variables

$$\rho = r/r_0, \qquad \tau = t/r_0^2 \qquad (48)$$

which reduce Eqs. (46) and (47) to

$$\frac{\partial \theta}{\partial \tau} = \rho^{1-m} \frac{\partial}{\partial \rho}\left(D\rho^{m-1} \frac{\partial \theta}{\partial \rho}\right) \qquad (49)$$

$$\begin{aligned} \tau = 0, & \quad \theta = \theta_0, \quad \rho > 1 \\ \tau \geq 0, & \quad \theta = \theta_1, \quad \rho = 1 \end{aligned} \qquad (50)$$

We seek solutions of Eq. (49) subject to conditions (50) of the form

$$\rho(\theta, \tau) = 1 + \phi_{*_1} \tau^{1/2} + \phi_{*_2} \tau + \phi_{*_3} \tau^{3/2} + \cdots \qquad (51)$$

where each ϕ_* is a function of θ. *A priori* justification of Eq. (51) may be established along lines similar to those by which Philip [87] originally developed the series solution of the one-dimensional infiltration problem (Sect. V).

The identity

$$(\partial \theta / \partial \tau)_\rho \cdot (\partial \rho / \partial \theta)_\tau = -(\partial \rho / \partial \tau)_\theta \qquad (52)[1]$$

converts Eq. (49) to a form with ρ as the dependent variable

$$-\frac{1}{m} \frac{\partial \rho^m}{\partial \tau} = \frac{\partial}{\partial \theta}\left(D\rho^{m-1} \frac{\partial \theta}{\partial \rho}\right) \qquad (53)$$

---

[1] Here, and at a later point where it is also necessary for clarity, we use the notation $(\partial a/\partial b)_c$ to denote the partial derivative of variable a with respect to variable b with variable c held constant.

Theory of Infiltration

and an integration with respect to θ yields

$$-\frac{1}{m}\frac{\partial}{\partial \tau}\int_{\theta_0}^{\theta} \rho^m \, d\theta = D\rho^{m-1}\frac{\partial \theta}{\partial \rho} \tag{54}$$

Equation (54) depends on the result that $\lim_{\theta \to \theta_0} D\rho^{m-1} \partial\theta/\partial\rho = 0$ (physically that the flow velocity at infinity is zero for all finite t).

Substituting Eq. (51) in Eq. (54), and assuming that the expansion in $\tau^{1/2}$ of each side of Eq. (54) is convergent, we may equate coefficients of powers of $\tau^{1/2}$ and so obtain the following set of ordinary integrodifferential equations:

$$\int_{\theta_0}^{\theta} \phi_{*_1} \, d\theta = -2(D/\phi'_{*_1}) \tag{55}$$

$$\int_{\theta_0}^{\theta} \phi_{*_2} \, d\theta = [D\phi'_{*_2}/(\phi'_{*_1})^2] - (m-1)\int_{\theta_0}^{\theta} D \, d\theta \tag{56}$$

$$\int_{\theta_0}^{\theta} \phi_{*_3} \, d\theta = -\frac{2}{3}\left[\frac{D[(\phi'_{*_2})^2 - \phi'_{*_1}\phi'_{*_3}]}{(\phi'_{*_1})^3} - (m-1)\int_{\theta_0}^{\theta} D\phi_{*_1} \, d\theta\right] \tag{57}$$

and so forth. The primes in Eqs. (55)–(57) signify differentiation with respect to θ. In view of the second of conditions (50), the condition on the ϕ_*'s is

$$\theta = \theta_1, \quad \phi_{*n} = 0, \quad n = 1, 2, 3, 4, \ldots \tag{58}$$

Now, apart from a change of symbolism, Eqs. (55) and (58) are identical to Eqs. (37) and (38), so that we have at once that

$$\phi_{*_1}(\theta) \equiv \phi(\theta) \tag{59}$$

That is, the leading term of the series solution for m-dimensional absorption is exactly the solution for one-dimensional absorption. The physical explanation for this is simply that, when the depth of penetration of the m-dimensional absorption process is small compared to r_0, the process is essentially one-dimensional [88]. We may put the matter mathematically as follows: Rewriting Eq. (46) as

$$\frac{\partial \theta}{\partial t} = \frac{\partial}{\partial r}\left(D\frac{\partial \theta}{\partial r}\right) + \frac{(m-1)D}{r}\frac{\partial \theta}{\partial r} \tag{60}$$

we may show that, for t and $r - r_0$ small enough, the second term of the right side of Eq. (60) is negligibly small compared with the first. Its neglect, of course, reduces Eq. (60) to the one-dimensional form of Eq. (31).

It is evident, then, that [when we do not already know $\phi(\theta)$] Eq. (55) subject to conditions (58) may be solved by the methods of Sects. IV.A and IV.B.

Equations (56), (57), etc., may then be solved in turn. These are all linear equations and are readily amenable to rapid and accurate numerical solution.

The methods devised by Philip [87] to solve the rather similar set of linear equations which arise in the series solution of the one-dimensional infiltration equation may be used here also. We notice that the form of Eq. (56) is such that

$$\phi_{*2(m=3)} = 2\phi_{*2(m=2)} \tag{61}$$

with a consequent economy of labor of calculation if solutions are required for both $m = 2$ and $m = 3$.

The influence of r_0 on the dynamics of (at least the early stages of) the absorption process is implicit in Eq. (51), and is made explicit if we rewrite Eq. (51) in terms of r and t:

$$r(\theta, t) = r_0 + \phi_{*_1} t^{1/2} + \phi_{*_2}(t/r_0) + \phi_{*_3}(t^{3/2}/r_0^2) + \cdots \tag{62}$$

We may develop series expansions for the absorption rate and for the cumulative absorption in terms either of τ or t. It is convenient here to treat the matter in terms of absorption rate $v_0(t)$ and cumulative absorption $i(t)$, each based on unit area of the water-supply surface $r = r_0$. Then

$$i = \frac{\int_{r_0}^{\infty} (\theta - \theta_0) r^{m-1} \, dr}{r_0^{m-1}} = \frac{\int_{\theta_0}^{\theta_1} r^m \, d\theta}{m r_0^{m-1}} \tag{63}$$

Putting Eq. (62) in Eq. (63), expanding r^m, and integrating term by term, we obtain $i(t)$ as a power series in $t^{1/2}$. The leading terms are

$$i = St^{1/2} + \frac{A_{*_2} t}{r_0} + \frac{A_{*_3} t^{3/2}}{r_0^2} + \cdots \tag{64}$$

S is, of course, the sorptivity defined in Sect. IV.C, and the A_{*_n} are appropriate integrals of the ϕ_{*_n}. Specifically,

$$A_{*_2} = \int_{\theta_0}^{\theta_1} \left[\frac{m-1}{2} \phi_{*_1}^2 + \phi_{*_2} \right] d\theta \tag{65}$$

and

$$A_{*_3} = \int_{\theta_0}^{\theta_1} \left[\frac{(m-1)(m-2)}{2 \cdot 3} \phi_{*_1}^3 + (m-1)\phi_{*_1}\phi_{*_2} + \phi_{*_3} \right] d\theta \tag{66}$$

It follows from Eqs. (61) and (65) that

$$A_{*_2(m=3)} = 2A_{*_2(m=2)} \tag{67}$$

Differentiating Eq. (64) with respect to t yields the expansion for $v_0(t)$:

$$v_0 = \tfrac{1}{2} S t^{-1/2} + \frac{A_{*_2}}{r_0} + \frac{3 A_{*_3} t^{1/2}}{2 r_0^2} + \cdots \tag{68}$$

Theory of Infiltration

It must be understood that the time range of applicability of these expansions is definitely limited. We shall not offer here a detailed discussion of the question of their convergence. It seems sufficient to point out that the characteristic time[2] t_{geom}, based on S, $(\theta_1 - \theta_0)$, and the mean curvature (in three dimensions) of the supply surface, r_*^{-1},

$$t_{geom} = \left(\frac{r_*(\theta_1 - \theta_0)}{S}\right)^2 \tag{69}$$

gives the order of magnitude of t at which the m-dimensional geometry can be expected to swamp the initially one-dimensional character of the process. Note that r_* is $2r_0$ for $m = 2$ and is r_0 for $m = 3$. One expects *a priori* that the useful t range of these expansions will be rather less than t_{geom}; and, in fact, empirical experience suggests that this range is about $0.12 t_{geom}$.

For the case $m = 3$, we may supplement this form of solution with the asymptotic (large-time) solution of Sect. IV.F.

F. Three-Dimensional Quasi-Analytical Large-Time Solution

It is readily shown that, in the limit as $t \to \infty$, the solution of Eq. (49) subject to conditions (50) for $m = 3$ is

$$\rho(\theta, \infty) = \int_{\theta_0}^{\theta_1} D\, d\theta \bigg/ \int_{\theta_0}^{\theta} D\, d\theta, \quad \text{i.e.,} \quad r(\theta, \infty) = r_0 \int_{\theta_0}^{\theta_1} D\, d\theta \bigg/ \int_{\theta_0}^{\theta} D\, d\theta \tag{70}$$

It follows that, in the limit of t large, $v_0(t)$ approaches the steady value $v_0(\infty)$ given by Eq. (71):

$$v_0(\infty) = \int_{\theta_0}^{\theta_1} D\, d\theta / r_0 \tag{71}$$

This result [86] is a particular form of the more general results of Sect. IV.H. Its physical importance is that it exemplifies the existence of a large-time steady phase of three-dimensional absorption *even in the absence of gravity*; and it is useful as an illustration of the falsity of certain notions in the literature about the use of similarity solutions in m-dimensional systems (Sect. IV.I).

G. Illustrative Example: Two- and Three-Dimensional Absorption

Figures 5 and 6 present for our illustrative example the sequence of moisture profiles during two- and three-dimensional absorption as calculated from Eq. (51) (first three terms) and Eq. (70). To this accuracy of calculation, we have that, because of Eq. (61), each profile for $m = 2$ represents, for the

---

[2] Discussion of physical processes in terms of characteristic time and length scales is a well-established procedure, although it has not been used much in the soil-water context [43, 89].

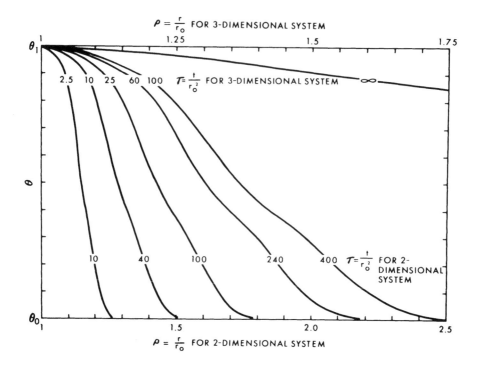

FIG. 5. Moisture profiles during two- and three-dimensional absorption, calculated for illustrative example of Yolo light clay of Figs. 1–3 with $\theta_0 = 0.2376$ and $\theta_1 = 0.4950$ [86]. Profiles computed from first three terms of Eq. (51), except for three-dimensional $\tau = \infty$ profile calculated from Eq. (71). Numerals on each profile represent time at which profile is realized in the form $\tau \, (= t/r_0^2)$.

appropriately adjusted scale of ρ and value of τ, a profile for $m = 3$ also. The $m = 3$ profile for $\tau = \infty$, computed from Eq. (70), is shown in part in Fig. 5, and is compared with the $\tau = 100$ profile in Fig. 6. It will be seen that, whereas the corresponding one-dimensional absorption profiles preserve similarity, the profiles for $m = 2$ and 3 grow *relatively* flatter as τ increases. The flattening proceeds more rapidly for $m = 3$ than for $m = 2$.

H. Two- and Three-Dimensional Arbitrary Geometry: Quasi-Analytical Large-Time and Steady Solutions

Of the whole class of absorption processes described by Eq. (14), a large subclass has boundary conditions such that, in the limit as $t \to \infty$, the phenomenon approaches a steady state. For such steady states, Eq. (14) becomes

$$\nabla \cdot (D \, \nabla \theta) = 0 \qquad (72)$$

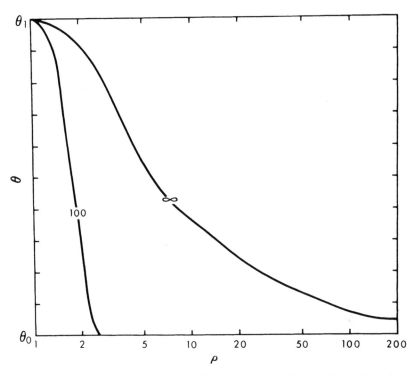

FIG. 6. As Fig. 5, showing only $\tau = 100$ and $\tau = \infty$ profiles for three-dimensional absorption [86].

The transformation [23, p. 11; 90–93]

$$\theta_* = \int_{\theta_0}^{\theta} D \, d\theta \tag{73}$$

reduces Eq. (72) to Laplace's equation

$$\nabla^2 \theta_* = 0 \tag{74}$$

For certain absorption phenomena (e.g., where water is under positive hydrostatic pressure in part of the flow region), the appropriate flow equation is that with Ψ, not θ, as the dependent variable [Eq. (13) with the second term of the right side omitted]. For steady states, this equation reduces to

$$\nabla \cdot (K \nabla \Psi) = 0 \tag{75}$$

and the transformation

$$\Psi_* = \int_{\Psi_0}^{\Psi} K \, d\Psi \tag{76}$$

again leads to Laplace's equation, now in the form

$$\nabla^2 \Psi_* = 0 \tag{77}$$

The whole corpus of established analytical methods of solution of Laplace's equation in two and three dimensions is thus available for the establishment of quasi-analytical large-time (or steady) solutions of appropriate two- and three-dimensional absorption problems.

It remains for us to specify these problems. It is necessary that their boundary conditions have definite limits as $t \to \infty$, and that these "limiting" conditions be such that a solution of Laplace's equation exists for them which is neither singular nor trivial.

Turning, in particular, to absorption in systems that are infinite in two or three dimensions, *but in which the supply surface is finite*, we remark that it can be shown that the steady state always exists in three dimensions, but never in two. This result holds whether the boundary condition at the supply surface is of the concentration type (as for most of the processes considered in this paper) or of the flux type.

It is not our purpose here to make an exhaustive exploration of these solutions, but we indicate two examples. As we have noted already, the solution of Sect. IV.F is of this class. A second example is that of absorption into a three-dimensional semi-infinite region from a plane circular region of radius r_0 at the surface of which the moisture content is maintained at the value θ_1. The required large-time solution follows at once from the relevant solution of Laplace's equation [23, p. 215]. It suffices here to remark that, with v_0 based on unit area of the supply surface,

$$v_0(\infty) = (4/\pi r_0) \int_{\theta_0}^{\theta_1} D \, d\theta \tag{78}$$

The large-time *total* absorption rate is then

$$q(\infty) = 4r_0 \int_{\theta_0}^{\theta_1} D \, d\theta \tag{79}$$

These results should be compared with those for absorption from a *hemispherical* supply surface of radius r_0 into a three-dimensional semi-infinite region (which follow at once from Sect. IV.F). In that case,

$$v_0(\infty) = (r_0)^{-1} \cdot \int_{\theta_0}^{\theta_1} D \, d\theta, \qquad q(\infty) = 2\pi r_0 \int_{\theta_0}^{\theta_1} D \, d\theta \tag{80}$$

I. SIMILARITY SOLUTIONS IN ABSORPTION PROBLEMS

Similarity methods form an important avenue of mathematical attack on nonlinear problems; as we have seen, the one-dimensional absorption solution is a similarity solution. These methods are better known in more conven-

Theory of Infiltration

tional branches of fluid mechanics than they seem to be in soil-water studies, and several confusions in the mathematical study of absorption have arisen from the misuse of similarity methods.

Miller and Miller [65] correctly observed that the cases $m = 2, 3, r_0 = 0$ lead to a similarity reduction of Eq. (46) subject to conditions (47), but that, for both values of m, a singularity at $r = 0$ implies a trivial "no-flow" solution in the physically meaningful situation where θ_1 is required to be finite. They went on, however, to suggest that experiments with an impressed flow rate proportional to $t^{1/2}$ and a small but nonzero r_0 would yield similarity. [It appears that they, in fact, intended to propose a constant (t^0) flow rate for $m = 2$ and a flow as $t^{1/2}$ for $m = 3$.] Their proposal depends on the heuristic idea that, if something is true for r_0 zero, it will be more or less true for r_0 "small": in fact, the singular and degenerate nature of the solution for $r_0 = 0$ and θ_1 finite completely invalidates the argument. It will be clear from the relevant preceding solutions (and from later supplementary results) that absorption solutions for $m = 2, 3$ with r_0 nonzero are never, even approximately, of the form

$$r(\theta, t) = f(\theta) t^{1/2}$$

as suggested by Miller and Miller [65]. For $m = 2$, in fact, v_0 varies initially as $t^{-1/2}$ and finally as $(\log t)^{-1}$; and, for $m = 3$, v_0 varies initially as $t^{-1/2}$ and finally as t^0: compare the "similarity predictions" that the behavior is as t^0 for $m = 2$ and as $t^{1/2}$ for $m = 3$.

The misuse of similarity methods seems to arise often from neglect of the following basic principle [86, 94]. A similarity substitution usefully reduces the number of independent variables in a partial differential equation only when the variables removed from the equation are removed also from all the governing conditions by the same substitution. Singh [95] and Singh and Franzini [96] attempted to study absorption for $m = 2$ by the use of similarity methods, but the disregard of this principle vitiates the results [86, 97]. Swartzendruber [98] and Philip [91] pointed out that the additive-variables substitution used by Kobayashi [99, 100] fails for essentially the same reason.

It appears that the *Boltzmann transformation* (33) is the one useful similarity substitution in studies of absorption with concentration-type boundary conditions. Kobayashi [100] claimed to have developed a diversity of substitutions, but they include nothing new that is both correct and useful [91]. The "three-dimensional" substitutions are simply the one-dimensional Boltzmann transformation in an inappropriate coordinate system, and other substitutions claimed as new are simply members of the indefinitely large class of "pseudotransformations" of the form $f(\phi)$ (with f a monotonic function of ϕ) which are totally equivalent to the Boltzmann transformation.

We see later in Sect. IX that other forms of similarity substitution are useful in connection with absorption processes involving a flux-type boundary condition.

V. Exact Solutions of Infiltration Equations

A. One-Dimensional Series Solution

Infiltration into an effectively semi-infinite, homogeneous, one-dimensional system of uniform initial moisture content is described by Eq. (81) subject to conditions (82):

$$\frac{\partial \theta}{\partial t} = \frac{\partial}{\partial z}\left(D \frac{\partial \theta}{\partial z}\right) - \frac{dK}{d\theta} \cdot \frac{\partial \theta}{\partial z} \quad (81) \text{ [also Eq. (22)]}$$

$$\begin{aligned} t = 0, & \quad z > 0, \quad \theta = \theta_0 \\ t \geq 0, & \quad z = 0, \quad \theta = \theta_1 \end{aligned} \quad (82)$$

Philip [33, 34, 87] developed the following method of solution of Eq. (81) subject to conditions (82). The solution is sought in the form

$$z(\theta, t) = \phi_1 t^{1/2} + \phi_2 t + \phi_3 t^{3/2} + \phi_4 t^2 + \cdots \quad (83)$$

where each of the ϕ's is a function of θ. The form of Eq. (83) was found by Philip [87] by a lengthy process of successive approximation. The reader who seeks a full *a priori* justification is referred to that paper. Equation (83) is, in fact, the immediate outcome of considering the solution of Eq. (81) subject to conditions (82) as a perturbation of the solution of Eq. (31) subject to conditions (32) (with z substituted for x). It will be noted that, here, as in Sect. IV.E, the method involves taking the space coordinate as a dependent variable, with θ as an independent variable.

Accordingly, we use the identity

$$(\partial \theta / \partial t)_z \cdot (\partial z / \partial \theta)_t = -(\partial z / \partial t)_\theta \quad (84)$$

to reduce Eq. (81) to the form

$$-\frac{\partial z}{\partial t} = \frac{\partial}{\partial \theta}\left(D \frac{\partial \theta}{\partial z}\right) - \frac{dK}{d\theta} \quad (85)$$

Integrating with respect to θ, we obtain

$$-\frac{\partial}{\partial t} \int_{\theta_0}^{\theta} z \, d\theta = D \frac{\partial \theta}{\partial z} - (K - K_0) \quad (86)$$

We write K_0 for $K(\theta_0)$. Equation (86) uses the result $\lim_{\theta \to \theta_0} D \, \partial \theta / \partial z = 0$ for all t, which is implicit in the first of conditions (82).

Theory of Infiltration

Putting Eq. (83) in Eq. (86), and assuming that the expansion in $t^{1/2}$ of each side of Eq. (86) is convergent, we may equate coefficients of powers of $t^{1/2}$ and so obtain the following set of ordinary integrodifferential equations:

$$\int_{\theta_0}^{\theta} \phi_1 \, d\theta = -2 \frac{D}{\phi_1'}, \tag{87}$$

$$\int_{\theta_0}^{\theta} \phi_2 \, d\theta = \frac{D\phi_2'}{(\phi_1')^2} + (K - K_0) \tag{88}$$

$$\int_{\theta_0}^{\theta} \phi_3 \, d\theta = \frac{2D}{3} \left[\frac{\phi_3'}{(\phi_1')^2} - \frac{(\phi_2')^2}{(\phi_1')^3} \right] \tag{89}$$

$$\int_{\theta_0}^{\theta} \phi_4 \, d\theta = \frac{D}{2} \left[\frac{\phi_4'}{(\phi_1')^2} - \frac{(\phi_2')^2}{(\phi_1')^3} \left\{ 2\frac{\phi_3'}{\phi_2'} - \frac{\phi_2'}{\phi_1'} \right\} \right] \tag{90}$$

and so on, the equation for ϕ_n ($n \geq 3$) being of the form

$$\int_{\theta_0}^{\theta} \phi_n \, d\theta = \frac{2D}{n} \left[\frac{\phi_n'}{(\phi_1')^2} - R_n(\theta) \right] \tag{91}$$

where R_n may be determined from $\phi_1, \ldots, \phi_{n-1}$. It will be understood that the prime in Eqs. (87)–(91) signifies differentiation with respect to θ. In view of the second of conditions (82), the condition on the ϕ's is

$$\theta = \theta_1, \quad \phi_n = 0, \quad n = 1, 2, 3, 4, \ldots \tag{92}$$

Now, apart from a difference of symbolism, Eqs. (87) and (92) are identical with Eqs. (37) and (38), so that we have at once that

$$\phi_1(\theta) \equiv \phi(\theta) \tag{93}$$

Evidently, if we do not already know $\phi(\theta)$, we may solve Eq. (87) subject to conditions (92) by the methods of Sects. IV.A and IV.B.

We see that the leading term of the series solution for one-dimensional infiltration is exactly the solution for one-dimensional absorption. In physical terms, the explanation is that, for small t, the potential gradients in the flow region due to capillarity are much greater than unity, the value of the potential gradient due to gravity. In mathematical terms, we may show that, for t and z small enough, the second term of the right side of Eq. (81) is negligibly small compared with the first; and the neglect of this term reduces Eq. (81) to Eq. (31).

With ϕ_1 known, we may solve Eq. (88) subject to conditions (92) for ϕ_2; then, with ϕ_2 known, we may solve Eq. (89) subject to conditions (92) for ϕ_3; and so on. The various equations (88)–(91) are linear, and may be rapidly and accurately solved by the numerical method of Philip [87]. In brief, the method uses appropriate finite difference forms and forward integration initiated by an assumed value of $\int_{\theta_0}^{\theta_1} \phi_n \, d\theta$. A recent account of this work [101], missing the point that these equations are linear, implies that the method of solution is iterative; however, because the equations are linear, iteration is not needed. One simply performs the calculation for two arbitrary values of $\int_{\theta_0}^{\theta_1} \phi_n \, d\theta$, and linear interpolation then yields the "exact" solution. The precision of the method is improved by use of the Richardson h^2-extrapolation [80] and by the use of analytical methods in the neighborhood of the "infinite tail," which, in general, occurs as $\theta \to \theta_0$.

The total increase of moisture content in the semi-infinite region during the time interval 0 to t equals the difference between the time integrals of the flow velocity at $x = 0$ and that at infinity. That is,

$$\int_0^\infty (\theta - \theta_0) \, dx = \int_{\theta_0}^{\theta_1} x \, d\theta = i(t) - K_0 t \tag{94}$$

where $i(t)$ is the cumulative infiltration. Putting Eq. (83) in Eq. (94), we obtain the series for $i(t)$ [34]:

$$i(t) = St^{1/2} + (A_2 + K_0)t + A_3 t^{3/2} + A_4 t + \cdots \tag{95}$$

Here S is the sorptivity, and

$$A_n = \int_{\theta_0}^{\theta_1} \phi_n \, d\theta \tag{96}$$

Differentiating Eq. (95) with respect to t, we obtain the series for the infiltration rate, $v_0(t)$ [34]:

$$v_0(t) = \tfrac{1}{2} S t^{-1/2} + (A_2 + K_0) + \tfrac{3}{2} A_3 t^{1/2} + 2 A_4 t + \cdots \tag{97}$$

A rigorous treatment of the convergence of these series seems rather difficult, and we limit ourselves here to some general remarks. In this connection, it is useful to introduce a characteristic time of the infiltration process, t_{grav}, based on S and $K_1 - K_0$ [we write K_1 for $K(\theta_1)$],

$$t_{\text{grav}} = \left(\frac{S}{K_1 - K_0} \right)^2 \tag{98}$$

The order of magnitude of the t value at which the effect of gravity on the process can be expected to be as great as that of capillarity is given by t_{grav}; and one expects *a priori* that the t range of useful convergence for expansions of Eqs. (83), (95), and (97) will be of about this magnitude.

It is observed empirically in numerical examples that

$$n > 2, \quad A_n/S > (A_2/S)^{n-1} \tag{99}$$

so that series (95) may be shown, by comparison with the geometric series, to converge for

$$t < (S/A_2)^2 \tag{100}$$

Theory of Infiltration

Now, for the usual case with $K_0/K_1 \ll 1$, it is found both empirically and by analysis (cf. Sects. VI and VIII) that $A_2 \approx \frac{1}{2}(K_1 - K_0)$. Inequality (100) then becomes

$$t < 4t_{\text{grav}} \tag{101}$$

One expects the practical limit of convergence to be given by inequality (101) with 4 replaced by a smaller numerical coefficient. Empirical experience suggests, in fact, that a suitable criterion for *practical convergence* is

$$t \lesssim t_{\text{grav}} \tag{102}$$

This method applies in a slightly modified form [81] to the case of water supplied under positive hydrostatic pressure. Equation (81) is replaced by the relevant form of Eq. (13), at least in the region of positive Ψ.

B. One-Dimensional Asymptotic (Large-Time) Solution

We supplement the preceding solution of Eq. (81) subject to conditions (82) with the asymptotic solution valid for large t obtained by Philip [73]. We seek a solution of the form

$$z(\theta, t) = u(t - t_0) + \zeta(\theta) \tag{103}$$

Substituting Eq. (103) in Eq. (86), we obtain

$$u(\theta - \theta_0) = (K - K_0) - D(d\theta/d\zeta) \tag{104}$$

For t sufficiently large, we evaluate u in Eq. (104) with the aid of the result

$$\lim_{t \to \infty} \lim_{\theta \to \theta_1} \partial\theta/\partial z \equiv 0 \tag{105}$$

which is the consequence of theorems established by Philip [73]. Equation (105) implies that, if Eq. (103) holds for indefinitely large t, $\lim_{\theta \to \theta_1} d\theta/d\zeta = 0$. It follows that

$$u = (K_1 - K_0)/(\theta_1 - \theta_0) \tag{106}$$

Putting Eq. (106) in Eq. (104) and integrating, we obtain

$$\zeta(\theta) = (\theta_1 - \theta_0) \int_\theta^{\theta_a} \frac{D\,d\theta}{(K_1 - K_0)(\theta - \theta_0) - (K - K_0)(\theta_1 - \theta_0)} \tag{107}$$

where the zero of ζ is taken at $\theta = \theta_a$. Usually, θ_a is conveniently taken as $\theta_1 - \Delta\theta$, where $\Delta\theta$ is a definite small positive quantity. It is usually necessary to take the zero of ζ at θ values other than θ_1 and θ_0, because, in general, ζ has singularities for both these θ values [73]. Except in the special case where $D \to 0$ sufficiently rapidly as $\theta \to \theta_0$, which is treated by Philip [48], we have that

$$\lim_{\theta \to \theta_1} \zeta = -\infty, \quad \lim_{\theta \to \theta_0} \zeta = +\infty \tag{108}$$

The singular behavior of ζ is not recognized in several accounts of the integration of Eq. (107) [101–103]. Evidently, $\zeta(\theta)$ for $\theta_0 < \theta < \theta_1$ can be evaluated simply by numerical integration; Philip [73] has shown how the accuracy of the calculation can be improved near $\theta = \theta_0$ and θ_1 by the use of analytical methods.

Because $d^2K/d\theta^2$ is inherently nonnegative and is positive for (at least) almost all θ, the denominator of the integrand of Eq. (107) is positive for $\theta_0 < \theta < \theta_1$. It is of some interest that the phenomenon of the "profile at infinity"—i.e., that, asymptotically, the moisture profile becomes of the constant shape $\zeta(\theta)$, moving downward with constant velocity, u—is inherent in the nonlinearity of Eq. (81). For the corresponding linearized equation (with K a linear function of θ), the profile at infinity does not occur, the gradients of θ simply becoming progressively smaller everywhere in the wetted region as t increases (Sect. VI.C). Equation (107) applies also to the case of water supplied under positive hydrostatic pressure [81].

The similar and independent analysis by Youngs [102] takes Eq. (105) for granted and erroneously puts $u = K_1/\theta_1$. Irmay [104] recognized that the form of Eq. (103) would constitute an asymptotic large-time solution of Eq. (81). [He had derived an equation formally equivalent to Eq. (81) from a physically unacceptable model of the flow process.] Childs [103] gave an elegant restatement of the theorems on moisture profile development which underlay the original analysis [73]. His improved treatment of the possible influence of capillary hysteresis makes use of the independent domain theory [66], discussed in Sect. II.D.6.

It follows from the preceding analysis [and from Eq. (105), in particular] that, as t increases indefinitely, the infiltration rate decreases monotonically to its final asymptotic value, $v_0(\infty)$,

$$v_0(\infty) = K_1 \tag{109}$$

These various results provide means of estimating $z(\theta, t)$ and $i(t)$ for large t values. Fairly simple interpolation and matching techniques to connect the series and asymptotic solutions appear to be adequate in practice. The reader is referred to Philip [73] for details. Farrell [105] has suggested some refinements.

C. Illustrative Example: One-Dimensional Infiltration

Figure 7 presents for our illustrative example the functions $\phi_2(\theta)$, $\phi_3(\theta)$, $\phi_4(\theta)$. It will be understood that $\phi(\theta)$ of Fig. 4 is the relevant $\phi_1(\theta)$ function. The calculated values of A_2, A_3, A_4 are, respectively, 4.654×10^{-6} cm-sec^{-1}, 1.405×10^{-9} cm-sec$^{-3/2}$, and 8.961×10^{-14} cm-sec^{-2}. $(K_1 - K_0) = 12.288 \times 10^{-6}$ cm-sec^{-1}. $t_{\text{grav}} = 1.041 \times 10^6$ sec. Figure 8 gives the profile at infinity,

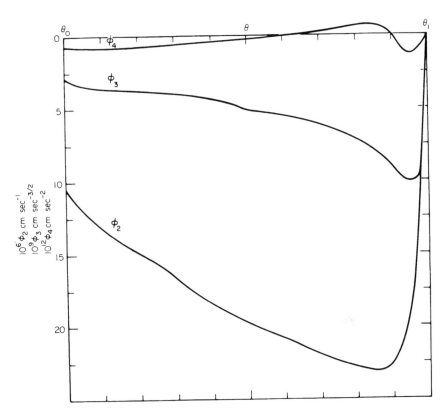

FIG. 7. The functions $\phi_2(\theta)$, $\phi_3(\theta)$, $\phi_4(\theta)$, calculated for one-dimensional infiltration in Yolo light clay of Figs. 1-3 with $\theta_0 = 0.2376$ and $\theta_1 = 0.4950$ [34].

$\zeta(\theta)$, for our example; θ_a is taken as $(0.95\theta_1 + 0.05\theta_0)$. $u = 47.74 \times 10^{-6}$ cm-sec^{-1}. Figure 9 presents the sequence of moisture profiles during one-dimensional infiltration. The profiles for $t \leqq 10^6$ sec are calculated from the first four terms of series (83); those for larger values of t are based on asymptotic solution (103).

D. Two- and Three-Dimensional Small-Time Solutions

The techniques of Sects. V.A and V.B cannot be extended satisfactorily to infiltration in two and three dimensions, which involves at least two space coordinates.

For the early stages of infiltration into such systems (with a concentration boundary condition at the absorbing surface), limited use can be made, however, of the absorption solutions of Sect. IV. We discuss the matter in terms of

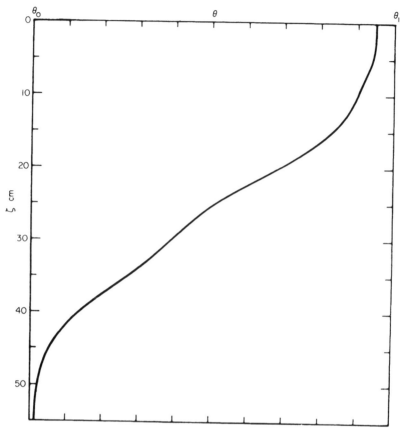

FIG. 8. The "profile at infinity," $\zeta(\theta)$, for illustrative example of one-dimensional infiltration [73]. $\theta_a = 0.95\theta_1 + 0.05\theta_0$.

the characteristic times t_{geom} [Eq. (69)] and t_{grav} [Eq. (98)]. When the corresponding absorption solution is known, it will provide a useful estimate of behavior for $t \leq ct_{\text{grav}}$, where c is perhaps about 0.01. For $t_{\text{geom}} \ll t_{\text{grav}}$, however, absorption solutions found as expansions in $t^{1/2}$ may cease to be useful at t values much smaller than ct_{grav}. On the other hand, for $t_{\text{grav}} \ll t_{\text{geom}}$, ct_{grav} will be so small that the one-dimensional absorption solution will be virtually indistinguishable from the absorption solution for the actual geometry. The one-dimensional solution is, in this limited sense, useful for infiltration in geometries for which no exact absorption solution is known. It should be noted that, when the supply surface is other than cylindrical or spherical, some care must be given to the choice of the r_*^{-1} used to define t_{geom}. The maximum mean curvature of the supply surface will often be the appropriate value.

Theory of Infiltration

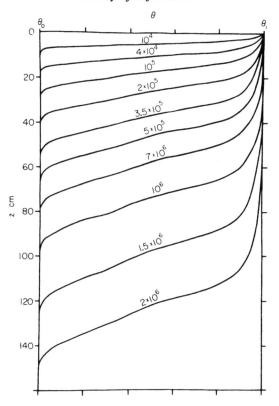

FIG. 9. Computed moisture profiles for illustrative example of one-dimensional infiltration. Numerals on each profile represent value of t (sec) at which profile is realized. Profiles for $t \leq 10^6$ calculated from first four terms of series (83); those for larger t are based on Eq. (103) [34, 73].

E. Two- and Three-Dimensional Arbitrary Geometry: Quasi-Analytical Large-Time and Steady Solutions

In this subsection, we discuss the subclass of infiltration processes described by Eq. (21) for which, in the limit as $t \to \infty$, the phenomenon approaches a steady state. It is significantly wider than the subclass of absorption processes with the same property; thus, infiltration from a finite supply surface into a region infinite in two dimensions asymptotically approaches a steady state at large t, whereas the corresponding absorption problems do not have steady solutions.

For steady states, Eq. (21) becomes

$$\nabla \cdot (D \, \nabla \theta) = \frac{dK}{d\theta} \cdot \frac{\partial \theta}{\partial z} \tag{110}$$

and transformation (73) reduces this to

$$\nabla^2 \theta_* = \frac{1}{D}\frac{dK}{d\theta} \cdot \frac{\partial \theta_*}{\partial z} \qquad (111)$$

Now, if

$$\frac{1}{D}\frac{dK}{d\theta} = \text{const} = \alpha \qquad (112)$$

important simplifications follow, so that it is of interest to examine the physical implications of Eq. (112). We find by combining Eqs. (11) and (112) that Eq. (112) is true when

$$K \propto e^{\alpha \Psi} \qquad (113)$$

Even though it cannot be claimed that Eq. (113) is universally exact, it does model in a reasonably convincing way the quite generally observed rapid and nonlinear decrease of K with Ψ. For many soils, Eq. (113) does, in fact, represent $K(\Psi)$ fairly well over Ψ ranges of interest in soil-water studies. α has the dimensions [length]$^{-1}$ and is conveniently expressed in cm^{-1}. Typically, α is about 0.01 cm^{-1}, and the range of values 0.05 to 0.002 cm^{-1} seems likely to cover most applications. α is a measure of the relative importance of gravity and capillarity for soil-water movement in the particular soil. Fine-textured soils, where capillarity tends to predominate, have small α values; and coarse-textured soils, where gravity effects manifest themselves most readily, have large α values.

Putting Eq. (112) in Eq. (111), we obtain the linear equation

$$\nabla^2 \theta_* = \alpha(\partial \theta_*/\partial z) \qquad (114)$$

It will be seen that transformation (73), supplemented by Eq. (112), enables us to reduce the highly nonlinear Eq. (110) to a linear form *while still retaining the basically nonlinear character of the flow process*. Evidently, the exploration of solutions of Eq. (114) can be expected to provide considerable illumination of the problems of steady infiltration in two and three dimensions.

Problems of this type (including seepage from trenches, canals, basins, etc.) have been treated classically as problems of saturated flow with a free surface (e.g., Muskat [106], Polubarinova-Kochina [107]). That mode of analysis may be relevant where a water table exists at shallow depth; but it is definitely inappropriate in situations, common in arid and semi-arid environments, where there is no shallow water table. In such cases, we seek to study the steady regime set up by prolonged infiltration (or seepage) into *an initially unsaturated system*. The phenomenon is essentially one of flow in unsaturated soil, and a formalism based on Eq. (114) promises to be both physically

Theory of Infiltration

apposite and mathematically amenable. It is of some interest that the unsaturated analysis, which might be expected *a priori* to be more complicated than the classical analysis, is the simpler in the sense that it is free of the difficulties connected with the free surface in the classical approach.

The pertinence of Eq. (114) has been recognized only recently [91–93], and the exploration of relevant solutions has not yet been taken very far. Here we limit specific discussion to two basic solutions, the source solutions in two and three dimensions, with θ_0 so small that K_0 is negligibly small.

1. Source Solution in Two Dimensions

The two-dimensional form of Eq. (114) is

$$\frac{\partial^2 \theta_*}{\partial x^2} + \frac{\partial^2 \theta_*}{\partial z^2} = \alpha \frac{\partial \theta_*}{\partial z} \tag{115}$$

It follows from Eq. (73) that, for infiltration into a two-dimensionally infinite region with initial moisture content θ_0, condition (116) must be satisfied:

$$\lim_{(x^2+z^2)^{1/2} \to \infty} \theta_* = 0 \tag{116}$$

The solution of Eq. (115) subject to condition (116) for the case of a continuous source of strength Q_2 cm^2-sec^{-1} at $x = 0$, $z = 0$ is well known [23, p. 267]. It is

$$\theta_* = (Q_2/2\pi)e^{\alpha z/2} \mathbf{K}_0[\alpha(x^2 + z^2)^{1/2}/2] \tag{117}$$

$\mathbf{K}_0$ is the modified Bessel function of the second kind of zero order. We may rewrite Eq. (117) as

$$\vartheta_2 = e^Z \mathbf{K}_0[(X^2 + Z^2)^{1/2}] \tag{118}$$

where

$$\vartheta_2 = 2\pi\theta_*/Q_2, \qquad X = \alpha|x|/2, \qquad Z = \alpha z/2 \tag{119}$$

This solution is presented in the form $\vartheta_2(X, Z)$ in Fig. 10.

This basic result can be used, as it stands, to develop solutions for infiltration from the surface of approximately circular cylindrical cavities in infinite regions; and appropriate source-sink distributions based on it yield solutions for infiltration from the surface of furrows of various shapes into semi-infinite regions bounded by an approximately horizontal upper surface.

2. Source Solution in Three Dimensions

The three-dimensional axisymmetric form of Eq. (114) is

$$\frac{1}{r}\frac{\partial}{\partial r}\left(r \frac{\partial \theta_*}{\partial r}\right) + \frac{\partial^2 \theta_*}{\partial z^2} = \alpha \frac{\partial \theta_*}{\partial z} \tag{120}$$

r is here $(x^2 + y^2)^{1/2}$. For axisymmetric infiltration into a three-dimensionally infinite region with initial moisture content θ_0, the final steady solution must satisfy the condition (121):

$$\lim_{(r^2+z^2)^{1/2}\to\infty} \theta_* = 0 \qquad (121)$$

The solution of Eq. (120) subject to condition (121) for the case of a continuous source of strength Q_3 cm^{3}sec^{-1} at $r = 0$, $z = 0$ is [23, p. 267]

$$\theta_* = \frac{Q_3}{4\pi(r^2+z^2)^{1/2}} \exp\left\{-\frac{\alpha}{2}[(r^2+z^2)^{1/2} - z]\right\} \qquad (122)$$

We may rewrite this as

$$\vartheta_3 = \frac{\exp[Z - (R^2 + Z^2)^{1/2}]}{(R^2 + Z^2)^{1/2}} \qquad (123)$$

where

$$\vartheta_3 = 8\pi\theta_*/\alpha Q_3, \qquad R = \alpha r/2 \qquad (124)$$

Figure 11 gives this solution in the form $\vartheta_3(R, Z)$.

Like Eq. (117), Eq. (122) leads to a diversity of solutions of interest. As it stands, it yields solutions for infiltration from the surface of approximately spherical cavities in infinite regions; and source-sink distributions based on it yield solutions for infiltration from the surface of basins of various shapes into semi-infinite regions bounded by an approximately horizontal upper surface.

3. Other Solutions of Eq. (114)

There remains much scope for the study of Eq. (114) by various methods. A boundary-layer technique neglecting $\partial^2\theta_*/\partial z^2$ {or, for the untransformed equation, $(\partial/\partial z)[D(\partial\theta/\partial z)]$}, similar to those employed in other contexts (e.g., Philip [94], Sutton [108], Wooding [109, 110]), is feasible and promises to give accurate results for the cases of infiltration from the surfaces of deep vertical slits and boreholes.

VI. Linearized Solutions

Apart from a limited range available for other classes of problem (see Sect. IX), we have exhausted the known solutions that may be applied to soils of arbitrary hydrologic characteristics. We go on to consider two classes of analytical solutions which are of some interest in their own right, and which are the basis of a comparison method for estimating integral properties of soils with arbitrary characteristics. We first discuss linearized solutions [74, 86, 111].

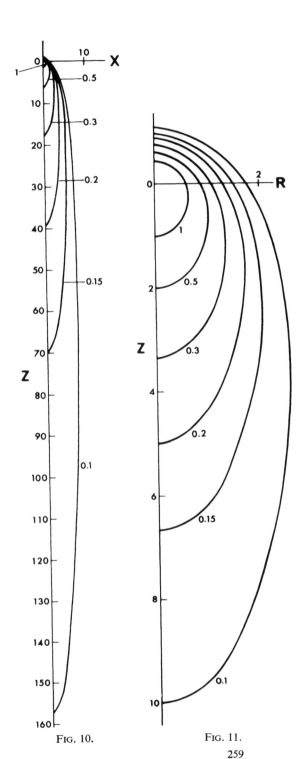

FIG. 10. Source solution for two-dimensional infiltration as $t \to \infty$, in the dimensionless form $\vartheta_2(X, Z)$. Numerals on the curves are values of ϑ_2.

FIG. 11. Source solution for three-dimensional infiltration as $t \to \infty$, in the dimensionless form $\vartheta_3(R, Z)$. Numerals on the curves are values of ϑ_3.

A. Principles of Method

The linearized form of Eq. (21) is

$$\partial \vartheta/\partial t = D_* \nabla^2 \vartheta - k(\partial \vartheta/\partial z) \qquad (125)$$

a linear Fokker–Planck equation. The equation is exact, and $\vartheta \equiv \theta$, for the case where

$$D(\theta) = \text{const} = D_* \qquad (126)$$

and

$$dK/d\theta = \text{const} = k \qquad (127)$$

As we have emphasized, D and $dK/d\theta$ are, for real media and soils, strongly varying functions of θ, so that solutions of Eq. (125) do not, in general, give an accurate detailed description of the phenomena that concern us here. We shall show, however, that Eq. (125) yields useful estimates of *integral properties* of these phenomena, $v_0(t)$ and $i(t)$. By choosing D_* so as to match the known integral behavior of absorption (and infiltration) processes in the limit as $t \to 0$, and by choosing k to match infiltration behavior in the limit as $t \to \infty$, we, essentially, use Eq. (125) as the basis of extrapolation and interpolation procedures that take known exact nonlinear solutions as reference points. In certain cases, which we shall identify, there exists a definite one-to-one relationship between θ and ϑ; and, in general, there is at least a loose ordinal relation between these two quantities.

B. Absorption Solutions

1. One-Dimensional Absorption

The linearized form of Eq. (31) subject to conditions (32) is

$$\frac{\partial \vartheta}{\partial t} = D_* \frac{\partial^2 \vartheta}{\partial x^2} \qquad (128)$$

subject to the conditions

$$t = 0, \quad x > 0, \quad \vartheta = \theta_0$$
$$t \geq 0, \quad x = 0, \quad \vartheta = \theta_1 \qquad (129)$$

The solution is

$$\vartheta = \theta_0 + (\theta_1 - \theta_0)\,\text{erfc}[\phi/2D_*^{1/2}] \qquad (130)$$

and

$$i = 2(D_*/\pi)^{1/2}(\theta_1 - \theta_0)t^{1/2} \qquad (131)$$

Theory of Infiltration

Matching Eq. (131) with Eq. (42), we have

$$D_* = \pi S^2/4(\theta_1 - \theta_0)^2 \tag{132}$$

The matching equation (132) is valid for all absorption and infiltration processes induced by a step-function change in θ (or Ψ) at the supply surface (since all such processes are indistinguishable from one-dimensional absorption in the limit as $t \to 0$).

Putting Eq. (132) in Eq. (130), we have

$$\vartheta(\theta) = \theta_0 + (\theta_1 - \theta_0) \operatorname{erfc} \frac{(\theta_1 - \theta_0)\phi(\theta)}{\pi^{1/2} S} \tag{133}$$

where $\phi(\theta)$ is the solution of Eq. (34) subject to conditions (35). Equation (133) defines the transformation between ϑ and θ which makes Eq. (130) *exact* for a medium with arbitrary $D(\theta)$. Applied solely to this problem, the linearization technique is exact, but yields no new information. It is useful only when we now apply it to more complicated problems.

2. Two- and Three-Dimensional Absorption

Linearizing Eqs. (46) and (47) gives

$$\frac{\partial \vartheta}{\partial t} = D_* r^{1-m} \frac{\partial}{\partial r}\left(r^{m-1} \frac{\partial \vartheta}{\partial r}\right) \tag{134}$$

$$\begin{aligned} t = 0, & \quad \vartheta = \theta_0, \quad r > r_0 \\ r \geq 0, & \quad \vartheta = \theta_1, \quad r = r_0 \end{aligned} \tag{135}$$

In the discussion that follows, we use the dimensionless variables

$$V = \frac{v_0 r_0}{D_*(\theta_1 - \theta_0)}$$

$$T = \frac{D_* t}{r_0^2} \tag{136}$$

$$I = \frac{i}{r_0(\theta_1 - \theta_0)}$$

It follows from Eq. (69) that

$$m = 2, \quad T = \pi t/t_{\text{geom}}; \quad m = 3, \quad T = \pi t/4 t_{\text{geom}} \tag{137}$$

For $m = 2$, the solution of Eq. (134) subject to conditions (135) [23, p. 335] is complicated. It suffices here to give the result for the absorption rate

$$V = \frac{4}{\pi^2} \int_0^\infty \frac{\exp(-w^2 T)\, dw}{w[J_0^2(w) + Y_0^2(w)]} \tag{138}$$

$\mathbf{J}_0$ and $\mathbf{Y}_0$ are the Bessel functions of the first and second kind of zero order. Numerical values of the integral are given by Jaeger and Clarke [112] and are graphed by Carslaw and Jaeger [23, p. 338]. For small T, the series expansion of Eq. (138)

$$V = (\pi T)^{-1/2} + \tfrac{1}{2} - \tfrac{1}{4}(T/\pi)^{1/2} + (T/8) - \cdots \qquad (139)$$

is useful. The asymptotic expansion (large T) is

$$V \sim 2\left[\frac{1}{\log(4T) - 2\gamma} - \frac{\gamma}{[\log(4T) - 2\gamma]^2} + \cdots\right] \qquad (140)$$

$\gamma = 0.57722 \cdots$ is Euler's constant.

For $m = 3$, the solution of Eq. (134) subject to conditions (135) [23, p. 247] is

$$\vartheta = \theta_0 + \frac{\theta_1 - \theta_0}{\rho}\,\text{erfc}\,\frac{\rho - 1}{2T^{1/2}} \qquad (141)$$

This yields the simple result [88]

$$V = (\pi T)^{-1/2} + 1 \qquad (142)$$

The leading terms of Eqs. (139) and (142) are identical; and the second term of Eq. (142) is exactly twice that of Eq. (139). The corresponding nonlinear results have the same features.

C. Infiltration Solutions

1. One-Dimensional Infiltration

The linearized form of Eqs. (81) and (82) is

$$\frac{\partial \vartheta}{\partial t} = D_* \frac{\partial^2 \vartheta}{\partial z^2} - k \frac{\partial \vartheta}{\partial z} \qquad (143)$$

$$\begin{aligned} t = 0, & \quad z > 0, \quad \vartheta = \theta_0 \\ t \geq 0, & \quad z = 0, \quad \vartheta = \theta_1 \end{aligned} \qquad (144)$$

We have fixed D_* already through Eq. (132). We now, similarly, fix k by matching values of $\lim_{t \to \infty} v_0$. We thus obtain the value

$$k = (K_1 - K_0)/(\theta_1 - \theta_0) \qquad (145)$$

so that k is identical with u, the velocity of the "profile at infinity" [Eq. (106)].

Theory of Infiltration

The solution of Eq. (143) subject to conditions (144) is [87]

$$\vartheta = \theta_0 + \frac{\theta_1 - \theta_0}{2}\left[\operatorname{erfc}\frac{z - kt}{2(D_* t)^{1/2}} + \exp\left(\frac{kz}{D_*}\right)\operatorname{erfc}\frac{z + kt}{2(D_* t)^{1/2}}\right] \quad (146)$$

As $t \to \infty$ in Eq. (146), the solution (at other than small z) approaches the form

$$z(\vartheta, t) = kt + t^{1/2}\omega(\vartheta) \quad (147)$$

The centroid of the profile tends to move downward with constant velocity $k(=u)$, and the profile tends to maintain its relative shape about the centroid, with the gradients of ϑ decreasing as $t^{-1/2}$. It is illuminating to compare Eq. (147) with the corresponding nonlinear result, Eq. (103). The constant absolute shape of the "profile at infinity," Eq. (103), is a demonstrable consequence of nonlinearity in $K(\theta)$.

We express the dynamics of one-dimensional infiltration in terms of the following dimensionless variables:

$$V_1 = \frac{v_0 - K_1}{K_1 - K_0}, \quad T_1 = \frac{k^2 t}{4D_*} = \frac{(K_1 - K_0)^2 t}{\pi S^2} = \frac{t}{\pi t_{\text{grav}}}$$

$$I_1 = \int_0^{T_1} V \, dT' = \frac{K_1 - K_0}{\pi S^2}(i - K_1 t) \quad (148)$$

Then, under the simplifying, and generally justified, assumption that K_0/K_1 is negligibly small, it follows from Eq. (146) that

$$V_1 = \tfrac{1}{2}[(\pi T_1)^{-1/2}\exp(-T_1) - \operatorname{erfc} T_1^{1/2}]$$

$$I_1 = \tfrac{1}{2}[\pi^{-1/2} T_1^{1/2}\exp(-T_1) + \tfrac{1}{2}\operatorname{erf} T_1^{1/2} - T_1 \operatorname{erfc} T_1^{1/2}] \quad (149)$$

The series expansions of Eq. (149), useful for small T_1, are

$$V_1 = \tfrac{1}{2}\left[(\pi T_1)^{-1/2} - 1 + \pi^{-1/2}\left\{T_1^{1/2} - \frac{T_1^{3/2}}{6} + \frac{T_1^{5/2}}{30} - \frac{T_1^{7/2}}{168} + \cdots\right\}\right]$$

$$I_1 = \pi^{-1/2}T_1^{1/2} - \frac{T_1}{2} + \pi^{-1/2}\left\{\frac{T_1^{3/2}}{3} - \frac{T_1^{5/2}}{30} + \frac{T_1^{7/2}}{210} - \frac{T_1^{9/2}}{1512} + \cdots\right\} \quad (150)$$

The asymptotic expansions for large T_1 are

$$V_1 \sim \tfrac{1}{4}\pi^{-1/2}T_1^{-3/2}\exp(-T_1)\left[1 - \frac{3}{4T_1} + \frac{15}{4T_1^2} - \cdots\right]$$

$$I_1 \sim \tfrac{1}{4}\left[1 - \pi^{-1/2}T_1^{-3/2}\exp(-T_1)\left\{1 - \frac{3}{T_1} + \cdots\right\}\right] \quad (151)$$

2. Two- and Three-Dimensional Infiltration

The linearized formalism describing infiltration from the surfaces of a cylindrical ($m = 2$) or spherical ($m = 3$) cavity consists of Eq. (125) subject to the conditions (135).

The substitution [87]

$$\vartheta_* = (\vartheta - \theta_0) \exp\left[-\frac{kz}{2D_*} + \frac{k^2 t}{4D_*}\right] \tag{152}$$

reduces Eq. (125) to

$$\partial \vartheta_*/\partial t = D_* \nabla^2 \vartheta_* \tag{153}$$

Now, *so long as* $kr_0/2D_*$ *is small*, we may replace conditions (135) by

$$\begin{aligned} t = 0, &\quad \vartheta_* = 0, &\quad r > r_0 \\ t \geq 0, &\quad \vartheta_* = (\theta_1 - \theta_0) \exp\frac{k^2 t}{4D_*}, &\quad r = r_0 \end{aligned} \tag{154}$$

We thus reduce these problems to ones of diffusion in cylindrical and spherically symmetrical systems with a boundary concentration increasing exponentially with time.

We discuss the solution in terms of T_1 and the new dimensionless quantities

$$R_* = \frac{kr}{4D_*}, \quad R_0 = \frac{kr_0}{4D_*}, \quad \sin\psi = \frac{z}{r}, \quad V_2 = \frac{\bar{v}_0}{k(\theta_1 - \theta_0)} \tag{155}$$

$\bar{v}_0$ is the average value of v_0 over the supply surface. The solutions we present appear to be useful for $R_0 \leq 0.25$, a range that includes many practical situations. The expressions for V_2 given below, and later, are simplified through the (usually justifiable) assumption that K_0/K_1 is negligibly small.

For $m = 2$, the solution of Eq. (153) subject to conditions (154) involves some complicated integrals. In the limit as $T_1 \to \infty$, it is, for R_0 small,

$$\vartheta = \theta_0 + (\theta_1 - \theta_0) \exp[2(R_* - R_0) \sin\psi] \frac{K_0(2R_*)}{K_0(2R_0)} \tag{156}$$

With suitable changes of symbolism, Eq. (156) for R_0 small is of the same form as Eq. (117), as it should be. Equation (156) gives, for R_0 small, the final infiltration rate

$$V_2(\infty) = [4R_0 K_0(2R_0)]^{-1} \tag{157}$$

By averaging v_0 over only the *lower semicircle*, we may use Eq. (156) to obtain a first estimate (demonstrably an *underestimate*) of the final infiltration rate from a semicircular furrow. We thus obtain

$$V_2(\infty) = \pi^{-1} + [4R_0 K_0(2R_0)]^{-1} \tag{158}$$

Theory of Infiltration

For $m = 3$, the solution of Eq. (153) subject to conditions (154) reduces, for R_0 small, to

$$\vartheta = \theta_0 + \frac{(\theta_1 - \theta_0)R_0}{2R_*} \exp[2(R_* - R_0)\sin\psi]$$

$$\times \left[\exp\{-2(R_* - R_0)\} \operatorname{erfc} \frac{R_* - R_0 - T_1}{T_1^{1/2}} \right.$$

$$\left. + \exp\{2(R_* - R_0)\} \operatorname{erfc} \frac{R_* - R_0 + T_1}{T_1^{1/2}} \right] \quad (159)$$

In the limit as $T_1 \to \infty$, Eq. (159) becomes

$$\vartheta = \theta_0 + \frac{(\theta_1 - \theta_0)R_0}{R_*} \exp[-2(R_* - R_0)(1 - \sin\psi)] \quad (160)$$

With the symbolism appropriately adjusted, Eq. (160) for R_0 small is of the same form as Eq. (122), as it should be. Equation (159) yields the infiltration rate

$$V_2 = \tfrac{1}{2}[(\pi T_1)^{-1/2} \exp(-T_1) + (2R_0)^{-1} + \operatorname{erf} T_1^{1/2}] \quad (161)$$

and

$$V_2(\infty) = \tfrac{1}{2} + (4R_0)^{-1} \quad (162)$$

Averaging v_0 over the lower hemisphere, we obtain the following (lower) estimates for infiltration from a hemispherical basin:

$$V_2 = \tfrac{1}{2}[(\pi T_1)^{-1/2} \exp(-T_1) + \tfrac{1}{2} + (2R_0)^{-1} + \operatorname{erf} T_1^{1/2}] \quad (163)$$

$$V_2(\infty) = \tfrac{3}{4} + (4R_0)^{-1} \quad (164)$$

D. Large-Time Solutions in Two and Three Dimensions

For two- and three-dimensional absorption problems that approach a steady state, the linearized Eq. (125) reduces, as $t \to \infty$, to

$$\nabla^2 \vartheta = 0 \quad (165)$$

which is identical in form to Eq. (74). Linearization solutions for this class of problem are thus exact in the limit as $t \to \infty$, when the one-to-one relation between ϑ and θ is

$$\vartheta = \theta_0 + (\theta_1 - \theta_0) \int_{\theta_0}^{\theta} D\, d\theta \bigg/ \int_{\theta_0}^{\theta_1} D\, d\theta \quad (166)$$

For infiltration problems of this type, Eq. (125) becomes, in the limit as $t \to \infty$,

$$\nabla^2 \vartheta = \frac{k}{D_*} \frac{\partial \vartheta}{\partial z} \quad (167)$$

This is identical in form to Eq. (114), so long as

$$\alpha = k/D_* \tag{168}$$

Linearization solutions of these problems are exact in the limit as $t \to \infty$ for media and soils satisfying Eq. (113). In such cases, also, Eq. (166) is the one-to-one relation between ϑ and θ in the limit as $t \to \infty$.

These results suggest that linearization should be a fairly reliable technique when applied to two- and three-dimensional problems at large t.

VII. Delta-Function Solutions

A. Principles of Method

Green and Ampt [113] were the first to propose a model of water movement during absorption and infiltration from free water (or water under positive hydrostatic pressure) which involved a definite sharp wet front, behind which θ and K are supposed to take the constant values $\bar{\theta}$ and $\bar{K}$, and ahead of which θ retains its initial value θ_0. The influence of capillarity (and of any constant nonzero hydrostatic pressure of the supplied water) is modeled by a constant difference in moisture potential between the supply surface and the wetting front, for which we shall use the symbol C with the convention that C is positive.

For systems involving only one space coordinate (one-dimensional, and cylindrically and spherically symmetrical systems) the resulting formalism replaces the basic partial differential equation considered in Sects. II–V by an ordinary differential equation in i and t. The parameters $\bar{\theta}$, $\bar{K}$, C enter equations for absorption only in the combination $\bar{K}C(\bar{\theta} - \theta_0)$; for infiltration, $\bar{K}$ also enters separately.

Lambe [114] recognized that, in general, $\bar{\theta} < \theta_1$ and $\bar{K} < K_1$. Philip [115] put the model on a somewhat firmer basis by showing that, in one-dimensional systems, the formalism gave $i(t)$ exactly, provided the moisture profiles, whatever their shape, preserved similarity. He showed that, under these conditions, the appropriate values of $\bar{\theta}$ and $\bar{K}$ are

$$\bar{\theta} = \int_0^{\eta_{wf}} \theta \, d\eta / \eta_{wf}, \qquad \bar{K} = v_0 \eta_{wf} \bigg/ \int_0^{\eta_{wf}} v/K \, d\eta \tag{169}$$

η is the space coordinate, with $\eta = 0$ at the supply surface and $\eta = \eta_{wf}$ at the wet front. v is the flow velocity.

Subsequently, Philip [84] related the model to the "diffusion" description of absorption and infiltration described in Sects. II–V by showing that the model is exact for a medium in which the moisture diffusivity $D(\theta)$ is defined as follows:

$$D = S^2(\theta_1 - \theta_0)\delta(\theta_1 - \theta)/2 \tag{170}$$

Theory of Infiltration

δ is the Dirac delta function. Equation (170) signifies that D is very large in a very small θ region near $\theta = \theta_1$, is negligibly small in the rest of the θ range, and is such that $\int_{\theta_0}^{\theta_1} D \, d\theta = S^2/2$. The model does not require any particular form of $K(\theta)$, although it is physically evident that media tending to satisfy Eq. (170) must also have a very sharp peak in K at $\theta = \theta_1$. The gradual steepening of moisture profiles as the peak in D becomes sharper is shown by Philip [48]. The delta-function medium embodies, essentially, the limiting result of this process. With the model interpreted thus, $\bar{\theta} = \theta_1$.

The similarity condition for the validity of the model in one-dimensional systems is rather less restrictive than Eq. (170); but the model can be given a precise meaning in two- and three-dimensional systems [for which moisture profiles in media not satisfying Eq. (170) do not satisfy similarity even approximately] only through Eq. (170).

So far as soils of arbitrary hydrological characteristics are concerned, delta-function solutions give no information about details of the moisture profiles, but they do offer estimates of integral properties such as $v_0(t)$ and $i(t)$. Just as for the linearized model, we may use this model also as a means of extrapolation and interpolation. We fix $\bar{K}C(\theta_1 - \theta_0)$ by matching known integral properties of the one-dimensional absorption solution [86, 111]; and we assign $\bar{K}$ in problems involving gravity by matching known behavior as $t \to \infty$ [74, 111].

B. Absorption Solutions

1. One-Dimensional Absorption

The delta-function equation for one-dimensional absorption is

$$v_0 = \frac{di}{dt} = \frac{\bar{K}C(\theta_1 - \theta_0)}{i} \qquad (171)$$

subject to the condition

$$t = 0, \quad i = 0 \qquad (172)$$

The solution is

$$i = [2\bar{K}C(\theta_1 - \theta_0)]^{1/2} t^{1/2} \qquad (173)$$

Matching this with Eq. (42), we have

$$\bar{K}C(\theta_1 - \theta_0) = S^2/2 \qquad (174)$$

Use of Eq. (174) for media of arbitrary properties ensures that delta-function solutions yield correct integral behavior, in the limit as $t \to 0$, for *all* absorption and infiltration processes induced by a step-function change in θ (or Ψ) at the supply surface.

2. Two- and Three-Dimensional Absorption

Substituting Eqs. (174) and (132) in Eq. (136), we see that we may use the dimensionless variables V, T, I to present delta-function solutions for the two- and three-dimensional absorption processes for which we have already considered the exact and linearized solutions.

For the two-dimensional case, the delta-function model yields the differential equation

$$V = dI/dT = (4/\pi)[\log(1 + 2I)]^{-1} \tag{175}$$

subject to the condition

$$T = 0, \quad I = 0 \tag{176}$$

Integration gives

$$T = (\pi/8)[(1 + 2I)\log(1 + 2I) - 2I] \tag{177}$$

Equations (175) and (177) may then be combined to give the relation between V and T:

$$T = \frac{\pi}{8}\left[\left(\frac{4}{\pi V} - 1\right)e^{4/\pi V} + 1\right] \tag{178}$$

In three dimensions, the differential equation is

$$V = dI/dT = (2/\pi)[1 - (1 + 3I)^{-1/3}]^{-1} \tag{179}$$

subject to condition (176). Integration yields

$$T = (\pi/2)[I - \tfrac{1}{2}\{(1 + 3I)^{2/3} - 1\}] \tag{180}$$

Combining Eqs. (179) and (180) gives the relation between V and T:

$$T = \frac{\pi}{12}\left[2\left(1 - \frac{2}{\pi V}\right)^{-3} - 3\left(1 - \frac{2}{\pi V}\right)^{-2} + 1\right] \tag{181}$$

C. Infiltration Solutions

1. One-Dimensional Infiltration

The delta-function equation for one-dimensional infiltration is

$$v_0 = \frac{di}{dt} = \bar{K}\left(1 + \frac{C(\theta_1 - \theta_0)}{i}\right) \tag{182}$$

According to Eq. (182),

$$v_0(\infty) = \bar{K} \tag{183}$$

so that, matching this with Eq. (109), we find

$$\overline{K} = K_1 \tag{184}$$

We therefore use matching equation (184), as well as Eq. (174), in the delta-function model of infiltration. The model thus agrees with integral properties of the exact solution both as $t \to 0$ and as $t \to \infty$.

Substituting Eqs. (174) and (184) (and using the simplifying, but generally justified, assumption that K_0/K_1 is negligibly small) in Eq. (148), we find that we may use the dimensionless variables V_1, T_1, I_1, to discuss the delta-function solution for one-dimensional infiltration. In terms of these variables, Eq. (182) becomes

$$V_1 = \frac{dI_1}{dT_1} = \frac{1}{2\pi(I_1 + T_1)} \tag{185}$$

subject to condition

$$T_1 = 0, \quad I_1 = 0 \tag{186}$$

The solution is

$$T_1 = (2\pi)^{-1}[\exp(2\pi I_1) - 2\pi I_1 - 1] \tag{187}$$

Combining Eqs. (185) and (187) gives $T_1(V_1)$:

$$T_1 = (2\pi)^{-1}[V_1^{-1} - \log(1 + V_1^{-1})] \tag{188}$$

2. Two- and Three-Dimensional Infiltration

For two- and three-dimensional infiltration, the delta-function model does not lead to an ordinary equation, but to a complicated formulation in which the wet front is a moving "free surface." The delta-function model appears, in any case, to be of doubtful accuracy for two- and three-dimensional processes, particularly for large t, since it fails to embody two important properties that may be established from exact solutions and from general physical arguments (and, somewhat more circumstantially, from linearized solutions). These are (i) that, for such processes, moisture profiles fail to preserve similarity and become progressively less steep as t increases; (ii) that, in systems with a steady state, the region of significant increase of moisture content (i.e., the region behind any "wet front") *remains definitely finite in both horizontal and vertical directions as $t \to \infty$*.

Vedernikov and Averjanov [cf. 107] have used delta-function models to study two-dimensional steady infiltration, but the results may be of limited value because of these physical shortcomings of the model.

VIII. Comparison Technique for Estimating Integral Properties of Solutions

A. Principles of Method

Figure 12 compares schematically the forms of $D(\theta)$ dependence for linearized and delta-function media with that for a typical real soil or porous

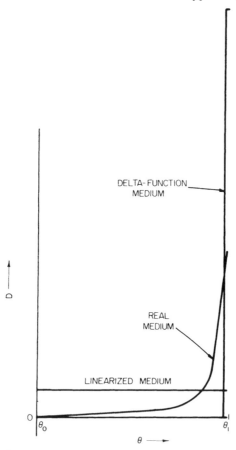

FIG. 12. Schematic figure comparing mode of variation of D with θ for (i) the delta-function medium, (ii) the linearized medium, and (iii) a typical real soil or porous medium.

medium. The latter, with a marked decrease of D with θ, is evidently intermediate in character between that of the two media treated in the preceding two sections.

We therefore expect integral forms of exact solutions for absorption processes [which depend only on $D(\theta)$] in real media to be intermediate in

Theory of Infiltration

character between those given by the (appropriately matched) linearized and delta-function solutions for the same process [86, 91–93, 111]. The linearized and delta-function solutions thus form *an envelope of possible behavior*.

For processes involving gravity, however, solutions are influenced, not only by $D(\theta)$, but also by the relative behavior of $D(\theta)$ and $dK/d\theta$. Both linearized and delta-function models are consistent with $D^{-1} dK/d\theta$ being constant, whereas, for media of arbitrary characteristics, this is not necessarily so. It follows that we cannot extend the envelope concept to such processes: we limit ourselves to the expectation that the difference between linear and delta-function solutions will indicate the order of magnitude of the deviation of either solution from the exact result for any real medium.

These notions have been confirmed by comparing exact nonlinear solutions for particular real media with linear and delta-function solutions. Some details are given below. The various differences between solutions are small, so that we have at our disposal a useful means of estimating integral behavior of media of arbitrary characteristics.

We are able to employ this approach at three different levels:

(i) For absorption phenomena with both linearized and delta-function solutions available, a close estimate can be made of the integral behavior of any real medium (and the possible error of the estimate is known).

(ii) For infiltration phenomena with both linearized and delta-function solutions available, either gives an estimate of the integral behavior of any real medium. An estimate of the order of magnitude of the error is provided by the difference between the two solutions.

(iii) For phenomena where only one form of solution (usually the linearized solution) is available, that solution offers a fairly reliable estimate of integral behavior.

B. Absorption Solutions

1. Two-Dimensional Absorption

Figure 13 compares $V(T)$ for two-dimensional absorption according to the linearized solution, the delta-function solution, and the nonlinear solution (68) (first two terms) calculated for our illustrative example of Yolo light clay. The three solutions evidently agree very closely. Consonant with the concepts developed in Sect. VIII.A, the linearized and delta-function solutions indicate upper and lower bounds to the behavior of real media.

We compare the leading terms of the various $V(T)$ expansions appropriate for small T:

linearized solution, Eq. (139), $\quad V = (\pi T)^{-1/2} + \tfrac{1}{2} - \cdots$
delta-function solution, Eq. (178), $\quad V = (\pi T)^{-1/2} + (4/3\pi) - \cdots \qquad (189)$
nonlinear example, Eq. (68), $\quad V = (\pi T)^{-1/2} + 0.44 - \cdots$
$4/3\pi = 0.424 \ldots$

The comparison in the limit of T large is as follows:

linearized solution, Eq. (140), $\quad V \sim 2(\log T)^{-1}$
delta-function solution, Eq. (178), $\quad V \sim (4/\pi)(\log T)^{-1} \qquad (190)$

The results of Eqs. (190) establish that, for media of arbitrary characteristics, the asymptotic behavior of V at large T is as $(\log T)^{-1}$. For real media, the numerical factor will be between the limits $4/\pi$ and 2.

2. Three-Dimensional Absorption

Figure 14 shows the similar comparison of $V(T)$ calculations for three-dimensional absorption. In this case, the comparison includes also the exact nonlinear solution for $V(\infty)$ [Eq. (71)] computed for our illustrative example. Again the solutions are in good agreement, and are consistent with the notion of the linearized and delta-function solutions as upper and lower envelopes of the possible behavior of real media.

The leading terms of the various expansions for $V(T)$ at small T are compared in Eqs. (191):

linearized solution, Eq. (142), $\quad V = (\pi T)^{-1/2} + 1$
delta-function solution, Eq. (181), $\quad V = (\pi T)^{-1/2} + (8/3\pi) - \cdots \qquad (191)$
nonlinear example, Eq. (68), $\quad V = (\pi T)^{-1/2} + 0.87 - \cdots$
$8/3\pi = 0.848 \ldots$

In the limit of T large, the comparison is

linearized solution, Eq. (142), $\quad V = 1 + (\pi T)^{-1/2}$
delta-function solution, Eq. (181), $\quad V \sim (2/\pi) + (4\pi^2/3)^{1/3} T^{-1/3} \qquad (192)$
nonlinear example, Eq. (71), $\quad V \sim 0.710$
$2/\pi \approx 0.636 \ldots$

FIG. 13. The dynamics of two-dimensional absorption in the dimensionless form $V(T)$ [86]. Solid curve: the linearized solution Eq. (139). Crosses: two-term nonlinear solution for Yolo light clay Eq. (68). Dotted curve: delta-function solution Eq. (178).

FIG. 14. The dynamics of three-dimensional absorption in the dimensionless form $V(T)$ [86]. Solid curve: the linearized solution Eq. (142). Crosses: two-term nonlinear solution for Yolo light clay Eq. (68). Dotted curve: delta-function solution Eq. (181). Values of $V(\infty)$ are shown for all three solutions. Nonlinear $V(\infty)$ solution calculated from Eq. (71).

Theory of Infiltration

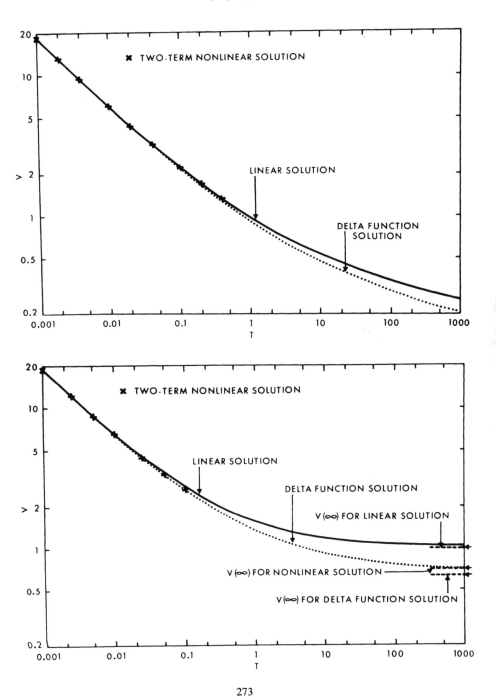

C. Infiltration Solutions

The various solutions for one-dimensional capillary rise [76] and infiltration also agree closely, although the "envelope" concept cannot be extended to these processes involving gravity. Figure 15 compares various $V_1(T_1)$ calculations for one-dimensional infiltration. The nonlinear solution (97) for our illustrative example was calculated from the first four terms.

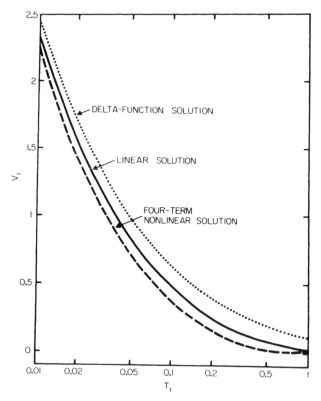

Fig. 15. The dynamics of one-dimensional infiltration in the dimensionless form $V_1(T_1)$ [111]. Solid curve: the linearized solution Eq. (150). Broken curve: nonlinear four-term solution for Yolo light clay Eq. (97). Dotted curve: delta-function solution Eq. (188).

The leading terms of the various $V_1(T_1)$ expansions for small T_1 are

linearized solution, Eq. (150), $\quad V_1 = \tfrac{1}{2}(\pi T_1)^{-1/2} - \tfrac{1}{2} + \cdots$

delta-function solution, Eq. (188), $\quad V_1 = \tfrac{1}{2}(\pi T_1)^{-1/2} - \tfrac{1}{3} + \cdots$ [84] (193)

nonlinear example, Eq. (97), $\quad V_1 = \tfrac{1}{2}(\pi T_1)^{-1/2} - 0.62 + \cdots$

For $T_1 \to \infty$, the asymptotic results are

linearized solution, Eq. (151), $V_1 \sim \tfrac{1}{4}\pi^{-1/2}T_1^{-3/2}\exp(-T_1)$

delta-function solution, Eq. (188), $V_1 \sim (2\pi T_1)^{-1}$ (194)

nonlinear example, Eq. (109), $V_1 \sim 0$

The nonlinear analysis of Sect. V.B is unable to give details of the approach of V_1 to 0 (i.e., of v_0 to K_1). Since the linearized model does not embody the "profile at infinity" behavior (Sect. VI.C), whereas the delta-function model does, the approach to the quasi-stationary state with $V_1 = 0$ (i.e., $v_0 = K_1$) for real media may resemble the T_1^{-1} behavior of the delta-function model, rather than the more rapid approach of the linearized model.

On the basis of the analysis of Sect. V.A, Philip [33, 84] suggested the two-parameter infiltration equations

$$i = St^{1/2} + At$$
$$v_0 = \tfrac{1}{2}St^{-1/2} + A \quad (195)$$

for use in applied hydrology when t is not too large. Clearly the relation

$$v_0 = K_1 \qquad (196) \text{ [also Eq. (109)]}$$

holds in the limit as $t \to \infty$.

It has been suggested recently [19] that this implies that A must be K_1. However, *the coefficients of corresponding terms in a series expansion with limited radius of convergence and in an asymptotic (large argument) expansion of the same function are not necessarily equal.* We have, in fact, from Eqs. (193) that $A = K_1/2$, $2K_1/3$, and $0.38K_1$ for the linearized model, the delta-function model, and for our nonlinear example.

Only linearized solutions are presently available for transient two- and three-dimensional infiltration. The comparison technique is, so far, useful for these problems only in the sense that it indicates that the process of linearization, combined with matching, yields results of useful accuracy.

D. Aggregated Media

A comparison technique was applied recently to the problem of absorption into aggregated media. The study used linearized and wet-front (modified delta-function) models of water movement in the macroporosity [35, 36].

IX. Exact Solutions of More Complicated Problems

Sections IV–VIII have dealt exclusively with flow processes subject to concentration boundary conditions in homogeneous systems. Before proceeding in the final sections to a discussion of the physical significance of the solutions of Sects. IV–VIII, we examine briefly the more difficult problems of flow processes subject to flux conditions and of flow in heterogeneous media. For these problems, we identify special subclasses for which exact solutions may be found by methods similar to those considered earlier in this article.

A. Processes Subject to Flux Boundary Conditions

The additional mathematical difficulty of transient flow processes subject to flux boundary conditions arises, essentially, because the θ range varies with t (so that θ cannot readily be made an independent variable). Purely numerical techniques, made less laborious by the use of high-speed computers, may be the only means of analyzing many such problems [116–118]. However, several subclasses of exact solutions are available. We discuss these below.

1. One-Dimensional Absorption

One-dimensional absorption with constant surface flux into a semi-infinite homogeneous medium of uniform initial moisture content θ_0 is described by Eq. (15) subject to conditions (197):

$$t = 0, \quad x > 0, \quad \theta = \theta_0$$
$$t \geq 0, \quad x = 0, \quad D\, \partial\theta/\partial x = -U_0 \tag{197}$$

When $D(\theta)$ may be represented by the form

$$D(\theta) = D_s \left(\frac{\theta - \theta_0}{\theta_s - \theta_0} \right)^n \quad \text{for} \quad n > -1 \tag{198}$$

there exists a similarity solution of Eq. (15) subject to conditions (197). Here θ_s is the value of θ corresponding to $\Psi = 0$, and D_s denotes $D(\theta_s)$. In general, Eq. (198) offers a reasonably accurate representation, but an important limitation is that it requires $D(\theta_0) = 0$ for $n > 0$. For this reason, Eq. (198) is best fitted to soils and media that are initially rather dry. The appropriate range of n seems to be from about 3 to 7, although smaller values would apply if the method were used with large values of θ_0.

The dimensionless quantities

$$\Theta = \frac{\theta - \theta_0}{\theta_s - \theta_0}, \quad \tau_1 = \frac{U_0^2 t}{D_s(\theta_s - \theta_0)^2}, \quad \chi = \frac{U_0 x}{D_s(\theta_s - \theta_0)} \tag{199}$$

reduce Eq. (15) subject to conditions (197), with D defined by Eq. (198), to the form

$$\frac{\partial \Theta}{\partial \tau_1} = \frac{\partial}{\partial \chi} \left(\Theta^n \frac{\partial \Theta}{\partial \chi} \right) \tag{200}$$

subject to the conditions

$$\tau_1 = 0, \quad \chi > 0, \quad \Theta = 0$$
$$\tau_1 \geq 0, \quad \chi = 0, \quad \Theta^n \partial\Theta/\partial\chi = -1 \qquad (201)$$

The similarity substitutions

$$\Theta_* = \Theta\tau^{-1/(n+2)}, \qquad \chi_* = \chi\tau^{-(n+1)/(n+2)} \qquad (202)$$

then enable the reduction of Eq. (200) subject to conditions (201) to an ordinary nonlinear differential equation in Θ_* and χ_*, subject to conditions expressed solely in terms of these quantities. Numerical methods yield the solution in the form $\Theta_*(\chi_*)$, i.e.,

$$\theta(x, t) = \theta_0 + t^{1/(n+2)} F(xt^{-(n+1)/(n+2)}) \qquad (203)$$

Clearly, this solution is valid only so long as $\theta(0) \leq \theta_s$, i.e., for

$$t \leq \left(\frac{\theta_s - \theta_0}{F(0)}\right)^{n+2} \qquad (204)$$

For larger t, the constant flux boundary condition can be maintained only by continuous increase of the hydrostatic pressure of the water supplied at $x = 0$.

2. One-Dimensional Infiltration

The preceding result is relevant also to the early stages of one-dimensional infiltration with a constant flux boundary condition. When $U_0 \leq K_s$, we may supplement this result with an asymptotic large-time solution of the same character as that described in Sect. V.B, except that we now write U_0 for K_1 and θ_1 is replaced by θ_U, where $K(\theta_U) = U_0$.

For $U_0 > K_s$, the constant flux condition cannot be maintained indefinitely unless water is supplied to the surface under a continuously increasing hydrostatic pressure. The required rate of increase of pressure is, asymptotically,

$$\frac{(U_0 - K_0)(U_0 - K_s)}{K_s(\theta_s - \theta_0)} \quad \text{cm-sec}^{-1}$$

3. Two- and Three-Dimensional Problems

The techniques of Sects. IV.H and V.E for large-time (steady) absorption and infiltration in two and three dimensions apply for a flux condition at the supply surface as well as for a concentration condition. These techniques are appropriate only if U_0 is nowhere greater than the steady flux density on the supply surface when it is maintained at moisture content θ_s.

B. Absorption and Infiltration in Heterogeneous Media

1. The Mathematical Problem

For unsteady unsaturated water movement in the simplest heterogeneous system, namely, a horizontal one-dimensional one, Eq. (9) reduces to

$$\frac{\partial \theta}{\partial t} = \frac{\partial}{\partial x}\left[K(\theta, x)\frac{\partial}{\partial x}\Psi(\theta, x)\right] \tag{205}$$

Equation (205) implies that, for each point x, there is a definite functional dependence of K and Ψ on θ. Superficial consideration might suggest that, in analogy with the diffusion formalism for homogeneous media, Eq. (205) might reduce to a diffusion equation with the diffusivity a function of both θ and x. In fact, Eq. (205) reduces to the complicated and generally intractable form

$$\frac{\partial \theta}{\partial t} = \frac{\partial}{\partial x}\left[D(\theta, x)\frac{\partial \theta}{\partial x}\right] + E(\theta, x)\frac{\partial \theta}{\partial x} + F(\theta, x) \tag{206}$$

The difficulty of the problem is aggravated by the fact that there is little hope of receiving a realistic and useful picture of the properties of equations such as Eq. (205), so long as $K(\theta, x)$ and $\Psi(\theta, x)$ are free to vary totally independently of each other. The computer use of numerical methods (e.g., Hanks and Bowers [119]) is clearly helpful, but offers no panacea. The concept of scale-heterogeneity, referred to in Sect. IX.B.3, offers means of reducing the problem to more manageable proportions by using a physically based connection between the x variation of K and Ψ.

2. Delta-Function Approach

Childs [101] has recently carried further an early unpublished attempt by the present author to study one-dimensional infiltration in heterogeneous systems by a delta-function model. The model appears to be quite unsuited to profiles where the soil texture becomes coarser in depth. In this case, the lower parts of the profile will tend to be permanently at relatively large negative Ψ; moisture profiles in that region will tend to be gradual, unlike the delta-function model as used by Childs. The model, superficially, appears better fitted to the opposite case where coarse-textured soil overlies finer; but even here the model may lead to such paradoxes as an infiltration rate that increases when the putative wet front encounters the finer-textured material.

Further to these quite serious shortcomings, we observe that $\overline{K}$ and C must reflect in a mutually consistent way the heterogeneity of the medium. Manipulations in which these parameters vary with depth quite independently of each other [101] must, in this author's opinion, be treated with great caution.

3. Exact Solutions for Scale-Heterogeneous Media

Philip [120] has considered a limited class of heterogeneity called *scale-heterogeneity*. A scale-heterogeneous medium is one in which the internal geometry is everywhere geometrically similar, but in which the characteristic internal length λ is free to vary spatially. The properties of the medium are then defined by λ, and the reduced conductivity and potential functions $K_{**}(\theta)$ and $\Psi_{**}(\theta)$. In a one-dimensional system, for example,

$$K(\theta, x) = [\lambda(x)]^2 \cdot K_{**}(\theta)$$
$$\Psi(\theta, x) = \Psi_{**}(\theta)/\lambda(x) \tag{207}$$

Equations (207) depend on the assumption that the flow process is essentially viscous flow of water of normal viscosity, and that Ψ is capillary in nature. See Miller and Miller [65] for a discussion of these concepts, and Klute and Wilkinson [121, 122] for experimental work verifying them.

Philip showed that for the special functional forms

$$K_{**} \propto \Psi_{**}^{-2}$$
$$\Psi_{**} \propto e^{-\beta\theta} \tag{208}$$

Eq. (207) reduces Eq. (206) to a nonlinear diffusion equation in $\log(-\Psi)$. Eq. (208) at least embodies the general experience that K decreases rapidly as $|\Psi|$ increases; and it offers a range of modes of variation of Ψ with θ, the representation being exact where the *pF curve* is linear. An important limitation is that these relations cannot apply satisfactorily when Ψ exceeds the air-entry value anywhere in the medium.

The method of Sect. IV.A yields the solution of the problem of one-dimensional absorption into scale-heterogeneous media satisfying Eq. (208). The solution is obtained in terms of Ψ; the further calculation of the solution for θ is a simple matter. The approach has been extended to one-dimensional infiltration. An interesting aspect of this study is that the solution for Ψ is independent of the mode of variation of λ. This determines only the conversion of the solution for Ψ to the solution for θ.

X. Physics of One-Dimensional Infiltration

We return here to the implications of the exact solutions for one-dimensional infiltration into a homogeneous soil of uniform initial moisture content. These solutions constitute a striking example of interaction between capillary and gravity. As we have seen, capillary potential gradients are dominant at small t; and the final quasi-stationary state is an expression of dynamic equilibrium between capillary and gravitational influences, brought about by the nonlinearity of $K(\theta)$. The existence of relatively steep moisture gradients ("wetting zone" and "wet front") are manifestations of the nonlinearity of the flow equation: specifically, they arise whenever D decreases strongly with θ.

A. Moisture Profile

We have presented in Fig. 9 the sequence of moisture profiles calculated for our illustrative example. The salient features indicated by the solutions are as follows:

(i) For small t, the moisture profile maintains a nearly constant *relative* shape, z ordinates for fixed θ increasing at $t^{1/2}$. This is, of course, the phase when capillarity dominates, and the behavior is indistinguishable from that for one-dimensional absorption.

(ii) At intermediate t, the z ordinates gradually depart from their proportionality to $t^{1/2}$, and the profile shape changes slowly. The moisture gradient near $\theta = \theta_1$ becomes flatter, but it becomes *relatively* steeper over the rest of the θ range. The profile as a whole (except at $\theta = \theta_1$) moves down at a velocity that decreases continuously.

(iii) Finally, at very large t, these tendencies reach their full expression. The profile of nearly constant *absolute* shape attached to a lengthening near-straight close to $\theta = \theta_1$ moves down at a velocity indistinguishable from $(K_1 - K_0)/(\theta_1 - \theta_0)$.

B. Comparison with Experiment

The exact solutions for one-dimensional infiltration are in excellent agreement with the relevant experimental data [102, 123–130]. For work earlier than about 1957, the agreement cannot be confirmed in full detail, since, before then, the experimental soils were not, in general, characterized by their $D(\theta)$ and $K(\theta)$ functions. The later work tests the theory more rigorously.

The observed moisture profiles during infiltration are found to exhibit four zones, arranged as follows from the surface $z = 0$ downward:

(i) The saturated zone: a surface zone of presumed saturation extending (in experiments on infiltration from shallow ponded water) to a depth of the order of 1 cm.

(ii) The transmission zone: an upper region in which θ changes quite slowly with both z and t, which lengthens as infiltration proceeds.

(iii) The wetting zone: a region of fairly rapid change of θ with respect to both z and t.

(iv) The wet front: a region of very steep moisture gradient which represents the visible limit of moisture penetration.

Theory of Infiltration

These various zones are predicted correctly by the exact solutions. Philip [48] has given a detailed discussion of how the shape of the $D(\theta)$ function influences the shape of the moisture profiles, and so leads to the observed characteristics. The "saturated zone" is the region in which $\Psi \geq \Psi_e$, and tends to become more significant as the water depth over the soil (the hydrostatic pressure at the supply surface) increases [22, 81].

Bodman and Colman [123, 124], and some later experimenters, observed a "transition zone" between the saturated zone and the transmission zone. This is a zone, up to a few centimeters deep, through which the moisture content decreases rapidly from "saturation" to a value characteristic of the transmission zone. It is noteworthy that this zone has been observed by some investigators, but not by others. An explanation in terms of air-entrapment has been offered, but it is possible that the transition zone is simply a

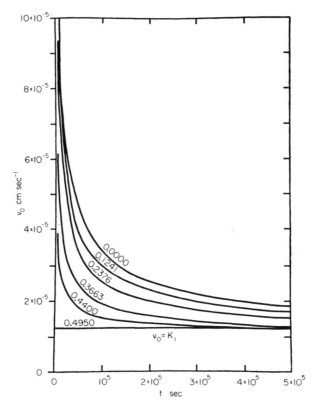

FIG. 16. Computed influence of the initial moisture content, θ_0, on one-dimensional infiltration rate v_0 for the soil of Figs. 1–3 with $\theta_1 = 0.4950$ [85]. Numerals on each curve denote values of θ_0. Note that curve for $\theta_0 = \theta_1$ is the line $v_0 = K_1$.

C. Effect of Initial Moisture Content

Figure 16 shows, for our illustrative example, the effect of the initial moisture content, θ_0, on the time-course of the infiltration rate, $v_0(t)$, as calculated from the exact solutions. Increasing θ_0 decreases v_0 at small t, but the effect decreases as t increases, and there is no influence on the final infiltration rate. On the other hand, increasing θ_0 *increases* the velocity of wet front advance at all t. These results are primarily the consequence of the decrease in "storage capacity" as θ_0 increases. The decrease of capillary potential differences is demonstrably a secondary influence. Philip [85] gives a detailed discussion of the influence of θ_0 on the shape and general behavior of the moisture profile.

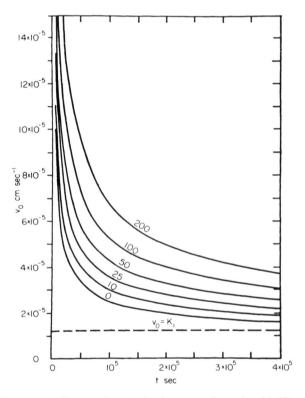

Fig. 17. Computed influence of water depth on one-dimensional infiltration rate v_0 for the soil of Figs. 1–3 with $\theta_0 = 0.2376$ [81]. Numerals on each curve denote values of water depth in centimeters. All curves are asymptotic to line $v_0 = K_1$.

D. Effect of Water Depth

Figure 17 presents, for our illustrative Yolo light clay, the effect of water depth over the soil on $v_0(t)$, as calculated from the exact solutions. Increasing the water depth increases v_0 at small t, but the effect becomes less important as t increases, and there is no effect in the limit as $t \to \infty$. The influence on wet front velocity is essentially similar. We have noticed already, in Sect. VII.A, that the parameter C of the delta-function model combines additively the magnitude of a putative wet front capillary potential and the hydrostatic pressure at the supply surface, so that C can be indifferently made up of various relative hydrostatic and capillary contributions. Exact analysis of the problem [81] shows that, in fact, the main effect of water depth on infiltration dynamics is that *an increase in water depth gives an increase in the sorptivity*, S. S is increased typically by about 2%/cm of water depth. The delta-function model is found to give a reliable prediction of the influence of water depth on S [22]. Increasing the water depth increases the steepness of moisture gradients in the wetting zone and the wet front (for small and moderate values of t) [81].

E. Algebraic Infiltration Equations

It is useful in applied hydrology to be able to characterize the dynamics of infiltration by a small number of parameters. Certain forms of algebraic infiltration equation cannot be related directly to the physical processes involved in infiltration. The Horton [131] exponential equation is particularly ill-fitted to describe the dynamics of one-dimensional infiltration into a uniform soil, despite its three disposable parameters, because it cannot adequately represent the $t^{1/2}$ behavior of i at small t. The Kostiakov [132]–Lewis [133] formulation is rather better, but, for infiltration into a uniform soil, its power law exponent for i must be $\tfrac{1}{2}$ at small t, and it must tend to increase gradually as data are fitted to larger t ranges.

The delta-function model yields an equation of the form

$$t = Y_1[i - Z_1 \log(1 + i/Z_1)] \tag{209}$$

where Y_1 and Z_1 have definite meanings in relation to either the similarity or delta-function interpretations of the model. [Despite its different appearance, Eq. (209) is exactly Eq. (187).] It will be evident from Sects. VIII and IX that Eq. (209) offers a good representation of one-dimensional infiltration into a uniform soil. Its implicit form is, however, inconvenient. It follows from Sects. VII and IX that the linearized solution offers an alternative two-parameter equation that will be equally reliable. Equations (149) are the $v_0(t)$ and $i(t)$ forms of this equation in dimensionless form. The linearized equation is at least explicit, but it seems far too complicated to be useful.

The representation

$$v_0 = \tfrac{1}{2}St^{-1/2} + A$$
$$i = St^{1/2} + At \quad (210)$$

based on the analysis of Sect. V.A, has the merit of being both physically well founded and simple. It appears to offer the most suitable two-parameter representation of the dynamics of infiltration. The limits on the validity of Eq. (210) at large t are discussed in Sect. IX.C. As we show there, when A is fixed by the behavior of v_0 or i at small t, it is definitely not equal to K_1. The form of Eq. (210) should, however, fit v_0 data over the whole t range fairly well, and a "best fit" over such a range will tend to give $A = K_1$. The matter of algebraic infiltration equations is discussed further by Philip [33, 84].

XI. Effects of Geometry and Gravity on Infiltration

The relevant two- and three-dimensional solutions of the flow equation lead to some fairly general insights into the interactions of capillarity, gravity, and geometry.

In the following discussion, two- (or three-) dimensional absorption and infiltration mean absorption and infiltration in a two- (or three-) dimensional region that is infinite (or semi-infinite) in the two (or three) dimensions from a supply surface that is finite in both (or all three) dimensions. The geometry we use for specific illustrations is as follows: Two-dimensional absorption: absorption in a cylindrically symmetrical infinite system from a supply surface of radius r_0. Three-dimensional absorption: absorption in a spherically symmetrical infinite system from a supply surface of radius r_0. Two-dimensional infiltration: infiltration in a two-dimensional system (infinite in the horizontal and semi-infinite in the vertical) from a semicircular furrow in its upper surface. Three-dimensional infiltration: infiltration into a three-dimensional region (infinite in both horizontal directions and semi-infinite in the vertical) from a hemispherical basin in its upper surface.

Strictly, t_{grav} will never be infinite, although it may be very large for some media. It follows that gravity must ultimately become nonnegligible in three-dimensional systems, even in the finest-textured media. Nevertheless, it remains useful and illuminating to include in our discussion (as we have done earlier) reference to the large t solutions for three-dimensional absorption.

A. Moisture Distribution

We have already observed that one-dimensional absorption moisture profiles preserve similarity. A well-defined wetting region and wet front are produced as a consequence of the shape of the $D(\theta)$ curve, but gradients (at fixed θ) everywhere decrease as $t^{-1/2}$, so that there is finally no real persistence of a wet front. In one-dimensional infiltration, on the other hand, nonlinearity in $K(\theta)$ produces the profile at infinity at large t, and a relatively sharp front persists for all t.

In two-dimensional absorption, profiles become less steep more rapidly than in one dimension, and gradients are always very small at large r. This can be regarded as a consequence of the fact that flow velocity must always decrease spatially more rapidly than r^{-1} in two-dimensional systems. In three-dimensional absorption, gradients decrease even more rapidly than in two dimensions: in this case, the flow velocity cannot decrease spatially less rapidly than r^{-2}. In three dimensions, however, there exists a final steady state, so that the gradients do not decrease to zero everywhere in the limit as $t \to \infty$, as they do in two dimensions.

There is no persistence of a wet front in two- and three-dimensional infiltration comparable to that in one dimension. In both cases, there is a final steady moisture distribution, and, everywhere within a finite distance of the supply surface, gradients tend to a small finite value.

The effect of gravity is, of course, to distort the symmetrical moisture distributions produced in the comparable symmetrical absorption problems. This distortion is very much stronger for two-dimensional systems than for three-dimensional ones [Figs. 9 and 10]. It also increases with increasing r_0 for both systems [86, 91]. The variation of the properties of large t moisture distributions during infiltration with the number of dimensions is a striking illustration of the influence of geometry on the effect of gravity. In one dimension, we have a quasi-stationary profile; in two dimensions, a very strongly distorted, but stationary, distribution; in three dimensions, a relatively weakly distorted, stationary distribution.

B. Existence of Steady State and Character of Ultimate Wetting

Table I summarizes the large-time behavior of these various systems. By *ultimate wetted region* (UWR), we mean the region that undergoes a definite, *nonzero and noninfinitesimal* increase of moisture content if the infiltration process is continued indefinitely. By *complete ultimate wetting*, we mean that, if infiltration continues indefinitely, all points at a finite distance from the supply surface reach a moisture content arbitrarily close to θ_1; by *incomplete wetting*, we mean that, even when infiltration continues indefinitely, all parts of the system a definite distance from the supply surface remain at moisture contents definitely less than that at the supply surface. Some modification of the *wetting* definitions is needed when the processes are complicated by

J. R. PHILIP

TABLE I

EFFECT OF GEOMETRY AND GRAVITY ON INFILTRATION

No. of dimensions m	Absorption (without gravity)			Infiltration (with gravity)		
	Ultimate wetted region	Ultimate wetting	Steady state?	Ultimate wetted region	Ultimate wetting	Steady state?
1	Infinite	Complete	No	Infinite	Complete	Quasi-steady
2	Infinite	Complete	No	Finite	Incomplete	Yes
3	Finite	Incomplete	Yes	Finite	Incomplete	Yes

air-entrapment. Also, where water is supplied at a potential greater than Ψ_e, it will be understood that there is a finite lower bound to the "definite distance from the supply surface" [22].

We note the following:

(i) The existence of a finite UWR, incomplete ultimate wetting, and a steady state are intimately connected. The one-dimensional gravity case is, in a sense, singular in that it has a quasi-steady state with an infinite UWR.

(ii) Gravity has a stabilizing effect. Two-dimensional UWR's are infinite without gravity, but are finite with gravity.

(iii) Increasing the number of dimensions has a stabilizing effect. No one-dimensional UWR's are finite; but two-dimensional UWR's with gravity are finite, and *all* three-dimensional UWR's are finite.

These considerations are obviously of great importance in connection with techniques of surface (and sub-) irrigation.

C. INFILTRATION RATE

The effects of geometry and gravity on the time course of the infiltration rate v_0 are illustrated by Figs. 18–20.

Figure 18 shows $V(T)$ for linearized and delta-function absorption solutions for $m = 1, 2,$ and 3. Figure 19 brings together linearized results for

FIG. 18. The dynamics of absorption in one, two, and three dimensions in the dimensionless form $V(T)$ [86]. Shaded areas indicate range of variation of $V(T)$ as soil properties vary between the linearized and delta-function extremes.

FIG. 19. The dynamics of absorption and infiltration in one, two, and three dimensions (linearized solutions) in the dimensionless form $V_2(T_1)$ [86]. Solutions for $m = 2$ and 3 are shown for $R_0 = 0.05$ and 0.25. Full curves: infiltration. Broken curves: absorption.

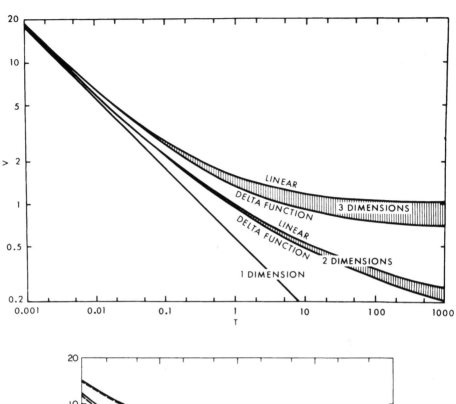
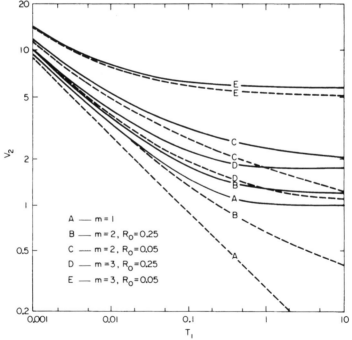

absorption and infiltration for $m = 1, 2$, and 3 with R_0 values 0.05 and 0.25 for the cases $m = 2, 3$.

Figure 20 shows the dependence of $V_2(\infty)$ on R_0 for $m = 2$ and 3 as indicated by Eqs. (158) and (164). The value $V_2(\infty) = 1$ for $m = 1$ is shown for comparison. Note that, in view of the definition of R_0 in Eq. (155), these curves can be interpreted as showing, for fixed r_0 and k, *the variation of the final infiltration rate with the capillary properties of the soil.*

We see that final infiltration rates for $m = 2$ and 3 depend in a complicated way on both the hydraulic and the capillary properties of the soil. It follows that final infiltration rates measured in the field by (necessarily finite) infiltrometers are *not*, as is often supposed, simply measures of the hydraulic properties of the soil. They are not directly related to (and are not necessarily well correlated with) the final infiltration rate for one-dimensional infiltration over large areas, which (for a uniform soil) depends only on hydraulic properties.

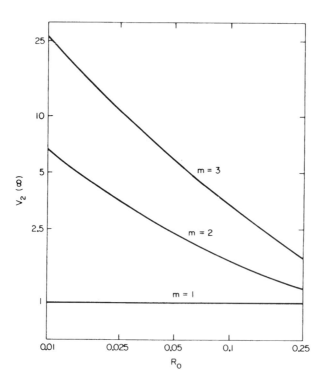

FIG. 20. The influence of geometry on final infiltration rate in the dimensionless form $V_2(\infty)$ as a function of R_0 and m. $V_2(\infty)$ for $m = 2$ and 3 are from Eqs. (158) and (164), respectively.

Theory of Infiltration

Symbols

Some symbols have been used in more than one way, but this should not lead to confusion. We indicate consistent convenient units where these are relevant.

A $(A_2 + K_0)$, coefficient of second term of series in Eq. (95) and of infiltration equation (195) (cm-sec^{-1})

A_n $\int_{\theta_0}^{\theta_1} \phi_n \, d\theta$ (cm-sec$^{-n/2}$)

A_{*n} Coefficient of nth term of series in Eq. (64) (cm^n-sec$^{-n/2}$)

C Capillary and hydrostatic potential difference between supply surface and (putative) wet front in delta-function model, taken as positive (cm)

c A small numerical factor

D Moisture diffusivity (cm^2-sec^{-1})

D_s Value of D for $\theta = \theta_s$ (cm^2-sec^{-1})

D_* Diffusivity in linearized model (cm^2-sec^{-1})

$E(\theta, x)$ Coefficient of $\partial\theta/\partial x$ in horizontal one-dimensional flow equation in heterogeneous media (cm-sec^{-1})

erfc   The complementary error function

F Function entering solution for one-dimensional absorption with constant flux condition (sec$^{-1/(n+2)}$)

$F(\theta)$ Function entering analytical solution for one-dimensional absorption (cm-sec$^{-1/2}$)

$F(\theta, x)$ Zeroth-order term in horizontal one-dimensional flow equation in heterogeneous media (sec^{-1})

$f(\theta)$ Any function of θ

g Acceleration due to gravity, ≈ 980 cm-sec^{-2}

h Relative humidity

I Dimensionless form of i defined in Eq. (136)

I_1 Dimensionless form of cumulative infiltration defined in Eq. (148)

i Cumulative absorption or infiltration (cm)

J_0 Bessel function of the first kind of zero order

K Hydraulic conductivity (cm-sec^{-1})

K_0 Value of K for $\theta = \theta_0$ (cm-sec^{-1})

K_1 Value of K for $\theta = \theta_1$ (cm-sec^{-1})

$\bar{K}$ Value of K adopted for region between supply surface and (putative) wet front in delta-function model (cm-sec^{-1})

K_{**} Reduced hydraulic conductivity (cm^{-1}-sec^{-1})

$\mathbf{K}_0$ Modified Bessel function of the second kind of zero order

k Constant coefficient corresponding to $dK/d\theta$ in linearized model (cm-sec^{-1})

m Number of space dimensions

n Generally, a positive integer; in Sect. IX.A.1, the index of the power law dependence of D on $(\theta - \theta_0)$

P Pressure (dyne-cm^{-2})

Q_2 Two-dimensional source strength (cm^2-sec^{-1})

Q_3 Three-dimensional source strength (cm^3-sec^{-1})

q Total absorption rate from finite supply surface in three-dimensional systems (cm^2-sec^{-1})

R Dimensionless form of $(x^2 + y^2)^{1/2}$ in three-dimensional systems, defined in Eq. (124)

R_0 Dimensionless form of r_0 defined in Eq. (155)

R_* Dimensionless form of r defined in Eq. (155)

$\mathbf{R}$ Gas constant for water vapor $\approx 4.615 \times 10^6$ erg-gm^{-1}-°K^{-1}

r Radius (cm)

r_0 Radius of supply surface (cm)

r_*^{-1} Mean curvative of supply surface (cm^{-1})

S Sorptivity (cm-sec$^{-1/2}$)

$\mathscr{S}$ Intrinsic sorptivity (cm$^{1/2}$)

T Dimensionless form of t defined in Eq. (136)

T_1 Dimensionless form of t defined in Eq. (148)

T	Absolute temperature (°K)	Z	Dimensionless form of z defined in Eq. (119)
t	Time, normally with $t = 0$ at beginning of infiltration process (sec)	Z_1	Coefficient in delta-function infiltration equation (209) (cm)
t_0	Arbitrary constant in Eq. (103) (sec)	z	Vertical Cartesian coordinate; positive upward in Sect. II, positive downward in treatment of infiltration in later sections (cm)
t_{geom}	Characteristic time based on capillary properties of medium and geometry of supply surface, defined by Eq. (69) (sec)		
t_{grav}	Characteristic time during which gravity becomes significant during infiltration, defined by Eq. (98) (sec)	α	In Sect. II, the "wetting potential" of a volume element of the porosity of a medium (cm); also the constant defined by Eq. (112), which enters the representation of K as an exponential function of Ψ; see Eq. (113) (cm^{-1})
U_0	Flow velocity at supply surface in processes subject to flux conditions (cm-sec^{-1})		
U	Vector flow velocity (cm-sec^{-1})	β	In Sect. II, the "draining potential" of a volume element of the porosity of a medium (cm); also a constant entering the second of Eqs. (208)
u	Velocity of the "profile at infinity" in one-dimensional infiltration; see Eq. (106) (cm-sec^{-1})		
UWR	Ultimate wetted region; see Sect. XI	γ	Euler's constant $= 0.57722\ldots$
V	Dimensionless form of v_0 defined in Eq. (136)	δ	Dirac delta function
		$\zeta(\theta)$	The ordinate of the "profile at infinity" in one-dimensional infiltration; see Eq. (107) (cm)
V_1	Dimensionless form of infiltration rate defined in Eq. (148)		
V_2	Dimensionless form of $\bar{v}_0$ defined in Eq. (155)	η	A space coordinate (cm)
		η_{wf}	Distance from supply surface to wet-front in delta-function model (cm)
$\mathcal{V}$	The volume of the medium		
v	Flow velocity (cm-sec^{-1})	Θ	Reduced form of θ defined in Eq. (199)
v_0	Flow velocity at the supply surface; the absorption or infiltration rate based on unit area of supply surface (cm-sec^{-1})	Θ_*	Reduced form of Θ defined in Eq. (202)
		θ	Volumetric moisture content
$\bar{v}_0$	v_0 averaged over supply surface (cm-sec^{-1})	θ_0	Initial value of θ
		θ_1	Value of θ at the supply surface for $t \geq 0$
W	The supply surface		
w	Dummy variable of integration	θ_a	Value of θ at which zero of ζ is taken
X	Dimensionless form of $|x|$ in two-dimensional systems defined in Eq. (119)	θ_s	Value of θ corresponding to $\Psi = \theta$
x	Horizontal Cartesian coordinate (cm)	θ_U	Asymptotic (large t) value of θ at $z = 0$ for one-dimensional infiltration with constant flux condition
Y_1	Coefficient in delta-function infiltration equation (209) (sec-cm^{-1})		
		θ_*	$\int_{\theta_0}^{\theta} D\, d\theta$ (cm^2-sec^{-1})
Y_0	Bessel function of the second kind of zero order	ϑ	θ-like variable used in linearized model
y	Horizontal Cartesian coordinate (cm)	ϑ_2	Dimensionless form of θ_*, defined in Eq. (119)

ϑ_3 Dimensionless form of θ_*, defined in Eq. (124)
ϑ_* Transformed form of ϑ defined by Eq. (152)
κ Intrinsic permeability (cm^2)
λ Characteristic length scale of internal geometry of medium (cm)
μ Dynamic viscosity (gm-cm^{-1}-sec^{-1})
ρ In Sect. II, the density of liquid water ≈ 1 gm-cm^{-3}; dimensionless radius ($= r/r_0$)
σ Surface tension (dyne-cm^{-1})
τ Reduced form of t, defined in Eq. (48) (sec-cm^{-2})
τ_1 Dimensionless form of t, defined in Eq. (199)
Φ Total potential (cm)
ϕ Similarity variable for one-dimensional absorption, defined by Eq. (33) (cm-sec$^{-1/2}$)
ϕ_n Coefficient of nth term of series in Eq. (83) (cm-sec$^{-n/2}$)
ϕ_{*n} Coefficient of $(n+1)$th term of series in Eq. (51) (cm^n-sec$^{-n/2}$)
χ Dimensionless form of x, defined in Eq. (199)
χ_* Reduced form of χ, defined in Eq. (202)
Ψ' Moisture potential (cm)
Ψ'_0 Initial value of Ψ' (cm)
Ψ'_1 Value of Ψ' at the supply surface for $t \geq 0$ (cm)
Ψ'_e "Air-entry value" of moisture potential (cm)
Ψ'_* $\int_{\Psi_0}^{\Psi} K\,d\Psi'$ (cm^2-sec^{-1})
Ψ'_{**} Reduced moisture potential (cm^2)
ψ Angle defined in Eq. (155) (rad)
Ω Potential of external forces (cm)
$\omega(\vartheta)$ Function entering Eq. (147) (cm-sec$^{-1/2}$)
∇ Gradient of a scalar
$\nabla \cdot$ Divergence of a vector
∇^2 $\nabla \cdot \nabla$

References

1. Darcy, H. P. G., "Les Fontaines Publiques de la Ville de Dijon." Dalmont, Paris, 1856.
2. Lamb, H., "Hydrodynamics," 6th ed., p. 577. Cambridge Univ. Press, London and New York, 1932.
3. Richards, L. A., Capillary conduction of liquids through porous mediums. *Physics* **1**, 318–333 (1931).
4. Childs, E. C., and Collis-George, N., The permeability of porous materials. *Proc. Roy. Soc.* **A201**, 392–405 (1950).
5. Philip, J. R., Remarks on the analytical derivation of the Darcy equation. *Trans. Am. Geophys. Union* **38**, 782–784 (1957).
6. Moore, R. E., Water conduction from shallow water tables. *Hilgardia* **12**, 383–426 (1939).
7. Philip, J. R., The physical principles of soil water movement during the irrigation cycle. *Proc. Intern. Congr. Irrigation Drainage, 3rd, San Francisco, 1957*, pp. 8.125–8.154.
8. Philip, J. R., and de Vries, D. A., Moisture movement in porous materials under temperature gradients. *Trans. Am. Geophys. Union* **38**, 222–232 (1957).
9. Gardner, W. R., Dynamic aspects of water availability to plants. *Soil Sci.* **89**, 63–73 (1960).
10. Nielsen, D. R., and Biggar, J. W., Measuring capillary conductivity. *Soil Sci.* **92**, 192–193 (1961).
11. Green, R. E., Hanks, R. J., and Larson, W. E., Estimates of field infiltration by numerical solution of the moisture flow equation. *Soil Sci. Soc. Am. Proc.* **26**, 530–535 (1962).

12. Elrick, D. E., and Bowman, D. H., Note on an improved apparatus for soil moisture flow measurements. *Soil Sci. Soc. Am. Proc.* **28**, 450–453 (1964).
13. Topp, G. C., and Miller, E. E., Hysteretic moisture characteristics and hydraulic conductivities for glass-bead media. *Soil Sci. Soc. Am. Proc.* **30**, 156–162 (1966).
14. Edlefsen, N. E., and Anderson, A. B. C., Thermodynamics of soil moisture. *Hilgardia* **15**, 31–298 (1943).
15. Richards, L. A., Methods of measuring soil moisture tension. *Soil Sci.* **68**, 95–112 (1949).
16. Richards, L. A., Water conducting and retaining properties of soil in relation to irrigation. Desert Research. *Res. Council Israel Spec. Publ.* **2**, 523–546 (1953).
17. Childs, E. C., and Collis-George, N., The control of soil water. *Advan. Agron.* **2**, 233–272 (1950).
18. Croney, D., Coleman, J. D., and Bridge, P. M., The suction of moisture held in soil and other porous materials. *Road Res. Tech. Paper* **24** (1952).
19. Miller, E. E., and Klute, A., The dynamics of soil water. Part I—mechanical forces. *In* "Irrigation of Agricultural Lands" (R. M. Hagan, H. R. Haise, and T. W. Edminster, eds.), pp. 209–244. Am. Soc. Agron., Madison, Wisconsin, 1967.
20. Luthin, J. N., and Miller, R. D., Pressure distribution in soil columns draining into the atmosphere. *Soil Sci. Soc. Am. Proc.* **17**, 329–333 (1953).
21. Philip, J. R., The concept of diffusion applied to soil water. *Proc. Natl. Acad. Sci. India Sect. A* **24**, 93–104 (1955).
22. Philip, J. R., The theory of infiltration: 7. *Soil Sci.* **85**, 333–337 (1958).
23. Carslaw, H. S., and Jaeger, J. C., "Conduction of Heat in Solids," 2nd ed. Oxford Univ. Press (Clarendon), London and New York, 1959.
24. Chandrasekhar, S., Stochastic problems in physics and astronomy. *Rev. Mod. Phys.* **15**, 1–89 (1943).
25. Bailey, N. T. J., "The Elements of Stochastic Processes." Wiley, New York, 1964.
26. Philip, J. R., Diffusion by continuous movements. *Phys. Fluids* **11**, 38–42 (1968).
27. Mahony, J. J., and Philip, J. R., Equations modeling spatially variable stochastic processes. *Phys. Fluids* **10**, 1403–1405 (1967).
28. Boltzmann, L., Zur Integration der Diffusionsgleichung bei variabeln Diffusionscoefficienten. *Ann. Physik* **53**, 959–964 (1894).
29. Buckingham, E., Studies on the movement of soil moisture. *U. S. Dept. Agr. Bur. Soils Bull.* **38** (1907).
30. Childs, E. C., and George, N. C., Soil geometry and soil-water equilibria. *Discussions Faraday Soc.* **3**, 78–85 (1948).
31. Klute, A., A numerical method for solving the flow equation for water in unsaturated materials. *Soil Sci.* **73**, 105–116 (1952).
32. Klute, A., Some theoretical aspects of the flow of water in unsaturated soils. *Soil Sci. Soc. Am. Proc.* **16**, 144–148 (1952).
33. Philip, J. R., Some recent advances in hydrologic physics. *J. Inst. Engrs. Australia* **26**, 255–259 (1954).
34. Philip, J. R., The theory of infiltration: 1. The infiltration equation and its solution. *Soil Sci.* **83**, 345–357 (1957).
35. Philip, J. R., The theory of absorption in aggregated media. *Australian J. Soil Res.* **6**, 1–19 (1968).
36. Philip, J. R., Diffusion, dead-end pores, and linearized absorption in aggregated media. *Australian J. Soil Res.* **6**, 21–30 (1968).
37. Philip, J. R., Effect of aquifer turbulence on well drawdown—Discussion. *J. Hydraulics Div. Am. Soc. Civil Engrs.* **86**, HY5, 179–181 (1960).

38. Philip, J. R., Physics of water movement in porous solids. *Highway Res. Board Spec. Rept.* **40**, 147–163 (1958).
39. Philip, J. R., Transient fluid motions in saturated porous media. *Australian J. Phys.* **10**, 43–53 (1957).
40. Philip, J. R., The early stages of absorption and infiltration. *Soil Sci.* **88**, 91–97 (1959).
41. Olsen, H. W., Deviations from Darcy's law in saturated clays. *Soil Sci. Soc. Am. Proc.* **29**, 135–140 (1965).
42. Swartzendruber, D., Soil-water behavior as described by transport coefficients and functions. *Advan. Agron.* **18**, 327–370 (1966).
43. Quirk, J. P., and Schofield, R. K., The effect of electrolyte concentrations on soil permeability. *J. Soil Sci.* **6**, 163–178 (1955).
44. Elrick, D. E., Transient two-phase capillary flow in porous media. *Phys. Fluids* **4**, 572–575 (1961).
45. Youngs, E. G., and Peck, A. J., Moisture profile development and air compression during water uptake by bounded porous bodies: 1. Theoretical introduction. *Soil Sci.* **98**, 290–294 (1964).
46. Peck, A. J., Moisture profile development and air compression during water uptake by bounded porous bodies: 2. Horizontal columns. *Soil Sci.* **99**, 327–334 (1965).
47. Peck, A. J., Moisture profile development and air compression during water uptake by bounded porous bodies: 3. Vertical columns. *Soil Sci.* **100**, 44–51 (1965).
48. Philip, J. R., The theory of infiltration: 3. Moisture profiles and relation to experiment. *Soil Sci.* **84**, 163–178 (1957).
49. Peck, A. J., The diffusivity of water in a porous material. *Australian J. Soil Res.* **2**, 1–7 (1964).
50. Nielsen, D. R., Biggar, J. W., and Davidson, J. M., Experimental consideration of diffusion analysis in unsaturated flow problems. *Soil Sci. Soc. Am. Proc.* **26**, 107–112 (1962).
51. Philip, J. R., Evaporation, and moisture and heat fields in the soil. *J. Meteorol.* **14**, 354–366 (1957).
52. de Vries, D. A., Simultaneous transfer of heat and moisture in porous media. *Trans. Am. Geophys. Union* **39**, 909–916 (1958).
53. de Vries, D. A., and Philip, J. R., Temperature distribution and moisture transfer in porous materials. *J. Geophys. Res.* **64**, 386–388 (1959).
54. Cary, J. W., Onsager's relation and the non-isothermal diffusion of water vapor. *J. Phys. Chem.* **67**, 126–129 (1963).
55. Anderson, D. M., and Linville, A., Temperature fluctuations at a wetting front: I. Characteristic temperature time curves. *Soil Sci. Soc. Am. Proc.* **26**, 14–18 (1962).
56. Anderson, D. M., Sposita, G., and Linville, A., Temperature fluctuations at a wetting front: II. The effect of initial water content of the medium on the magnitude of the temperature fluctuations. *Soil Sci. Soc. Am. Proc.* **27**, 367–369 (1963).
57. Philip, J. R., Energy dissipation during absorption and infiltration: 1. *Soil Sci.* **89**, 132–136 (1960).
58. Philip, J. R., Energy dissipation during absorption and infiltration: 2. *Soil Sci.* **89**, 353–358 (1960).
59. Haines, W. B., Studies in the physical properties of soil. V. The hysteresis effect in capillary properties, and the modes of moisture distribution associated therewith. *J. Agr. Sci.* **20**, 97–116 (1930).
60. Youngs, E. G., Redistribution of moisture in porous materials after infiltration: 1. *Soil Sci.* **84**, 283–290 (1958).

61. Collis-George, N., Hysteresis in moisture content-suction relationships in soils. *Proc. Natl. Acad. Sci. India Sect. A* **24**, 80–85 (1955).
62. Everett, D. H., and Whitton, W. I., A general approach to hysteresis. *Trans. Faraday Soc.* **48**, 749–757 (1952).
63. Everett, D. H., and Smith, F. W., A general approach to hysteresis, 2. *Trans. Faraday Soc.* **50**, 187–197 (1954).
64. Everett, D. H., A general approach to hysteresis, 3. *Trans. Faraday Soc.* **51**, 1077–1096 (1954).
65. Miller, E. E., and Miller, R. D., Physical theory for capillary flow phenomena. *J. Appl. Phys.* **27**, 324–332 (1956).
66. Poulovassilis, A., Hysteresis of pore water, an application of the concept of independent domains. *Soil Sci.* **93**, 405–412 (1962).
67. Poulovassilis, A., An investigation of some problems of hydrostatics and dynamics of water in porous media. Ph.D. Thesis, Univ. Cambridge, Cambridge, England, 1962.
68. Enderby, J. A., The domain model of hysteresis, 1. *Trans. Faraday Soc.* **51**, 835–848 (1955).
69. Enderby, J. A., The domain model of hysteresis, 2. *Trans. Faraday Soc.* **52**, 106–120 (1956).
70. Everett, D. A., A general approach to hysteresis, 4. *Trans. Faraday Soc.* **51**, 1551–1557 (1955).
71. Philip, J. R., The gain, transfer, and loss of soil-water. "Water Resources, Use and Management," pp. 257–275. Melbourne Univ. Press, Melbourne, Australia, 1964.
72. Philip, J. R., Similarity hypothesis for capillary hysteresis in porous materials. *J. Geophys. Res.* **69**, 1553–1562 (1964).
73. Philip, J. R., The theory of infiltration: 2. The profile at infinity. *Soil Sci.* **83**, 435–448 (1957).
74. Philip, J. R., The dynamics of capillary rise. *Proc. UNESCO Netherlands Symp. Water Unsaturated Zone, Wageningen, 1966.*
75. Van Dyke, M., "Perturbation Methods in Fluid Mechanics." Academic Press, New York, 1964.
76. Philip, J. R., General method of exact solution of the concentration-dependent diffusion equation. *Australian J. Phys.* **13**, 1–12 (1960).
77. Crank, J., and Henry, M. E., Diffusion in media with variable properties: I. *Trans. Faraday Soc.* **45**, 636–642 (1949).
78. Crank, J., and Henry, M. E., Diffusion in media with variable properties: II. *Trans. Faraday Soc.* **45**, 1119–1128 (1949).
79. Philip, J. R., Numerical solution of equations of the diffusion type with diffusivity concentration-dependent. *Trans. Faraday Soc.* **51**, 885–892 (1955).
80. Richardson, L. F., The deferred approach to the limit, part 1. *Phil. Trans. Roy. Soc. London Ser. A* **226**, 299–349 (1927).
81. Philip, J. R., The theory of infiltration: 6. Effect of water depth over soil. *Soil Sci.* **85**, 278–286 (1958).
82. Philip, J. R., A very general class of exact solutions in concentration-dependent diffusion. *Nature* **185**, 233 (1960).
83. Philip, J. R., The function inverfc θ. *Australian J. Phys.* **13**, 13–20 (1960).
84. Philip, J. R., The theory of infiltration: 4. Sorptivity and algebraic infiltration equations. *Soil Sci.* **84**, 257–264 (1957).
85. Philip, J. R., The theory of infiltration: 5. The influence of the initial moisture content. *Soil Sci.* **84**, 329–339 (1957).
86. Philip, J. R., Absorption and infiltration in two- and three-dimensional systems. *Proc.*

UNESCO Netherlands Symp. Water Unsaturated Zone, Wageningen, 1966.
87. Philip, J. R., Numerical solution of equations of the diffusion type with diffusivity concentration-dependent. II. *Australian J. Phys.* **10**, 29–42 (1957).
88. Philip, J. R., Kinetics of growth and evaporation of droplets and ice crystals. *J. Atmospheric Sci.* **22**, 196–206 (1965).
89. Philip, J. R., and Farrell, D. A., General solution of the infiltration-advance problem in irrigation hydraulics. *J. Geophys. Res.* **69**, 621–631 (1964).
90. Irmay, S., Solution of the non-linear diffusion equation with a gravity term in hydrology. *Proc. UNESCO Netherlands Symp. Water Unsaturated Zone, Wageningen, 1966.*
91. Philip, J. R., Discussion to Session IIa. *Proc. UNESCO Netherlands Symp. Water Unsaturated Zone, Wageningen, 1966.*
92. Philip, J. R., Mathematical-physical approach to water movement in unsaturated soils. *Proc. Intern. Soil Water Symp., Prague, 1967*, pp. 309–319.
93. Philip, J. R., Extended techniques of calculation of soil-water movement, with some physical consequences. *Proc. Intern. Congr. Soil Sci., 9th, Adelaide, 1968*, **1**, 1–9.
94. Philip, J. R., The theory of local advection: 1 *J. Meteorol.* **16**, 535–547 (1959).
95. Singh, R., Unsteady and unsaturated flow in soil in two dimensions. *Dept. Civil Eng., Stanford Univ. Tech. Rept. 54.* Stanford Univ., Stanford, California, 1965.
96. Singh, R., and Franzini, J. B., Unsteady flow in unsaturated soils from a cylindrical source of finite radius. *J. Geophys. Res.* **72**, 1207–1215 (1967).
97. Philip, J. R., Unsaturated soil-water movement from a cylindrical source. *J. Geophys. Res.* **73**, 3968–3970 (1968).
98. Swartzendruber, D., Variables-separable solution of the horizontal flow equation with nonconstant diffusivity. *Soil Sci. Soc. Am. Proc.* **30**, 7–11 (1967).
99. Kobayashi, H., Study on the theoretical analysis for unsaturated flow in the soil. [In Japanese.] *Gifu Daigaku Nogakubu Kenkyu Hokoku* **12**, 283–294 (1960).
100. Kobayashi, H., A theoretical analysis and numerical solutions of unsaturated flow in soil. *Proc. UNESCO Netherlands Symp. Water Unsaturated Zone, Wageningen, 1966.*
101. Childs, E. C., Soil moisture theory. *Advan. Hydrosci.* **4**, 73–117 (1967).
102. Youngs, E. G., Moisture profiles during vertical infiltration. *Soil Sci.* **84**, 283–290 (1957).
103. Childs, E. C., The ultimate moisture profile during infiltration in a uniform soil. *Soil Sci.* **97**, 173–178 (1964).
104. Irmay, S., Extension of Darcy's law to unsteady, unsaturated flow through porous media. *Symp. Darcy Intern. Assoc. Sci. Hydrol.*, **2**, 57–66, Dijon, France, 1956.
105. Farrell, D. A., The hydraulics of surface irrigation. Ph.D. Thesis, Melbourne Univ., Melbourne, Australia, 1963.
106. Muskat, M., "The Flow of Homogeneous Fluids through Porous Media," Chapter VI. McGraw-Hill, New York, 1937.
107. Polubarinova-Kochina, P. Ya., "Theory of Ground Water Movement." Princeton Univ. Press, Princeton, New Jersey, 1962.
108. Sutton, O. G., Wind structure and evaporation in the lower atmosphere. *Proc. Roy. Soc.* **A146**, 701–722 (1934).
109. Wooding, R. A., Convection in a saturated porous medium at large Rayleigh number or Péclet number. *J. Fluid Mech.* **15**, 527–544 (1963).
110. Wooding, R. A., Mixing-layer flows in a saturated porous medium. *J. Fluid Mech.* **19**, 103–112 (1964).
111. Philip, J. R., A linearization technique for the study of infiltration. *Proc. UNESCO Netherlands Symp. Water Unsaturated Zone, Wageningen, 1966.*

112. Jaeger, J. C., and Clarke, M. E., A short table of $\int_0^\infty \frac{e^{-xu^2}}{J_0^2(u) + Y_0^2(u)} \frac{du}{u}$. *Proc. Roy. Soc. Edinburgh* **A61**, 229–230 (1942).
113. Green, W. H., and Ampt, G. A., Studies in soil physics. I. The flow of air and water through soils. *J. Agr. Sci.* **4**, 1–24 (1911).
114. Lambe, T. W., Capillary phenomena in cohesionless soils. *Trans. Am. Soc. Civil Engrs.* **116**, 401–432 (1951).
115. Philip, J. R., An infiltration equation with physical significance. *Soil Sci.* **77**, 153–157 (1954).
116. Rubin, J., and Steinhardt, R., Soil water relations during rain infiltration: I. Theory. *Soil Sci. Soc. Am. Proc.* **27**, 246–250 (1963).
117. Rubin, J., Steinhardt, R., and Reiniger, P., Soil water relations during rain infiltration: II. Moisture content profiles during rains of low intensities. *Soil Sci. Soc. Am. Proc.* **28**, 1–5 (1964).
118. Rubin, J., and Steinhardt, R., Soil water relations during rain infiltration: III. Water uptake at incipient ponding. *Soil Sci. Soc. Am. Proc.* **28**, 614–620 (1964).
119. Hanks, R. J., and Bowers, S. A., Numerical solution of the moisture flow equation for infiltration into layered soils. *Soil Sci. Soc. Am. Proc.* **26**, 530–535 (1962).
120. Philip, J. R., Sorption and infiltration in heterogeneous media. *Australian J. Soil Res.* **5**, 1–10 (1967).
121. Klute, A., and Wilkinson, G. E., Some test of the similar media concept of capillary flow: I. Reduced capillary conductivity and moisture characteristic data. *Soil Sci. Soc. Am. Proc.* **22**, 278–280 (1958).
122. Wilkinson, G. E., and Klute, A., Some tests of the similar media concept of capillary flow: II. Flow systems data. *Soil Sci. Soc. Am. Proc.* **23**, 434–437 (1959).
123. Bodman, G. B., and Coleman, E. A., Moisture and energy conditions during downward entry of water into soils. *Soil Sci. Soc. Am. Proc.* **8**, 116–122 (1944).
124. Coleman, E. A., and Bodman, G. B., Moisture and energy conditions during downward entry of water into moist and layered soils. *Soil Sci. Soc. Am. Proc.* **9**, 3–11 (1945).
125. Marshall, T. J., and Stirk, G. B., Pressure potential of water moving downward into soil. *Soil Sci.* **68**, 359–370 (1949).
126. Miller, R. D., and Richard, F., Hydraulic gradients during infiltration in soils. *Soil Sci. Soc. Am. Proc.* **16**, 33–38 (1952).
127. Nielsen, D. R., Kirkham, D., and van Wijk, W. R., Diffusion equation calculations of field soil water infiltration profiles. *Soil Sci. Soc. Am. Proc.* **25**, 165–168 (1961).
128. Davidson, J. M., Nielsen, D. R., and Biggar, J. W., The measurement and description of water flow through Columbia silt loam and Hesperia sandy loam. *Hilgardia* **34**, 601–616 (1963).
129. Gupta, R. P., and Staple, W. J., Infiltration in vertical columns of soil under a small positive head. *Soil Sci. Soc. Am. Proc.* **28**, 729–732 (1964).
130. Nielsen, D. R., and Vachaud, G., Infiltration of water into vertical and horizontal soil columns. *J. Indian Soc. Soil Sci.* **13**, 15–23 (1965).
131. Horton, R. E., Approach toward a physical interpretation of infiltration capacity. *Soil Sci. Soc. Am. Proc.* **5**, 399–417 (1940).
132. Kostiakov, A. N., On the dynamics of the coefficient of water-percolation in soils and on the necessity for studying it from a dynamic point of view for purposes of amelioration. *Trans. 6th Comm. Intern. Soc. Soil Sci., Moscow, 1932*, Russian Part A, pp. 17–21.
133. Lewis, M. R., The rate of infiltration of water in irrigation practice. *Trans. Am. Geophys. Union* **18**, 361–368 (1937).

AUTHOR INDEX

Numbers in parentheses are reference numbers and indicate that an author's work is referred to, although his name is not cited in the text. Numbers in italic show the page on which the complete reference is listed.

A

Alexander, G. N., 18(77), 21, *115*, *116*
Algert, J. A., 30(114), 56, *117*
Amorocho, J., 3, 5, 7, 8, 11, 57(4), 109, *112*, *113*
Amos, D. E., 48, 50(135), 64, 65, 82, *118*
Ampt, G. A., 157, *172*, 266, *296*
Anderson, A. B. C., 220(14), *292*
Anderson, D. M., 228(55, 56), *293*
Anderson, R. L., 79, *119*
Angell, J. K., 34(117), *117*
Aseltine, J. A., 44, *117*
Averjanov, S. F., 157(31), *171*

B

Bader, H., 174, *213*
Bagley, J. H., 7(6), 26, 27, *112*
Bailey, N. T. J., 222, *292*
Bartlett, M. S., 70(158), *119*
Battin, R. H., 44, *117*
Bayazit, M., 68, *119*
Bear, J., 145, *171*
Beard, L. R., 27, 33(106), *117*
Bendat, J. S., 29, 30, 41(109), 43, 44, 50, 64, 65, 69(109), 70, 71, 72, 73(109), 74(109), 75, 84(109), 86, 109, *117*
Benson, M. A., 10, *113*
Berg, M., 182, *213*
Biggar, J. W., 219(10), 227(50), 280(128), *291*, *293*, *296*
Black, R. P., 18(75), *115*
Blackman, R. B., 10(25), 41(25), 44, 73, 74(25), *113*
Bodman, G. B., 280(123, 124), 281, *296*
Boersma, L., 165, 167(40), *172*
Bogdanoff, J. L., 30(113), *117*
Boltzmann, L., 222, 235, *292*

Bouwer, H., 123, 124(2), 126, 130, 137(1), 140(13), 141, 142, 144(1), 145(1, 7, 19), 145(23, 24a), 148, 151(1, 13, 27), 152(7, 13, 19), 154(13, 27), 155(1, 13, 28), 156, 157(1), 158(27), 160(1), 163(2), 164(27), 165(27, 41), 166(27), 167(23, 41, 43, 44, 45), 168(46), *170*, *171*, *172*
Bowers, S. A., 278, *296*
Bowman, D. H., 219(12), *292*
Bratranek, A., 51(143), *118*
Bridge, P. M., 220(18), *292*
Brier, G. W., 10(29), 15, 34(115, 116), 54(146), 84(146), *113*, *117*, *118*
Brittan, M. R., 27, *116*
Brooks, C. E. P., 100, *119*
Brown, R. G., 51(140), *118*
Buckingham, E., 223, *292*
Bunge, M., 9, 11, 25, 29(20), 102, *113*
Busch, W. F., 15(64), 22(64), *115*

C

Caffey, J. E., 15, *115*
Carruthers, N., 100, *119*
Carslaw, H. S., 222(23), 245(23), 246(23), 257(23), 258(23), 261(23), 262, *292*
Cary, J. W., 227(54), *293*
Chandrasekhar, S., 222, *292*
Chang, S. S. L., 44, *117*
Cheng, D. K., 7, *113*
Childs, E. C., 218, 220(17), 223, 250(101), 252(101, 103), 278, *291*, *292*, *295*
Chow, V. T., 5, 7, 8, 13, 14(17), 26, 27, 33(102), 104, *112*, *113*, *116*
Clarke, M. E., 262, *296*
Coleman, E. A., 280(123, 124), 281, *296*
Coleman, J. D., 220(18), *292*
Collis-George, N., 162, *172*, 218, 220(17), 228, *291*, *292*, *294*
Cornell, C. A., 11, 26, *114*

AUTHOR INDEX

Cote, L. G., 30(113), *117*
Cox, D. R., 26, *116*
Cramer, H., 9, 11, *113*
Crank, J., 235, *294*
Crawford, N. H., 7(5), 11(5), *112*
Croney, D., 220(18), *292*
Cunnyngham, J., 51(137), 75(137), 80, *118*

D

Dachler, R., 126, 145, *170*
Dalrymple, T., 15(63), 20(63), 22(63), *115*
Darcy, H. P. G., 217, *291*
Davenport, W. B., 5, 44, *113*
Davidson, J. M., 227(50), 280(128), *293, 296*
Dawdy, D. R., 8, 11(52), 81, 64, *113, 114*
de Quervain, M. R., 176, 177(2), 179(2), *213*
de Vries, D. A., 219(8), 227, *291, 293*
Doering, E. J., 164(39), *172*
Doob, J. L., 11, *114*
Dorsey, N. E., 176, *213*

E

Eagleson, P. S., 45, 61(132), 63, 65(153, 154), 66, *118, 119*
Edlefsen, N. E., 220(14), *292*
Eichberger, L. C., 62(150), *118*
Elrick, D. E., 219(12), 226, *292, 293*
Enderby, J. A., 228(68, 69), *294*
Ernst, L. F., 126, 128, 147, *170*
Everett, D. A., 228(70), *294*
Everett, D. H., 228, *294*

F

Fair, G., 10(22), *113*
Faraday, M., 180, *213*
Farrell, D. A., 243(89), 252, *295*
Feller, W., 18(68), 101, *115*
Fiering, M. B., 18(71, 73), 20(71, 73), 22(88), 23(90), 26, 27, 29, 33(71, 73), 91(88), 102, 104, 105, 106, 107, 108), *115, 116, 117*
Franzini, J., 13(57), *115*, 247, *295*
Frenkiel, F. N., 10, *114*
Fritts, H. C., 40(119), *117*

G

Gardner, W. R., 219(9), *291*
Gargett, A., 84(167), *119*
George, N. C., 223, *292*
Gervais, J. G., 52(145), 84(145), 95(145), 97(145), 99(145), *118*
Geyer, J., 10(22), *113*
Giddings, J. C., 209, *214*
Gilman, C. S., 14(60), 15(61), *115*
Goldshlak, L., 180(9), 181(9), *213*
Goodman, N. R., 64, 82, *118, 119*
Gould, B. W., 18(70), 20(70), *115*
Gould, L. A., 44(122), *117*
Gracie, G., 51, 84(141), *118*
Granger, C. W. J., 10(28), 11, 12(28), 29, 41(28), 51(28), 74(28), 82, 86, *113*
Green, R. E., 219(11), *291*
Green, W. H., 157, *172*, 266, *296*
Gunnerson, C. G., 10, 32, 33(46), 38, 72, 74, 75, *114*
Gupta, R. P., 280(129), *296*

H

Haines, W. B., 228, *293*
Hald, A., 81, *119*
Hammad, H. Y., 136, 138, *170*
Hamon, B. V., 10(36), 29, 49(36), 91, 108, *114*
Hanks, R. J., 219(11), 278, *291, 296*
Hannan, E. J., 10(27, 36), 29, 49(36), 91(27, 36), 105, 108, *113, 114*
Harr, M. E., 134, 136, 156, *170*
Hart, W. E., 3, 5, 7, 11, 57(4), 109, *112*
Hays, J. R., 62(151), *118*
Hely, A. G., 15, 22(65), *115*
Henry, M. E., 235, *294*
Hershfield, D. M., 15(62), *115*
Hobbs, P. V., 182, 183, *213*
Holloway, J. Leith, Jr., 10(30), 84(30), *113*
Horowitz, I. M., 44, *117*
Horton, R. E., 283, *296*
Hosler, C. L., 180, 181(9), *213*
Hurst, H. E., 18(69, 75), 21, 101, *115, 116*

I

Irmay, S., 245(90), 252, *295*

AUTHOR INDEX

Ishihara, T., 7, 8, *113*
Itagaki, K., 183, *213*
Iwai, H., 184, 187, *213*

J

Jaeger, J. C., 222(23), 245(23), 246(23), 257(23), 258(23), 261(23), 262, *292, 296*
Jellinek, H. H. G., 181, *213*
Jenkins, E., 74(160), 84, *119*
Jenkins, G. M., 82, *119*
Jensen, D. C., 180(9), 181(9), *213*
Johnson, N. L., 19, *116*
Jones, R. H., 29, 84(110), 91(110), 109, *117*
Julian, P. R., 10, 21, 27, 33(43), 47(133), 83, 84(133), 85(133), 86, 88, 89, 90, 91(133), 93, 94, 97, 109, *114, 118*

K

Kaiser, J. F., 44(122), *117*
Karoly, A., 18(77), *115*
Kazmann, R., 21, *116*
Keeler, C. M., 182, *213*
Kendall, M. B., 50, 53, *118*
Kingery, W. D., 181, 182, *213*
Kinsman, B., 11, 24, *114*
Kirkham, D., 164, *172*, 280(127), *296*
Klute, A., 157(32), *171*, 220(19), 222(19), 223, 235, 275(19), 279, *292, 296*
Kobayashi, H., 247, *295*
Kohler, M., 13(56), 51, *115, 118*
Koopmans, L. H., 48, 50(135), 64, 65, 82, *118*
Korn, G. A., 6, 42, 44, *112, 117*
Korn, T. M., 42, 44, *117*
Korshover, J., 34(117), *117*
Kostiakov, A. N., 283, *296*
Kotz, S., 18(76), 100, *115, 119*
Kozeny, J., 134, *170*
Kozin, F., 30(113), *117*
Krenkel, P. A., 62(151), *118*
Kuczynski, G. C., 182, *213*
Kuroiwa, D., 181, *213*
Kuznetsov, P. I., 57(148), *118*

L

La Chapelle, E., 209, *214*
Lamb, H., 218, *291*

Lambe, T. W., 266, *296*
Langbein, W. B., 26, *116*
Laning, J. H., 44, *117*
Larson, W. E., 219(11), *291*
Laurenson, E. M., 8, 9, *113*
Lee, Y. W., 5, 7, 20(10), 30, 34, 44, 60, *113*
Leone, F, C., 19, *116*
Leopold, L. B., 21, *116*
Lewis, M. R., 283, *296*
Linsley, R. K., 7(5), 11(5), 13(56, 57), *112, 115*
Linville, A., 228(55, 56), *293*
Little, W. C., 145(19), 152(19), *171*
Lloyd, E. H., 26, *116*
Loucks, D. P., 11(53), *114*
Luthin, J. N., 145(18), *171*, 222(20), *292*

M

Maasland, M., 147(24), *171*
Mahony, J. J., 222(27), *292*
March, F., 66(154), *119*
Marshall, J. L., 5, 7, 8, 57(13), *113*
Marshall, T. J., 280(125), *296*
Mason, B. J., 182, *213*
Matalas, N. C., 8, 10, 21, 22(89), 28, 40, 53, 55, 56, 64, 65, 79, 81, 86, 91(89), 100, 102, 106, 107, *113, 114, 116*
Matsumoto, A., 180, *213*
Maxwell, J. C., 188, *214*
Mejia, R., 66(154), *119*
Mellor, M., 188, 199, 201(22), 203, 204(22), 205(31), 206(31), 207, 208(22, 31), *213, 214*
Melton, M. A., 33(112), *117*
Middleton, D., 44, *117*
Miller, E. E., 219(13), 220(19), 222(19), 228, 229, 247, 275(19), 279, *292, 294*
Miller, R. D., 222(20), 228, 247, 279, 280(126) *292, 294, 296*
Minshall, N. E., 8, *113*
Mitchell, J. M., Jr., 10(33, 38), 12(33), 35, 55(38), 58, 59, 60, 79, 82, 83, *114*
Moore, R. E., 218, 219(6), 221(6), 239, *291*
Moran, P. A. P., 26, 102, *116*
Morel-Seytoux, H. J., 148, *171*
Muller, F. B., 52(145), 84(145), 95(145), 97, 99, *118*
Munk, W. H., 10(37), *114*
Muskat, M., 126, 134, 136, *170*, 256, 269(107), *295*
Muster, E. L., 62(150), *118*

299

AUTHOR INDEX

N

Nakaya, U., 180, *213*
Nash, J. E., 26, *116*
Nettheim, N. F., 51(138), 75(138), *118*
Neumann, J., 18(76), 100, *115*, *119*
Newton, G. C., 44, *117*
Nielson, D. R., 219(10), 227, 280(127, 128, 130), *291*, *293*, *296*

O

O'Donnell, T., 11(52), 64, 68, *114*, *119*
Olsen, H. W., 225, *293*
Orlob, G. T., 7, 8, *113*

P

Panofsky, H. A., 10(29), 15, *113*
Papoulis, A., 23, 52(91), *116*
Parzen, E., 10(26), 19, 26, 52(78), 74, *113*, *115*, *119*
Paskusz, G. F., 62(150), *118*
Pattison, A., 27, 33(100), *116*
Paulhus, J. L., 13(56), 15(61), *115*
Pavlovsky, N. N., 134, *170*
Peck, A. J., 226, 227, *293*
Philip, J. R., 157(33), *171*, 215(48, 73, 81), 218(5, 7), 219(8), 222(21, 22, 26, 27), 223, 224, 225(37, 38, 39, 40), 227, 228(7, 21, 57, 58, 71, 72), 229, 232(34, 58, 73), 233(34, 51, 74), 234, 235, 236(48, 76, 79, 81, 82, 83), 238, 239(34), 240, 241(88), 242, 243(86, 89), 245(86, 91, 92, 93), 247(86, 91, 94, 97), 248, 249, 250(34), 252(73, 81), 253(34), 254(73), 255(34, 73), 257(91, 92, 93), 258(74, 86, 94, 111), 262(88), 264(87), 266, 267(48, 74, 86, 111), 271(86, 91, 92, 93, 111), 272(86), 274(76, 84, 111), 275(33, 35, 36, 84), 279, 281(22, 48, 81, 85), 282(81, 85), 283(22, 81), 284(33, 84), 285(86, 91), 286(22, 86), *291*, *292*, *293*, *294*, *295*, *296*
Piersol, A. G., 29, 30, 41(109), 43, 44, 50, 64, 65, 69(109), 70, 71, 72, 73(109), 74(109), 75, 84(109), 86, 109, *117*
Platzman, G. W., 34(118), 49(118), *117*
Polubarinova-Kochina, P. Ya., 134, 136, 156, 157(10), *170*, 256, *295*
Poulovassilis, A., 228, 252(66), *294*
Prabhu, N. U., 19, *116*
Preissmann, A., 149, *171*

Q

Quick, J. P., 226(43), 243(43), *293*

R

Ramaseshan, S., 27, 33(102), 104, *116*
Ramseier, R. O., 182, *213*
Rao, D. B., 34(118), 49(118), *117*
Reiniger, R., 276(117), *296*
Reisenauer, A. E., 145(16), 152(16), *171*
Restrepo, J. C. O., 65(153), *118*
Rice, R. C., 165, 167(41, 45), 168(46), *172*
Richard, F., 280(126), *296*
Richards, L. A., 218, 220(15, 16), 223, *291*, *292*
Richardson, L. F., 236, 250, *294*
Risenkampf, B. K., 155, 156, *171*
Rockwood, D. M., 9, *113*
Roden, G. I., 92, 95, *119*
Root, W. L., 5, 44, *113*
Rosenblatt, H. M., 51(139), 75(139), *118*
Rubin, J., 276(116, 117, 118), *296*

S

Saaty, T. L., 57(147), *118*
Sander, G. W., 182, *213*
Santing, G., 145(21), *171*
Schneck, H. Jr., 34(92), *116*
Schnelle, K. B., Jr., 62(151), *118*
Schofield, R. K., 226(43), 243(43), *293*
Schuster, A., 70(157), *119*
Schwartzchild, M., 10, *114*
Schweppe, J. L., 62(150), *118*
Schwerdtfeger, P., 188, 190(23), 191(23), 193(23), *214*
Scott, V. H., 145(22), *171*
Searcy, J. K., 14(59), 22(59), 91(59), *115*
Sellers, W. D., 22, 87, *116*
Sewell, J. I., 145(17), 152, *171*
Shack, W. J., 45, 61(132), 63, *118*
Shapiro, R., 11, *114*
Shaw, L. C., 15(64), 22(64), *115*
Shaw, R. W., 52(145), 84(145), 95(145), 97(145), 99(145), *118*
Shen, J., 61, *118*
Simaika, Y. M., 18(75), *115*
Singh, R., 247, *295*

AUTHOR INDEX

Smiles, D. E., 162, *172*
Smith, F. W., 228(63), *294*
Smith, O. J. M., 44, *117*
Smith, W. L., 26, *116*
Snodgrass, F. E.., 10(37), *114*
Snyder, W. M., 11, *114*
Solodovnikov, V. V., 44, *117*
Southwell, R. V., 145(15), *171*
Speight, J. G., 30(111), *117*
Sposita, G., 228(56), *293*
Staple, W. J., 280(129), *296*
Steinhardt, R., 276(116, 117, 118), *296*
Sterberg, Y. M., 145(22), *171*
Stirk, G. B., 280(125), *296*
Stommel, H., 10(32), 74, *113*
Stratonovich, R. L., 57(148), *118*
Streets, R. B. Jr., 41, 42, 44, *117*
Stubberud, A. R., 7, 8, *113*
Sulakvelidze, G. K., 195, *214*
Susts, A. B., 18(77), *115*
Sutton, O. G., 258, *295*
Swartzendruber, D., 225, 247, *293, 295*

T

Takasao, T., 7, 8, *113*
Taylor, G. S., 145(18), *171*
Thomas, C. W., 207(32), *214*
Thomas, H. A., Jr., 13(58), 18(71), 20(71), 27, 33(71, 104), 102, 104, *115, 116*
Thomson, J., 180, *213*
Tikhonov, V. I., 57(148), *118*
Todd, D. K., 145, *171*
Topp, G. C., 219(13), 229, *292*
Truxal, J. G., 44, *117*
Tucker, M. J., 10(37), *114*
Tukey, J. W., 10(25, 34), 11, 41(25), 44, 57(34), 73, 74(25), 81, 84, 85, 109, *113, 114, 119*
Tyndall, J., 180, *213*

V

Vachaud, G., 280(130), *296*
Van Beers, W. F. J., 128, 147, *170*

Van Dyke, M., 234(75)k *294*
Van Isacker, J., 10(31), *113*
Van Schilfgaarde, J., 145(17), 152, *171, 172*
van Wijk, W. R., 186, *213*, 280(127), *296*
Vedernikov, V. V., 134, 136, 156, *170, 171*
von Schelling, H., 3, *112*

W

Wallis, J. R., 10, 11, *113*
Ward, F., 11, *114*
Wastler, T. A., 10, 31, 32, 34(44), 74, 93, 96, *114*
Watson, K. K., 157(34), *172*
Weiss, L. L., 27, *116*
Whisler, F. D., 157(32), *171*
Whittle, P., 11, *114*
Whitton, W. I., 228(62), *294*
Wiener, N., 48, *118*
Wilkinson, G. E., 279, *296*
Wilks, S. S., 19, *116*
Winsor, C. P., 81, *119*
Wold, H., 105, *119*
Wooding, R. A., 258, *295*
Wu, I. P., 12(55), *114*

Y

Yagil, S., 27, 33(105), *116*
Yen, Y. C., 196(26), 197(26, 27, 28, 29), 198(26), 199, 200(27, 28), 201(28), *214*
Yevjevich, V. M., 10, 18(72, 74), 21, 53, 54(39, 40), 79, 86, *114, 115*
Yosida, Z., 178, 184, 187, 192, 193(24), 194, 199, *213, 214*
Youngs, E. G., 226, 228, 252, 280(102), *293, 295*

Z

Zangar, C. N., 167(42), *172*

SUBJECT INDEX

Absorption, 217
Air-entry permeameter, 165
Air-entry value, 222, 232
Albedo, 207
American Society of Civil Engineers, 1
Angular frequency, 33
ANOVA, 4
ASCE, see American Society of Civil Engineers
ASCE Committee on Surface-Water Hydrology, 1
Autocorrelation, 68
Auto-covariance function, 18

Basin response, 8
Basin system, 2
Biochemical oxygen demand, 10, 93
B.O.D., see Biochemical oxygen demand
Boltzmann transformation, 247
Bouger–Lambert law, 203
Bulk water, 225

Capillarity, 224
Capillary fringe, 153
Capillary hysteresis, 228
Capillary potential, 220
Capillary pressure, 220
Central tendency values, 26
Cesptrum, 11
Circular frequency, 33
Coherence, 48
  squared, 47
Colorado River, 82, 83
Columbia River, 93
Concentration condition, 232
Confidence limits, 79
Convolution integral, 57
Correlation, 18, 68
  spurious, 10
Correlation kernel, 19, 38
Correlograms, 10, 86
Covariance kernel, 18
Covariance spectrum, 31, 77
Covariance stationarity, 20
Cross bispectra, 11
Cross correlation, 77

Cross-correlation coefficient, 38
Cross-covariance function, 38
Cross spectra, 11
Cross-spectral methods, 29
Cross-spectrum sampling theory, 82

Density function
  cospectral, 46
  quadrature spectral, 46
  spectral, 68
Depth hoar, 175
D–F methods, see Dupuit–Forchheimer method
Diffusion coefficient, 192
Dissolved oxygen, 93
D.O., see Dissolved oxygen
Double-tube method, 167
Dupuit–Forchheimer flow, 124
Dupuit–Forchheimer method, 124

Extinction coefficient, 203

Feedback, 2
Fick's second law of diffusion, 209
Flux conditions, 233
Fokker–Planck equation, 216, 233, 260
  nonlinear, 222
Fourier heat-conduction equation, 195
Fourier transforms, 61
French Broad River Basin, 104
Frequency response function, 61

Great Lakes, 84

Hanning lag window weighting function, 73
Harmonic analysis, 10, 30
Harmonic coefficients, 33
Hodograph techniques, 148
Homogeneity, 13, 91
  spatial, 14
  temporal, 13
Hydraulic conductivity, 122 217
  measurement of, 163
Hydraulic head, 217
Hydraulic impedance, 168
Hydrologic cycle, 2

SUBJECT INDEX

Hydrologic system, 7
Hydrology
 parametric, 3
 physical, 3
 stochastic, 1

Ice adhesion, 181
Ice bonding, 181
Impulse response function, 58
Indeterministic model, 102
Infiltration, 217
 fluid-mechanical theory of, 229
Infiltrometer, 165
International Association of Scientific Hydrology, 15
International Conference on Low temperature Science, 174

John Day River, 89, 94
Joint distribution function, 18

Kernel, 74

Lake Michigan–Huron, 50, 96, 97, 99
Lambert's cosine law, 207
Linear regressive model, 28
Linear reservoir, 45

Manning's formula, 150
Mean value function, 18
Metamorphism
 constructive, 175, 176
 melt, 175
 pressure, 175
 snow, 174
Model
 building, 2, 9
 deterministic, 9
 indeterministic, 102
 probabilistic, 103
 stochastic, 9
Moisture diffusivity, 223
Moisture potential, 220
Moisture suction, 220
Moisture tension, 220
Moving average, 50, 52
Moving average coefficients, 51
Multivariate analysis, 10

Nash's model, 59
Navier–Stokes equation, 218

Negative pressure, 220
Newtonian fluid, 225
Noise
 red, 54, 57, 82
 white, 42
Nonlinearity, 10
Null continuum, 81

Oscillation, 34

Parametric hydrology, 3
Periodicity, 13, 34
 hidden, 52
Periodogram, 70
Permeability, intrinsic, 218
Persistence, 13, 87, 88
 Markov, 82, 89
 model, 57
 simple, 28
Physical hydrology, 3
Piezometer, 164
PMP, see Probable maximum precipitation
Population spectrum, 81
Porosity, 190
Potomac River, 75, 93, 97
Power spectrum, 31, 86
Pressure head, 220
Probabilistic model, 103
Probable maximum precipitation, 15
Process, 12
 autoregressive, 28
 deterministic, 15
 ergotic, 22
 Gaussian, 19
 generating, 52
 hydrologic, 12
 lag-one, 57
 linear autoregressive, 52, 54
 linear cyclic, 52
 Markov, 28, 56, 57
 moving average, 28, 50, 52
 nonlinear, 8
 nonreactive, 9
 pure random, 15
 random, 15
 stochastic, 15
  classification of, 26
Public Health Service, 10
Pumping test, 164

SUBJECT INDEX

Radiation interaction, 202
Randomness, 9
Random walk model, 209
Reflectrance, 207
  spectral, 207
Regelation, 180
Regression
  lagged, 29
  multiple, 29
  simple, 29
Response
  basin, 8
  dynamic, 18
Routing constant, 59

Salt River, 89, 90, 92
Sample function, 17
Scale heterogeneity, 279
Schuster periodogram, 70
Schwartz–Christoffel transformations, 148
Seepage, 121
  hydrodynamics of, 122
Seepage flow, 122
  transient, 157
Sensitivity analysis, 11
Sequence, 13
  decoupled, 19
  pure random, 19
  random, 16
Serial correlation coefficients, 28
Shallow well pump-in method, 166
Signal-to-noise ratio, 108
Sintering, 180, 181
Skewness, coefficient, 104
Slutzky–Yule effect, 54
Snow
  metamorphism of, 174
  porosity, 190
  radiation of, 201
  reflection from, 206
  rotten, 175
  thermal conductivity, 183–190
  thermal diffusivity, 190
  vapor diffusion in, 192
Snow metamorphism, 174
Soil
  anisotropic, 145
  heterogeneous, 145
  nonuniform, 145
  uniform, 131
  unsaturated, 216
Soil–air, 226
Sorptivity, 237
  intrinsic, 238
Spectral function,
  variance, 68
Spectral transmittance, 203
Stationarity, 27
  covariance, 20
Station-year method, 22
Stefan–Boltzmann law, 201
Stochastic forcing function, 18
Stochastic hydrology, 1, 3
Stochastic variable, 2
Storage time delay, 59
Sublimation transfer, 180
Surface viscosity, 225
System
  basin, 2
  linear, 57
  nonlinear, 8
  reactive, 9
  transformation, 2, 5
System function, 61
System time response function, 58
System weighing function, 58

Thermal conductivity, 183
Thermal diffusivity, 190
  effective, 195
Time series, 10, 12
  continuous, 12
  discrete, 12
  hydrologic, 12
  multivariate, 29
  nonstationary, 21
  periodic, 33
  random, 40
  stationary, 20
  transient, 39
Transformation system, 2, 5
Transient flow, 157
Transient function, 39
Transition zone, 227
Trend, 13, 28
Trispectrum, 11
Tukey sampling theory, 81

UHG, *see* Unit hydrograph

SUBJECT INDEX

Ultimate wetted region, 285
Umpqua River, 89, 90, 94
Unit hydrograph, 4
  instantaneous, 45, 49
Unsaturated flow, 151
U.S. Geological Survey, 15, 20
UWR, *see* Ultimate wetted region

Variance spectra, 10, 31

Water table, 122
Wet front, 280
Wetting, 285
Wetting zone, 280
Wiener–Hopf equation, 65

Printed in Poland
by Amazon Fulfillment
Poland Sp. z o.o., Wrocław

35957253R00181